Maxim Olegovich Korpusov, Alexey Vital'evich Ovchinnikov,
Alexey Georgievich Sveshnikov, and Egor Vladislavovich Yushkov
Blow-Up in Nonlinear Equations of Mathematical Physics

De Gruyter Series in Nonlinear Analysis and Applications

Volume 27

Maxim Olegovich Korpusov, Alexey Vital'evich Ovchinnikov,
Alexey Georgievich Sveshnikov, and Egor Vladislavovich Yushkov

Blow-Up in Nonlinear Equations of Mathematical Physics

Theory and Methods

DE GRUYTER

Mathematics Subject Classification 2010
Primary: 35B44, 35Qxx; Secondary: 46Txx, 35A15, 35C06

Authors
Prof. Dr Maxim Olegovich Korpusov
Lomonosov Moscow State University
Department of Mathematics
Faculty of Physics
Leninskie Gory
Moscow 119992
Russian Federation
korpusov@gmail.com

Prof. Dr Alexey Georgievich Sveshnikov
Lomonosov Moscow State University
Department of Mathematics
Faculty of Physics
Leninskie Gory
Moscow 119992
Russian Federation
tfl10@mail.ru

Prof. Dr Alexey Vital'evich Ovchinnikov
Lomonosov Moscow State University
Department of Mathematics
Faculty of Physics
Leninskie Gory
Moscow 119992
Russian Federation
ovchinnikov@viniti.ru

Prof. Dr Egor Vladislavovich Yushkov
Lomonosov Moscow State University
Department of Mathematics
Faculty of Physics
Leninskie Gory
Moscow 119992
Russian Federation
yushkov.msu@mail.ru

ISBN 978-3-11-060108-4
e-ISBN (PDF) 978-3-11-060207-4
e-ISBN (EPUB) 978-3-11-059900-8
ISSN 0941-813X

Bibliographic information published by the Deutsche Nationalbibliothek
The Deutsche Nationalbibliothek lists this publication in the Deutsche Nationalbibliografie; detailed bibliographic data are available on the Internet at http://dnb.dnb.de.

Typesetting: VTeX UAB, Lithuania
Printing and binding: CPI books GmbH, Leck

www.degruyter.com

Dedicated to the Shiny Memory
of Professor I. A. Shishmarev
and Professor S. I. Pokhozhaev

Contents

Introduction

In this monograph, we collected all methods used in the study of the blow-up of solutions to mathematical models of realistic physical phenomena. The blow-up effect occurs, for example, when a sea wave tumbles to the shore, when a computer breaks down as a result of electrical breakdown, or when a nuclear bomb explodes, and in a number of other interesting physical phenomena.

There exist several well-known methods of the study of the blow-up effect, which have their specific domain of applicability to corresponding problems of mathematical physics. In this monograph, all these methods are illustrated by numerous examples of problems for nonlinear equations.

Despite the fact that the title of the monograph contains the words "blow-up of solutions to *nonlinear equations*," this does not mean that the blow-up occurs only in nonlinear problems. One of examples is the *linear Schrödinger equation* for which the so-called *dispersion blow-up* is observed. Other examples of the dispersion blow-up are represented by the linear heat equation with an initial function that *does not belong to the Tikhonov–Tacklind class* and the *inversely parabolic linear equation*. A detailed study of the dispersion blow-up of solutions to linear problems is beyond the scope of the monograph; however, certain results are presented in the Appendix.

The authors are sincerely grateful to S. I. Pokhozhaev for useful discussions of the subject matter of this book and a number of valuable remarks and to V. A. Il'in, who took the trouble to get acquainted with the monograph. Also, we are grateful to A. A. Panin, who carefully read the manuscript and made valuable comments.

The research was partially supported by the Russian Foundation for Basic Research (project No 11-01-12018-ofi-m-2011) and the program "Scientific and Pedagogical Staff of Innovative Russia" (project No. 8215).

https://doi.org/10.1515/9783110602074-201

Notation

$\{a\} \times \{b\}$	the Cartesian product of the sets $\{a\}$ and $\{b\}$;
$A \otimes B$	the Cartesian product of the topological spaces A and B;
$\mathbb{N}$	the set of natural numbers;
$\mathbb{Z}$	the set of integer numbers;
$\mathbb{Z}^n$	the set consisting of ordered sets of integers of the form $(z_1, z_2, \ldots, z_n)$, $z_m \in \mathbb{Z}$, $m = \overline{1, n}$;
$\mathbb{Z}^n_+$	the set consisting of ordered sets of nonnegative integers;
$\mathbb{R}^{\mathbb{N}}$	the $\mathbb{N}$-dimensional Euclidean space;
$\lvert \cdot \rvert$	the norm in the Euclidean space $\mathbb{R}^N$;
$\mathbb{R}^1_+$	the set of nonnegative real numbers;
$\overline{\alpha}$	the multi-index $\overline{\alpha} = (\alpha_1, \ldots, \alpha_n)$, $\alpha_i \in \mathbb{R}^1_+$;
$\lvert\overline{\alpha}\rvert$	is the modulus of the multi-index: $\lvert\overline{\alpha}\rvert = \alpha_1 + \alpha_2 + \cdots + \alpha_n$;
D_x	the gradient operator with respect to the variable $x \in \mathbb{R}^{\mathbb{N}}$;
D^m_x	the operator of the form

$$D^m_x \equiv \left(\frac{\partial}{\partial x_1}\right)^{m_1} \left(\frac{\partial}{\partial x_2}\right)^{m_2} \cdots \left(\frac{\partial}{\partial x_N}\right)^{m_N},$$

	where $m = m_1 \times m_2 \times \cdots \times m_{\mathrm{N}}$;
$f^{(m)}(x)$	$f^{(m)}(x) = D^{m_1}_{x_1} \ldots D^{m_N}_{x_N} f(x)$, where $x = (x_1, \ldots, x_N) \in \mathbb{R}^N$ and $m = (m_1, \ldots, m_N)$;
∇_x	the gradient with respect to the variable $x \in \mathbb{R}^N$;
$\frac{\partial u}{\partial \mathbf{n}}$	the derivative in the direction of the outer normal $\mathbf{n}$ of the smooth boundary $\partial\Omega \in \mathbb{C}^{(1,\delta)}$ of a bounded domain $\Omega \subset \mathbb{R}^N$;
∂_t	the partial derivative with respect to t;
$\partial^2_{x_i x_j}$	the mixed partial derivative with respect to x_i and x_j;
$\partial^k_{x_i}$	the partial derivative of order k with respect to the variable $x_i \in \mathbb{R}^1$;
$\Delta_p u$	the pseudo-Laplacian (p-Laplacian), $\Delta_p u \equiv \mathrm{div}(\lvert\nabla u\rvert^{p-2}\nabla u)$;
Δ	the Laplace operator;
Δ^2	the biharmonic operator;
$\Delta_2 u$	the two-dimensional Laplace operator, $\Delta_2 u \equiv \partial^2_{x_1} u + \partial^2_{x_2} u$;
$\hat{u}$	the Fourier transform of a function u;
$\mathcal{L}(\mathbb{X}, \mathbb{Y})$	the set of linear continuous operators acting from $\mathbb{X}$ to $\mathbb{Y}$;
$\mathbb{A}'_u(\cdot)$	the Fréchet derivative of the operator $\mathbb{A}(u) : \mathbb{X} \to \mathbb{X}^*$, $\mathbb{A}'_u(\cdot) : \mathbb{X} \to \mathcal{L}(\mathbb{X}, \mathbb{X}^*)$;
$\mathbb{X}^*$	the dual space of a Banach space $\mathbb{X}$;
$\Omega \subset \mathbb{R}^{\mathbb{N}}$	a domain in $\mathbb{R}^{\mathbb{N}}$;
$\bar{\Omega}$	the closure of a domain Ω;
$\partial\Omega$	the boundary of a domain Ω;

https://doi.org/10.1515/9783110602074-202

$\partial\Omega \in \mathbb{C}^{(m,\delta)}$	the boundary $\partial\Omega$ of a domain $\Omega \in \mathbb{R}^{\mathbb{N}}$ in a neighborhood of each point $x \in \partial\Omega$ can be represented by local coordinates $$\zeta_i = \Phi_i(\xi_1, \xi_2, \dots, \xi_{N-1}, \eta), \quad i = \overline{1, N-1},$$ where the functions Φ_i are m times continuously differentiable with respect to their variables, and the functions $\Phi_i^{(m)}$, $m \in \mathbb{Z}_+^N$, are Hölder functions with exponent $\delta \in (0, 1]$;
$\|\cdot\|_{\mathbb{X}}$	the norm of a Banach space $\mathbb{X}$;
$\mathbb{C}^{(p)}(\Omega)$	the set of all functions on Ω that have $p \in \mathbb{N}$ continuous derivatives in Ω;
$\mathbb{C}_b^{(p)}(\overline{\Omega})$	the Banach space with the norm $$\|u\|_{\mathbb{C}^p} \equiv \sup_{x\in\Omega} \sum_{m=1}^{p} \|D_x^m u\|;$$
$\mathbb{C}_0^{\infty}(\Omega)$	the set of infinitely differentiable compactly supported functions;
$\operatorname{supp} u$	the support of a function u;
$\mathbb{C}^{(0,\delta)}(\Omega)$	the Banach space of Hölder functions with the norm $$\|u\|_{0,\delta} \equiv \sup_{x\in\Omega} \|u\| + \sup_{x,y\in\Omega} \frac{\|u(x) - u(y)\|}{\|x - y\|^{\delta}}, \quad \delta \in (0, 1];$$
$\mathbb{AC}(0, T)$	the space of absolutely continuous functions;
$\mathbb{BV}(0, T)$	the space of functions of finite variation;
$\mathbb{L}^p(\Omega)$	the Banach space of measurable functions that are summable with power $p \in [1, +\infty]$ in a domain Ω with the norm $$\|u\|_p \equiv \int_{\Omega} dx\, \|u\|^p;$$
$(\cdot, \cdot)$	the scalar product in $\mathbb{L}^2$ and in $\mathbb{R}^N$);
$\langle \cdot, \cdot \rangle$	the duality brackets between a reflexive Banach space $\mathbb{X}$ and its dual $\mathbb{X}^*$;
$\|\cdot\|_*$	the norm of the Banach space $\mathbb{X}^*$ if $\|\cdot\|$ is the norm of a Banach space $\mathbb{X}$;
$\mathbb{H}_0^m(\Omega)$	the Hilbert space of measurable functions with zero trace on the boundary of a domain Ω that possess $m \in \mathbb{N}$ generalized derivatives from $\mathbb{L}^2(\Omega)$ with the scalar product $$(u, v)_{\mathbb{H}_0^m} = \sum_{\|\alpha\|\leq m} (D^{\alpha}u, D^{\alpha}v);$$
$\mathbb{H}^{-m}(\Omega)$	the Hilbert space dual to $\mathbb{H}_0^m(\Omega)$; each element u of this space can be represented in the form $$u = \sum_{\|\alpha\|\leq m} D^{\alpha} g_{\alpha}, \quad g_{\alpha} \in \mathbb{L}^2(\Omega);$$

$\mathbb{H}^s(\Omega)$ the space defined by the real interpolation:

$$\mathbb{H}^s(\Omega) = \left[\mathbb{H}^m(\Omega), \mathbb{H}^0(\Omega)\right]_\theta, \quad (1-\theta)m = s, \quad m \in \mathbb{Z},\ 0 < \theta < 1;$$

$\mathbb{W}^{k,p}(\Omega)$ the Banach space of measurable functions that possess $k \in \mathbb{N}$ generalized derivatives summable with power $p \in \mathbb{R}^1_+$ in a domain Ω with the norm

$$\|u\|_{k,p} \equiv \sum_{m=1}^{k} \|D^m u\|_p;$$

$\mathbb{W}^{k,p}_0(\Omega)$ the Banach space consisting of elements of the Banach space $\mathbb{W}^{k,p}(\Omega)$ with zero trace on the boundary of the domain;

$\mathbb{W}^{-k,p'}(\Omega)$ the Banach space dual to the Banach space $\mathbb{W}^{k,p}_0(\Omega)$, $p' = p/(p-1)$; each element of this space can be represented in the form

$$u = \sum_{|\alpha|\le k} D^\alpha g_\alpha, \quad g_\alpha \in \mathbb{L}^{p'}(\Omega);$$

$\langle\cdot,\cdot\rangle_s$ the duality brackets between the Hilbert spaces $\mathbb{H}^s_0(\Omega)$ and $\mathbb{H}^{-s}(\Omega)$;

$\langle\cdot,\cdot\rangle_{k,p}$ the duality brackets between the Banach spaces $\mathbb{W}^{k,p}_0(\Omega)$ and $\mathbb{W}^{-k,p'}(\Omega)$, $p' = p/(p-1)$;

$\mathbb{L}^p(0,T;\mathbb{B})$ the Banach space of $\mathbb{B}$-valued functions that are strongly measurable on an interval $(0,T)$ and such that the Lebesgue integral

$$\int_0^T dt\, \|u\|^p_{\mathbb{B}}$$

is finite; the norm of this space is

$$\|u\| = \left(\int_0^T dt\, \|u\|^p_{\mathbb{B}}\right)^{1/p};$$

$\mathbb{C}^m(0,T;\mathbb{B})$ the space of m times continuously differentiable functions with values in a Banach space $\mathbb{B}$;

$\mathcal{D}(\Omega)$ the space of compactly supported test functions;

$\mathcal{D}'(\Omega)$ the space of generalized functions dual to $\mathcal{D}(\Omega)$.

List of nonlinear equations

1. $u_t = \Delta u + u^q$ (1.1);
2. $x'' \geq |x|^q$ (1.7);
3. $y'' + x^\sigma |y|^{k-1} y = 0$ (1.23);
4. $\xi^{-2}(\xi^2\theta')' + |\theta|^{k-1}\theta = 0$ (1.24);
5. $x'' \geq \beta t^\alpha |x|^q$ (1.25);
6. $-(|u'|^{p-2}u')' \geq a(x)u^q$ (1.46);
7. $-\Delta u \geq u^q$ (1.47);
8. $-|x|^2\Delta u \geq |u|^q$ (1.64);
9. $u_t - \Delta u \geq u^q$ (1.71);
10. $u_t - |x|^2\Delta u \geq |u|^q$ (1.86);
11. $u_{tt} - \Delta u \geq |u|^q$ (1.94);
12. $u_{tt} - a(\int_{\mathbb{R}^N} |\nabla u|^2\, dx)\Delta u \geq |u|^q$ (1.111);
13. $u_t = \nabla(u^\sigma \nabla u) + u^\beta$ (2.1), (2.30), (2.78);
14. $\theta''_A + 1/2\theta'_A\xi - m\theta_A = 0$ (2.7);
15. $u_t = B(u) = \nabla(k(u)\nabla u) + Q(u)$ (2.9);
16. $-\Delta u = |u|^{q-1}u$ (3.1), (10.68);
17. $u_t = \Delta u + u^{1+\alpha}$ (3.10);
18. $u_t - uu_x + u_{xxx} = 0$ (3.27);
19. $u_t = -(\partial_x^2 + \partial_y^2)^2 u + u^2$ (4.36);
20. $\lambda u_t - u_{xxt} = u_{xx} + u^2$ (4.39);
21. $u_t = u_{xx} + u(x,t)\int_0^\pi K(x,y)u^2(y,t)\, dy$ (4.41);
22. $u_t - \mu\Delta u_t = -\Delta^2 u + u^{1+\beta}$ (4.53);
23. $w_{tt} = -(\partial_x^2 + \partial_y^2)^2 w + F(w)$ (4.92);
24. $u_{tt} - \Delta u + au_t|u_t|^{m-1} = bu|u|^{p-1}$ (5.1);
25. $u_{tt} - \Delta u + \int_0^t g(t-s)\Delta u(x,s)\, ds + u_t|u_t|^{m-2} = |u|^{p-2}u$ (5.62);
26. $u_t = \Delta_p u + |u|^{\sigma-2}u$ (6.37);
27. $u_{tt} = \Delta u + |u|^{p-1}u$ (6.49);
28. $u_{tt} = -(-\Delta)^m u + |u|^{p-1}u$ (6.50), (6.53);
29. $u_{tt} = \operatorname{div}(|\nabla u|^{q-1}\nabla u) + |u|^{p-1}u$ (6.51), (6.68);
30. $u_{tt} = -(\int |\nabla^m u|^2)^q(-\Delta)^m u + |u|^{p-1}u$ (6.52), (6.72);
31. $u_{tt} = -(\int |\nabla^m u|^2)^q(-\Delta)^m u + |u|^{p-1}u$ (6.74);
32. $u_{tt} = \Delta u + b(x,t)|u|^p$ (6.76);
33. $u_{tt} = u_{xx} - c_1 u_{xxxx} + c_2(u^2)_{xx}$ (7.41);
34. $u_{tt} = \pm(u_x^m)_x$ (7.45);
35. $u_{tt} = -u_{xxxx} - u_{xx} \mp (u_x^2)_x$ (7.46);
36. $u_{tt} = -u_{xxxx} - u_{xx} - u + u^3$ (7.47);

https://doi.org/10.1515/9783110602074-203

37. $\partial_t\varphi(x,u) - \Delta(|u|^{q_1}u) = |u|^{q_2}u$ (8.3);
38. $\partial_t\varphi(x,u) - \operatorname{div}(h(x,|\nabla u|)\nabla u) + g(x,u) = f(x,u)$ (8.26);
39. $\partial_t\varphi(x,u) - \chi(\|\nabla u\|_2^2)\Delta u = f(x,u)$ (8.64);
40. $u_{tt} - \Delta u + u = u^2 + u^3$ (8.87);
41. $u_{tt} + \partial_x^4 u = \partial_x^4 u(\partial_x^2 u)^2 + 2(\partial_x^3 u)^2\partial_x^2 u$ (8.90);
42. $-u_{xxtt} + u_{tt} + u_{xxxx} - u_{xx} = -(u_x^2)_x$ (8.91);
43. $u_{tt} + \Delta^2 u = \Delta(\Delta u)^3$ (8.124);
44. $\mathbf{A}_{tt} + \mu\mathbf{A}_t - \Delta\mathbf{A} - h(x,|\mathbf{A}|)\mathbf{A} = \nabla\psi$ (8.144);
45. $\partial_t(\Delta u + \Delta_{p_1}u) + \Delta u - \Delta_{p_2}u = 0$ (8.194);
46. $\partial_t(\Delta u - \varphi(x,u)) + \operatorname{div}(h(x,|\nabla u|)\nabla u) - g(x,u) + f(x,u) = 0$ (8.209);
47. $\partial_t(-\Delta^2 u + \Delta u + \Delta_{p_1}u) + \int_0^t ds\, h(t-s)\Delta u(s) + \Delta u + \alpha_1\partial_{x_1}(\partial_{x_2}u\partial_{x_3}u) + \alpha_2\partial_{x_2}(\partial_{x_3}u\partial_{x_1}u) + \alpha_3\partial_{x_3}(\partial_{x_1}u\partial_{x_2}u) - \Delta_{p_2}u = 0$, $\alpha_1 + \alpha_2 + \alpha_3 = 0$ (8.241);
48. $\partial_t^2(\partial_x^2 u - u) + \partial_{tx}^2(\partial_x u - (\partial_x u)^2) + \partial_x^2 u = 0$ (8.257);
49. $\partial_t^2(\partial_x^2 u - u) + \partial_t(u - u^2) + \partial_x^2 u = 0$ (8.259);
50. $\partial_t^2(\Delta u - \gamma^2(x_3)u) + \omega^2(x_3)\Delta_2 u + f(x,u) = 0$ (8.282);
51. $\partial_t^2(\Delta u - u) + \partial_t(\Delta u - g(x,u)) + \Delta u + f(x,u) = 0$ (8.307);
52. $iu_t = -\Delta u - |u|^{p-1}u$ (9.1), (9.15);
53. $iu_t = \Delta u + F(|u|^2)u$ (9.5);
54. $u_t - u_{xxt} + 3uu_x = 2u_xu_{xx} + uu_{xxx}$ (11.1);
55. $u_t + uu_x + \int_{-\infty}^{+\infty} \mathrm{K}(x-s)u_s(s,t)\,ds = 0$ (11.7).

1 Nonlinear capacity method of S. I. Pokhozhaev

In this chapter, we consider one of the most powerful methods of nonlinear analysis developed in the last decades, which allowed one to solve numerous problems for nonlinear equations of modern mathematical physics. This method was developed by S. I. Pokhozhaev and E. L. Mitidieri (see [74] and the references therein).

1.1 Critical exponent

Before we present the nonlinear capacity method, we recall the notion of the critical exponent for nonlinear problems. Following Mitidieri, we consider the following Cauchy problem for the semilinear parabolic equation:

$$\frac{\partial u}{\partial t} = \Delta u + u^q, \quad u(x,0) = u_0(x) \geq 0, \quad q > 1. \tag{1.1}$$

Since the initial function $u_0(x)$ is nonnegative, a solution of the problem also is nonnegative due to the maximum principle. Consider the fundamental solution $\mathcal{E}(x,t)$ of the operator

$$\frac{\partial}{\partial t} - \Delta,$$

which, as is well known, satisfies the equation

$$\frac{\partial \mathcal{E}(x,t)}{\partial t} - \Delta \mathcal{E}(x,t) = \delta(x)\delta(t)$$

in the sense of generalized functions (distributions) from $\mathcal{D}'(\mathbb{R}_+^{N+1})$ and has the following explicit form:

$$\mathcal{E}(x,t) = \frac{\theta(t)}{(4\pi t)^{N/2}} \exp\left(-\frac{1}{4t}|x|^2\right), \quad \theta(t) = \begin{cases} 1 & \text{for } t \geq 0, \\ 0 & \text{for } t < 0. \end{cases} \tag{1.2}$$

Clearly, the fundamental solution decreases with rate $t^{-N/2}$ as $t \to +\infty$, and the solution of the Cauchy problem for the linear heat transfer equation

$$u(x,t) = \int_{\mathbb{R}^N} \mathcal{E}(x-y,t) u_0(y)\, dy$$

also decreases.

Now we consider the initial-value problem for the following ordinary differential equation:

$$\frac{dx}{dt} = x^q, \quad x(0) = x_0 > 0, \quad q > 1. \tag{1.3}$$

https://doi.org/10.1515/9783110602074-001

Its solution is

$$x(t) = \left(x_0^{1-q} - (q-1)t\right)^{-1/(q-1)};$$

obviously, this solution blows up in finite time:

$$\lim_{t\to t_0^-} x(t) = +\infty, \quad t_0 = \frac{x_0^{1-q}}{q-1}, \quad \text{as } (c_1 - c_2 t)^{-1/(q-1)}.$$

Note that the nonlinear partial differential equation (1.1) contains two competing factors, namely, the linear dissipative term

$$\frac{\partial u}{\partial t} - \Delta u$$

and the nonlinear blow-up term

$$\frac{\partial u}{\partial t} - u^q.$$

Qualitatively, the situation is as follows: if the linear term overcomes, then a solution exists globally in time, whereas if the nonlinear term overcomes, then a solution blows up in finite time. The quantitative description is as follows:

$$\frac{N}{2} > \frac{1}{q-1} \quad \Rightarrow \quad \text{global solvability,} \tag{1.4}$$

$$\frac{N}{2} < \frac{1}{q-1} \quad \Rightarrow \quad \text{blow-up.} \tag{1.5}$$

The value

$$\frac{1}{q-1} \tag{1.6}$$

is called the *critical exponent* for problem (1.1). The behavior of solutions for $N/2 = 1/(q-1)$ required an additional analysis. (Note that in problem (1.1) the blow-up occurs in this case.)

Remark. We further prove the following exact result. If inequality (1.5) holds, then the blow-up of any weak nontrivial solution for a finite time occurs. In inequality (1.4) holds, then the global solvability occurs for sufficiently "small" initial function $u_0(x)$. If the initial function $u_0(x)$ is sufficiently "large," then the solution blows up in finite time. The first result can be obtained by the Pokhozhaev nonlinear capacity method, the second – by the method of contraction mappings, and the third – by one of the modifications of the energy method presented in Chapters 6–8.

Fujita [23] was the first who obtained this result in 1966; so the critical exponents are also called the *Fujita critical exponents*.

We further examine critical exponents for all three fundamental classes of partial differential equations, namely, *elliptic, parabolic, and hyperbolic*, and also for the class of *Sobolev* equations.

1.2 Ordinary differential inequalities

1.2.1 Example

First, we consider the following problem:

$$\frac{d^2x}{dt^2} \geq |x|^q, \quad q > 1, \tag{1.7}$$

$$x(0) = x_0, \quad x'(0) = x_1. \tag{1.8}$$

We will examine weak solutions of problem (1.7)–(1.8) in the sense of the following definition.

Definition 1.1. A *weak solution* of problem (1.7)–(1.8) is a function

$$x(t) \in \mathbb{L}^q_{\mathrm{loc}}(0, +\infty)$$

satisfying the inequality

$$\int_0^T \varphi''(t)x(t)\,dt - x_1\varphi(0) + x_0\varphi'(0) \geq \int_0^T \varphi(t)|x|^q\,dt \tag{1.9}$$

for all functions $\varphi(t) \in \mathbb{C}^{(2)}([0, +\infty))$ and the conditions

$$\varphi(t) \geq 0 \quad \text{and} \quad \operatorname{supp}(\varphi) \subset [0, T)$$

for certain $T > 0$.

We have the following chain of inequalities:

$$\begin{aligned}
\left|\int_0^T \varphi''(t)x(t)\,dt\right| &\leq \int_0^T |\varphi''||x(t)|\,dt = \int_0^T \frac{|\varphi''|}{\varphi^{1/q}}\varphi^{1/q}|x(t)|\,dt \\
&\leq \left(\int_0^T \frac{|\varphi''|^{q'}}{\varphi^{q'/q}}\,dt\right)^{1/q'} \left(\int_0^T \varphi(t)|x|^q\,dt\right)^{1/q} \\
&\leq \frac{1}{q'}\int_0^T \frac{|\varphi''|^{q'}}{\varphi^{q'/q}}\,dt + \frac{1}{q}\int_0^T \varphi(t)|x|^q\,dt.
\end{aligned} \tag{1.10}$$

Here we applied the Hölder inequality and the arithmetic Hölder inequality

$$ab \leq \frac{a^q}{q} + \frac{b^{q'}}{q'}, \quad \frac{1}{q} + \frac{1}{q'} = 1, \quad a, b \geq 0.$$

Taking into account the chain of expressions (1.10) and inequality (1.9), we obtain the expression

$$\int_0^T \frac{|\varphi''|^{q'}}{\varphi^{q'/q}}\,dt - q'x_1\varphi(0) + q'x_0\varphi'(0) \geq \int_0^T \varphi(t)|x|^q\,dt. \tag{1.11}$$

Now we separately consider the nonlinear capacity, that is, the integral in the left-hand side of (1.11):

$$\int_0^T \frac{|\varphi''|^{q'}}{\varphi^{q'/q}}\,dt. \tag{1.12}$$

We choose the following test function $\varphi(t)$:

$$\varphi(t) = \varphi_0\left(\frac{t}{T}\right), \quad \varphi_0(s) = \begin{cases} 1 & \text{if } 0 \leq s \leq 1/2, \\ 0 & \text{if } s \geq 1. \end{cases} \tag{1.13}$$

We have the following equalities:

$$\int_0^T \frac{|\varphi''|^{q'}}{\varphi^{q'/q}}\,dt = \int_0^T \frac{|\varphi''|^{q'}}{\varphi^{1/(q-1)}}\,dt = T^{1-2q'} \int_{1/2\leq s\leq 1} \frac{|\varphi_0''(s)|^{q'}}{\varphi_0^{1/(q-1)}(s)}\,ds. \tag{1.14}$$

We prove that the integral

$$\int_{1/2\leq s\leq 1} \frac{|\varphi_0''(s)|^{q'}}{\varphi_0^{1/(q-1)}(s)}\,ds \tag{1.15}$$

converges at least for one test function $\varphi_0(s)$ of the class specified.

As a test function $\varphi_0(s) \in \mathbb{C}^{(2)}([0, +\infty))$, we choose a function of the form

$$\varphi_0(s) = \psi_0^\gamma(s) \quad \text{for } \gamma = 2q'. \tag{1.16}$$

The following expressions are valid:

$$\varphi_0'(s) = \gamma\psi_0^{\gamma-1}(s)\psi_0'(s), \tag{1.17}$$

$$\varphi_0''(s) = \gamma(\gamma-1)\psi_0^{\gamma-2}(\psi_0'(s))^2 + \gamma\psi_0^{\gamma-1}\psi_0''(s), \tag{1.18}$$

and we arrive at the inequality

$$|\varphi_0''(s)|^{q'} \leq c_1(q',\gamma)[\psi_0^{(\gamma-2)q'}|\psi_0'|^{2q'} + \psi_0^{(\gamma-1)q'}|\psi_0''|^{q'}]. \tag{1.19}$$

Thus, the integrand in (1.15) can be estimated as follows ($\gamma = 2q'$):

$$\frac{|\varphi_0''(s)|^{q'}}{\varphi_0^{1/(q-1)}(s)} \leq c_1(q',\gamma)[\psi_0^{(\gamma-2)q'-\gamma/(q-1)}|\psi_0'|^{2q'} + \psi_0^{(\gamma-1)q'-\gamma/(q-1)}|\psi_0''|^{q'}]$$

$$= c_1(q',\gamma)[|\psi_0'|^{2q'} + \psi_0^{q'}|\psi_0''|^{q'}] \le c_2(q') < +\infty. \tag{1.20}$$

Therefore, integral (1.15) converges for a sufficiently wide class of functions $\varphi_0(s)$. Then inequality (1.11) implies the inequality

$$c_0 T^{1-2q'} - q'x_1\varphi(0) + q'x_0\varphi'(0) \ge 0, \quad c_0 = \int\limits_{1/2\le s\le 1} \frac{|\varphi_0''(s)|^{q'}}{\varphi_0^{1/(q-1)}(s)}\,ds. \tag{1.21}$$

Now we impose the condition

$$x_1 > 0; \tag{1.22}$$

note that, due to our choice of the function $\varphi_0(s)$, we have the following equalities:

$$\varphi(0) = 1, \quad \varphi'(0) = 0.$$

Therefore, the initial condition $x(0) = x_0$ is irrelevant for the analysis of the blow-up of solutions of problem (1.7), (1.8). Since

$$2q' > 2 \quad \text{for } q > 1,$$

inequality (1.21) implies that it is valid only for

$$c_0 T^{1-2q'} > q'x_1 \quad \Rightarrow \quad T < T_\infty = \left(\frac{c_0}{q'x_1}\right)^{(q-1)/(q+1)}.$$

However, T_∞ is exactly the blow-up time of the weak solution $x(t)$ (in the sense of Definition 1.1).

1.2.2 Singular differential inequality of the Emden–Fowler type

The Emden–Fowler equation

$$y'' + x^\sigma |y|^{k-1}y = 0 \tag{1.23}$$

was first deduced by Emden [15] in connection with the model problem for a polytropic gas model used for description of equilibrium stellar configurations. This equation was obtained from the equation

$$\frac{1}{\xi^2}\frac{d}{d\xi}\left(\xi^2\frac{d\theta}{d\xi}\right) + |\theta|^{k-1}\theta = 0, \tag{1.24}$$

in which ξ is proportional to the distance from the center of a star, and the function $(\theta(\xi))^k$ is proportional to the density of the star (see [112]). Such equations also appear in plasma physics, gas dynamics, and also in the study of Kolmogorov diameters.

Asymptotic properties of the Emden–Fowler equation for various parameters σ and k were studied by Bellman [3], Sansone [95], Hartman [34], etc. Moreover, higher-order Emden–Fowler equations (of order $n > 2$) were examined by Kiguradze and Chanturia (see [42], Kondratiev and Samovol [43], and others).

Estimates with common domain are interesting due to their applications to qualitative characteristics of solutions; moreover, they allow us to prove the absence of global nontrivial solutions. In particular, such results for the equation

$$y^{(n)} \geq \beta |y|^q, \quad q > 1,$$

were obtained in [74]. In this section, we recall results of Hay [35] for differential inequalities of the Emden–Fowler-type inequalities.

Consider the following ordinary differential inequality:

$$\frac{d^2x}{dt^2} \geq \beta t^\alpha |x|^q, \quad \alpha \in (-\infty, +\infty),\ \beta > 0,\ q > 1,\ t > 0. \tag{1.25}$$

Definition 1.2. A *classical solution* of inequality (1.25) is a function

$$x(t) \in \mathbb{C}^{(2)}(0, t_0] \quad \text{for certain } t_0 > 0,$$

satisfying inequality (1.25).

We follow the method proposed by Hay [35]. Prove the following assertion: *Let either*

$$t_0 \frac{dx}{dt}(t_0) - x(t_0) \leq 0, \quad \alpha \leq -2,$$

or

$$\frac{dx}{dt}(t_0) \leq 0, \quad \alpha \leq -1 - q.$$

Then problem (1.25) *has no nontrivial local solutions for* $t \in (0, t_0]$.

In inequality (1.25), we perform the change of variables

$$\tau = \frac{1}{t}.$$

The derivatives take the form

$$\frac{dx}{dt} = \frac{dx}{d\tau}\frac{d\tau}{dt} = -\frac{dx}{d\tau}\frac{1}{t^2}, \quad \frac{d^2x}{dt^2} = 2\frac{dx}{d\tau}\frac{1}{t^3} + \frac{d^2x}{d\tau^2}\frac{1}{t^4} = 2\tau^3\frac{dx}{d\tau} + \tau^4\frac{d^2x}{d\tau^2}.$$

Then in the new variables inequality (1.25) has the form

$$\tau^4 \frac{d^2x}{d\tau^2} + 2\tau^3 \frac{dx}{d\tau} \geq \beta \tau^{-\alpha} |x|^q. \tag{1.26}$$

Take the following test functions:

$$\varphi(\tau) = \tau^{\delta}\varphi_1(\tau), \tag{1.27}$$

$$\varphi_1(\tau) = \begin{cases} 1 & \text{if } 0 < \tau_0 \le \tau \le T/2, \\ 0 & \text{if } \tau \ge T, \end{cases} \qquad \varphi_1(\tau) \in \mathbb{C}^{(2)}(0, T], \tag{1.28}$$

where the parameter δ will be chosen later.

Multiplying both sides of inequality (1.26) by the test function $\varphi(\tau)$ and integrating over $\tau \in (\tau_0, T]$, we arrive at the inequality

$$\int_{\tau_0}^{T} [\tau^{4+\delta}x'' + 2\tau^{3+\delta}x']\varphi_1(\tau)\,d\tau \ge \beta \int_{\tau_0}^{T} \tau^{-\alpha+\delta}\varphi_1(\tau)|x|^q\,d\tau. \tag{1.29}$$

Integrating by parts, we rewrite the integrands in the left-hand side of the last inequality, taking into account the fact that, due to the choice of the test function $\varphi_1(\tau)$, the following equalities are valid:

$$\varphi_1(\tau_0) = 1, \quad \varphi_1'(\tau_0) = 0, \quad \varphi_1(T) = 0, \quad \varphi_1'(T) = 0.$$

We verify the following expressions:

$$\begin{aligned}
\int_{\tau_0}^{T} \tau^{4+\delta}x''\varphi_1(\tau)\,d\tau &= \varphi_1(\tau)\tau^{4+\delta}x'(\tau)\Big|_{\tau=\tau_0}^{\tau=T} \\
&\quad - \int_{\tau_0}^{T} [(4+\delta)\tau^{3+\delta}\varphi_1(\tau)x'(\tau) + \varphi_1'(\tau)\tau^{4+\delta}x'(\tau)]\,d\tau \\
&= -\tau_0^{4+\delta}\frac{dx}{d\tau}(\tau_0) + (4+\delta)\tau_0^{3+\delta}x(\tau_0) \\
&\quad + \int_{\tau_0}^{T} [(4+\delta)(3+\delta)\tau^{2+\delta}\varphi_1(\tau)x(\tau) \\
&\quad + 2(4+\delta)\tau^{3+\delta}\varphi_1'(\tau)x(\tau) + \varphi_1''(\tau)\tau^{4+\delta}x(\tau)]\,d\tau,
\end{aligned} \tag{1.30}$$

$$\begin{aligned}
2\int_{\tau_0}^{T} \tau^{3+\delta}x'(\tau)\varphi_1(\tau)\,d\tau &= -2\tau_0^{3+\delta}x(\tau_0) \\
&\quad - 2\int_{\tau_0}^{T} [(3+\delta)\tau^{2+\delta}x(\tau)\varphi_1(\tau) + \tau^{3+\delta}x(\tau)\varphi_1'(\tau)]\,d\tau.
\end{aligned} \tag{1.31}$$

Taking into account (1.30) and (1.31), from inequality (1.29) we obtain the following estimate:

$$-\tau_0^{4+\delta}\frac{dx}{d\tau}(\tau_0) + (2+\delta)\tau_0^{3+\delta}x(\tau_0)$$

$$+ \int_{\tau_0}^{T} (([\varphi_1'' \tau^{4+\delta} + 2(3+\delta)\varphi_1'(\tau)\tau^{3+\delta}]x(\tau)$$
$$+ (2+\delta)(3+\delta)\varphi_1(\tau)\tau^{2+\delta}x(\tau))\,d\tau$$
$$\geq \beta \int_{\tau_0}^{T} \tau^{-\alpha+\delta}\varphi_1(\tau)|x|^q\,d\tau. \tag{1.32}$$

Now we choose the parameter $\delta > 0$ so that the “bad” term

$$\int_{\tau_0}^{T} (2+\delta)(3+\delta)\varphi_1(\tau)\tau^{2+\delta}x(\tau)\,d\tau$$

vanishes, that is, either $\delta = -2$ or $\delta = -3$. This choice is explained by the fact that the support of the function $\varphi_1(\tau)$ is contained in the interval (τ_0, T), whereas the support of the function $\varphi^{(k)}(\tau)$ is contained in the interval $(T/2, T)$.

Thus, let either $\delta = -2$ or $\delta = -3$. Then from (1.32) we obtain the following inequality:

$$I_0 + \int_{\tau_0}^{T} [\varphi_1'' \tau^{4+\delta} + 2(3+\delta)\varphi_1'(\tau)\tau^{3+\delta}]x(\tau)\,d\tau$$
$$\geq \beta \int_{\tau_0}^{T} \tau^{-\alpha+\delta}\varphi_1(\tau)|x|^q\,d\tau, \tag{1.33}$$

where

$$I_0 = -\tau_0^{4+\delta}\frac{dx}{d\tau}(\tau_0) + (2+\delta)\tau_0^{3+\delta}x(\tau_0). \tag{1.34}$$

In the first and second terms, we introduce the factors

$$(\varphi_1\beta c_1)^{1/q}\tau^{-(\alpha-\delta)/q}, \quad (\varphi_1\beta c_2)^{1/q}\tau^{-(\alpha-\delta)/q};$$

then the following inequality holds:

$$I_0 + \int_{\tau_0}^{T} (\varphi_1\beta c_1)^{1/q}\tau^{-(\alpha-\delta)/q}|x|\frac{|\varphi_1''|}{(\varphi_1\beta c_1)^{1/q}\tau^{(-\alpha+\delta-q(4+\delta))/q}}\,d\tau$$
$$+ 2\int_{\tau_0}^{T} (\varphi_1\beta c_2)^{1/q}\tau^{-(\alpha-\delta)/q}|x|\frac{|\varphi_1'(\tau)|}{(\varphi_1\beta c_2)^{1/q}\tau^{(-\alpha+\delta-q(3+\delta))/q}}\,d\tau$$
$$\geq \beta \int_{\tau_0}^{T} \tau^{-\alpha+\delta}\varphi_1(\tau)|x|^q\,d\tau \tag{1.35}$$

with constants $c_1 > 0$ and $c_2 > 0$ to be specified later. Now, applying the arithmetic Hölder inequality, we obtain the estimate

$$I_0 + \int_{\tau_0}^{T} \left[\frac{c_1}{q} \beta \varphi_1 \tau^{-\alpha+\delta} |x|^q + \frac{1}{q'} \frac{|\varphi_1''|^{q'}}{(\varphi_1 \beta c_1 \tau^{-\alpha+\delta-q(4+\delta)})^{q'-1}} \right] d\tau$$

$$+ \int_{\tau_0}^{T} \left[\frac{c_2}{q} \beta \varphi_1 \tau^{-\alpha+\delta} |x|^q + \frac{2^{q'}}{q'} \frac{|\varphi_1'|^{q'}}{(\varphi_1 \beta c_2 \tau^{-\alpha+\delta-q(3+\delta)})^{q'-1}} \right] d\tau$$

$$\geq \beta \int_{\tau_0}^{T} \tau^{-\alpha+\delta} \varphi_1(\tau) |x|^q \, d\tau. \tag{1.36}$$

Grouping the summands, we obtain the inequality

$$I_0 + \int_{\tau_0}^{T} \frac{1}{q'} \frac{|\varphi_1''|^{q'}}{(\varphi_1 \beta c_1 \tau^{-\alpha+\delta-q(4+\delta)})^{q'-1}} \, d\tau$$

$$+ \int_{\tau_0}^{T} \frac{2^{q'}}{q'} \frac{|\varphi_1'|^{q'}}{(\varphi_1 \beta c_2 \tau^{-\alpha+\delta-q(3+\delta)})^{q'-1}} \, d\tau$$

$$\geq c_0 \beta \int_{\tau_0}^{T} \tau^{-\alpha+\delta} \varphi_1(\tau) |x|^q \, d\tau, \tag{1.37}$$

where

$$c_0 = \frac{q - c_1 - c_2}{q} > 0 \quad \text{for small } c_1, c_2 > 0.$$

Since the supports of the functions $\varphi_1^{(k)}$, $k = 1, 2$, are contained in the interval $(T/2, T)$, from inequality (1.37) we obtain the following inequality:

$$I_0 + \int_{T/2}^{T} \frac{1}{q'} \frac{|\varphi_1''|^{q'}}{(\varphi_1 \beta c_1 \tau^{-\alpha+\delta-q(4+\delta)})^{q'-1}} \, d\tau$$

$$+ \int_{T/2}^{T} \frac{2^{q'}}{q'} \frac{|\varphi_1'|^{q'}}{(\varphi_1 \beta c_2 \tau^{-\alpha+\delta-q(3+\delta)})^{q'-1}} \, d\tau$$

$$\geq c_0 \beta \int_{\tau_0}^{T} \tau^{-\alpha+\delta} \varphi_1(\tau) |x|^q \, d\tau. \tag{1.38}$$

In (1.38), we change the variable by the rule $\tau = Ts$; then the inequality takes the form

$$I_0 + T^{(\alpha-\delta+q(4+\delta))(q'-1)-2q'+1} \int_{1/2}^{1} \frac{|\varphi_1''(s)|^{q'}}{q'(\varphi_1(s) \beta c_1 s^{-\alpha+\delta-q(4+\delta)})^{q'-1}} \, ds$$

$$+ T^{(\alpha-\delta+q(3+\delta))(q'-1)-q'+1} \int_{1/2}^{1} \frac{2^{q'} |\varphi_1'(s)|^{q'}}{q'(\varphi_1(s)\beta c_1 s^{-\alpha+\delta-q(3+\delta)})^{q'-1}} ds$$

$$\geq c_0\beta \int_{\tau_0}^{T} \tau^{-\alpha+\delta} \varphi_1(\tau)|x|^q \, d\tau. \tag{1.39}$$

As before, we can choose a test function $\varphi_1(s)$ and a constant c_3 such that

$$\int_{1/2}^{1} \frac{|\varphi_1''(s)|^{q'}}{q'(\varphi_1(s)\beta c_1 s^{-\alpha+\delta-q(4+\delta)})^{q'-1}} ds \leq c_3,$$

$$\int_{1/2}^{1} \frac{2^{q'} |\varphi_1'(s)|^{q'}}{q'(\varphi_1(s)\beta c_1 s^{-\alpha+\delta-q(3+\delta)})^{q'-1}} ds \leq c_3.$$

Then from inequality (1.39) we obtain the following inequality:

$$I_0 + T^{(\alpha-\delta+q(4+\delta))(q'-1)-2q'+1} c_3 + T^{(\alpha-\delta+q(3+\delta))(q'-1)-q'+1} c_3$$

$$\geq c_0\beta \int_{\tau_0}^{T} \tau^{-\alpha+\delta} \varphi_1(\tau)|x|^q \, d\tau. \tag{1.40}$$

Now we impose the following conditions:

$$(\alpha - \delta + q(4+\delta))(q'-1) - 2q' + 1 < 0,$$
$$(\alpha - \delta + q(3+\delta))(q'-1) - q' + 1 < 0.$$

Clearly, these conditions imply the inequality

$$\alpha < \delta - q(3+\delta) + 1. \tag{1.41}$$

Since either $\delta = -3$ or $\delta = -2$, we can easily verify that, for $\delta = -3$, the condition $I_0 \leq 0$ means

$$t_0 \frac{dx}{dt}(t_0) - x(t_0) \leq 0, \quad \alpha < -2, \tag{1.42}$$

whereas, for $\delta = -2$, the condition $I_0 \leq 0$ means that

$$\frac{dx}{dt}(t_0) \leq 0, \quad \alpha < -1 - q. \tag{1.43}$$

Thus, from inequality (1.40), passing to the limit as $T \to +\infty$, we obtain a contradiction; hence, under conditions (1.42) or under conditions (1.43), the blow-up occurs.

Now we prove that the blow-up also occurs for critical exponents $\alpha = -2$ or $\alpha = -1 - q$. For example, consider the case $\delta = -3$. Then from inequality (1.35) we obtain the following inequality:

$$I_0 + \int_{T/2}^{T} (\varphi_1 \beta c_1)^{1/q} \tau^{-(\alpha+3)/q} |x| \frac{|\varphi_1''|}{(\varphi_1 \beta c_1)^{1/q} \tau^{(-\alpha-3-q)/q}} \, d\tau$$
$$+ 2 \int_{T/2}^{T} (\varphi_1 \beta c_2)^{1/q} \tau^{-(\alpha+3)/q} |x| \frac{|\varphi_1'(\tau)|}{(\varphi_1 \beta c_1)^{1/q} \tau^{(-\alpha-3)/q}} \, d\tau$$
$$\geq \beta \int_{\tau_0}^{T} \tau^{-\alpha-3} \varphi_1(\tau) |x|^q \, d\tau. \tag{1.44}$$

Together with the Hölder inequality, it implies the inequality

$$I_0 + \left(\int_{T/2}^{T} \varphi_1 \beta c_1 \tau^{-3-\alpha} |x|^q \, d\tau \right)^{1/q} \left(\int_{T/2}^{T} \frac{|\varphi_1''|^{q'}}{(\varphi_1 \beta c_1 \tau^{-\alpha-3-q})^{q'-1}} \, d\tau \right)^{1/q'}$$
$$+ \left(2 \int_{T/2}^{T} \varphi_1 \beta c_2 \tau^{-3-\alpha} |x|^q \, d\tau \right)^{1/q} \left(\int_{T/2}^{T} \frac{|\varphi_1'|^{q'}}{(\varphi_1 \beta c_2 \tau^{-\alpha-3})^{q'-1}} \, d\tau \right)^{1/q'}$$
$$\geq \beta \int_{\tau_0}^{T} \varphi_1 \tau^{-\alpha-3} |x|^q \, d\tau. \tag{1.45}$$

For $\delta = -3$, inequality (1.39) implies that the integral

$$\int_{\tau_0}^{+\infty} \varphi_1 \tau^{-\alpha-3} |x|^q \, d\tau$$

converges; therefore, due to the absolute continuity of the Lebesgue integral, we obtain

$$\int_{T/2}^{T} \varphi_1 \beta c_1 \tau^{-3-\alpha} |x|^q \, d\tau \to +0 \quad \text{as } T \to +\infty.$$

The case $\delta = -2$ can be examined similarly. Thus, we have proved the following theorem.

Theorem 1.1 (Hay [35]). *Let either*

$$t_0 \frac{dx}{dt}(t_0) - x(t_0) \leq 0, \quad \alpha \leq -2,$$

or

$$\frac{dx}{dt}(t_0) \le 0, \quad \alpha \le -1 - q.$$

Then problem (1.25) *has no nontrivial local solutions for* $t \in (0, t_0]$.

Note that results obtained by S. I. Pokhozhaev, E. L. Mitidieri, and J. Hay were further developed by E. I. Galakhov. In particular, in [25] the following problem was considered:

$$\begin{gathered} -(|u'|^{p-2}u')' \ge a(x)u^q, \quad x \in (0, x_0], \\ u(x) \ge 0, \quad x \in (0, x_0], \\ u'(x_0) > 0, \end{gathered} \tag{1.46}$$

under the conditions $x_0 > 0$, $p > 1$, and $q > p - 1$.

1.3 Elliptic differential inequalities

1.3.1 Example

We start with the problem of the following classical form:

$$-\Delta u \ge u^q, \quad u \ge 0,\ q > 1,\ x \in \mathbb{R}^N. \tag{1.47}$$

We formulate the definition of a weak solution of the differential inequality (1.47).

Definition 1.3. A weak generalized solution of problem (1.47) is a function $u(x) \in \mathbb{L}^q_{\text{loc}}(\mathbb{R}^N)$ satisfying the inequality

$$-\int_{\mathbb{R}^N} \Delta\varphi(x)u(x)\,dx \ge \int_{\mathbb{R}^N} |u|^q \varphi(x)\,dx \tag{1.48}$$

for all $\varphi \in \mathbb{C}_0^\infty(\mathbb{R}^N)$, $\varphi(x) \ge 0$.

Now we deduce necessary and sufficient conditions under which inequality (1.47) has no nontrivial weak generalized solutions. Consider the following inequalities:

$$\begin{aligned} \left| \int_{\mathbb{R}^N} \Delta\varphi(x)u(x)\,dx \right| &\le \int_{\mathbb{R}^N} |u|\varphi^{1/q}\frac{|\Delta\varphi|}{\varphi^{1/q}}\,dx \\ &\le \left(\int_{\mathbb{R}^N} \varphi|u|^q\,dx \right)^{1/q} \left(\int_{\mathbb{R}^N} \frac{|\Delta\varphi|^{q'}}{\varphi^{q'-1}}\,dx \right)^{1/q'} \\ &\le \frac{1}{q}\int_{\mathbb{R}^N} \varphi|u|^q\,dx + \frac{1}{q'}\int_{\mathbb{R}^N} \frac{|\Delta\varphi|^{q'}}{\varphi^{q'-1}}\,dx, \quad q' = \frac{q}{q-1}. \end{aligned} \tag{1.49}$$

Inequalities (1.48) and (1.49) imply the following a priori estimate considered before:

$$\int_{\mathbb{R}^N} \frac{|\Delta\varphi|^{q'}}{\varphi^{q'-1}}\,dx \ge \int_{\mathbb{R}^N} |u|^q \varphi\,dx. \tag{1.50}$$

Choose the following test function $\varphi(x)$:

$$\varphi(x) = \varphi_0\left(\frac{|x|^2}{R^2}\right), \quad \varphi_0(s) = \begin{cases} 1 & \text{if } 0 \le s \le 1; \\ 0 & \text{if } s \ge 2. \end{cases} \tag{1.51}$$

Substitute this test function into inequality (1.50) and perform the change of variables $x = R\xi$ under the integral sign; we obtain the estimate

$$c_0 R^{N-2q'} \ge \int_{\mathbb{R}^N} |u|^q \varphi_0\left(\frac{|x|^2}{R^2}\right) dx, \quad c_0 = \int_{1 \le |\xi| \le \sqrt{2}} \frac{|\Delta_\xi \varphi_0(|\xi|)|^{q'}}{\varphi_0^{q'-1}(|\xi|)}\,d\xi. \tag{1.52}$$

Now we impose the conditions

$$N < 2q', \quad q' = \frac{q}{q-1}.$$

After this, we pass to the limit as $R \to +\infty$ in estimate (1.52). Note that using the same arguments as in Section 1.1, we can prove the existence of test functions $\varphi_0(s)$ for some finite nonlinear capacity $c_0 > 0$. Finally, from inequality (1.52) we obtain the expression

$$\int_{\mathbb{R}^N} |u|^q\,dx = 0, \tag{1.53}$$

which immediately implies the absence of nontrivial global solutions of the differential inequality (1.47).

We separately consider the critical case

$$2q' = N. \tag{1.54}$$

Let us prove that in this case there are no nontrivial global solutions. Note that due to inequality (1.52) the integral

$$\int_{\mathbb{R}^N} \varphi_0\left(\frac{|x|^2}{R^2}\right) |u|^q\,dx \tag{1.55}$$

converges for any function $\varphi_0(s)$. Consider inequalities (1.49). Note that the support of the function $|\Delta\varphi(x)|$ lies in the segment $|x| \in [\sqrt{R}, \sqrt{2R}]$, and therefore the following "clarified" inequalities hold:

$$\left| \int_{\sqrt{R} \le |x| \le \sqrt{2R}} \Delta\varphi(x) u(x)\,dx \right| \le \int_{\sqrt{R} \le |x| \le \sqrt{2R}} |u| \varphi^{1/q} \frac{|\Delta\varphi|}{\varphi^{1/q}}\,dx$$

$$\leq \Bigg(\int\limits_{\sqrt{R}\leq|x|\leq\sqrt{2R}} \varphi|u|^q \, dx \Bigg)^{1/q} \Bigg(\int\limits_{\sqrt{R}\leq|x|\leq\sqrt{2R}} \frac{|\Delta\varphi|^{q'}}{\varphi^{q'-1}} \, dx \Bigg)^{1/q'} . \tag{1.56}$$

Taking into account inequality (1.48), we arrive at the inequality

$$\begin{aligned} &\Bigg(\int\limits_{\sqrt{R}\leq|x|\leq\sqrt{2R}} \varphi|u|^q \, dx \Bigg)^{1/q} \Bigg(\int\limits_{\sqrt{R}\leq|x|\leq\sqrt{2R}} \frac{|\Delta\varphi|^{q'}}{\varphi^{q'-1}} \, dx \Bigg)^{1/q'} \\ &\quad \geq \int\limits_{\mathbb{R}^N} |u|^q \varphi(x) \, dx. \end{aligned} \tag{1.57}$$

Therefore, due to the convergence of the integral (1.55), we obtain the limit equality

$$\lim_{R\to+\infty} \int\limits_{\sqrt{R}\leq|x|\leq\sqrt{2R}} \varphi|u|^q \, dx = 0.$$

Hence, by inequality (1.57), passing to the limit as $R \to +\infty$, we have

$$\int\limits_{\mathbb{R}^N} |u|^q \, dx = 0. \tag{1.58}$$

Therefore, for the critical exponent (1.54), there are no nontrivial global solutions.

Finally, we show that the result obtained cannot be strengthened. Indeed, consider the function

$$u_0(x) = \frac{\varepsilon}{(1+|x|^2)^p}, \quad p = \frac{1}{q-1}, \; q > 1. \tag{1.59}$$

Let us prove that under the condition

$$N > 2q', \quad q' = \frac{q}{q-1}, \tag{1.60}$$

the function $u_0(x)$ is a solution of inequality (1.47) for sufficiently small $\varepsilon > 0$. Since $u_0 = u_0(r)$ for $r = |x|$, inequality (1.47) takes the form

$$-u_0'' - \frac{N-1}{r} u_0' \geq u_0^q. \tag{1.61}$$

Calculating the derivatives

$$u_0'(r) = -2p\varepsilon \frac{r}{(1+r^2)^{p+1}},$$

$$u_0''(r) = -2p\varepsilon \frac{1}{(1+r^2)^{p+1}} + p(p+1)2\varepsilon \frac{2r^2}{(1+r^2)^{p+2}},$$

we obtain the following expressions:

$$\begin{aligned} -u_0'' - \frac{N-1}{r}u_0' &= \frac{2\varepsilon p}{(1+r^2)^{p+1}}\left[N - 2(p+1)\frac{r^2}{1+r^2}\right] \ge \frac{2\varepsilon p}{(1+r^2)^{p+1}}[N-2(p+1)] \\ &= \frac{2\varepsilon p}{(1+r^2)^{p+1}}[N-2q']. \end{aligned} \tag{1.62}$$

Moreover, note that

$$u_0^q = \frac{\varepsilon^q}{(1+|x|^2)^{p+1}}. \tag{1.63}$$

Comparing equation (1.63) with inequality (1.62), we see that, under condition (1.60) for sufficiently small $\varepsilon > 0$, inequality (1.61) is valid.

1.3.2 Singular Emden–Fowler-type differential inequality

Now we consider the following differential inequality with critical degeneration in $\mathbb{R}^N \backslash \{0\}$:

$$-|x|^2 \Delta u \ge |u|^q, \tag{1.64}$$

where $q > 0$ and $N > 2$. We follow the paper [74]. First, we state the definition of a weak solution of the differential inequality (1.64).

Definition 1.4. Let $q > 1$. A function $u(x)$ is called a *weak solution* of problem (1.64) if
(i) $u \in \mathbb{L}^q_{\mathrm{loc}}(\mathbb{R}^N \backslash \{0\})$;
(ii) $u \in \mathbb{L}^1_{\mathrm{loc}}(\mathbb{R}^N \backslash \{0\})$ and

$$\int_{\mathbb{R}^N} \frac{|u|^q}{|x|^N} \varphi \, dx \le -\int_{\mathbb{R}^N} u \Delta(|x|^{2-N}\varphi) \, dx \tag{1.65}$$

for any nonnegative function $\varphi \in \mathbb{C}_0^\infty(\mathbb{R}^N \backslash \{0\})$.

Theorem 1.2. *Let* $q > 1$. *Then problem* (1.64) *has no weak nontrivial solutions.*

Proof. Assume the contrary; let u be a nontrivial weak solution of problem (1.64). From inequality (1.65) we obtain

$$\int_{\mathbb{R}^N} \frac{|u|^q}{|x|^N} \varphi \, dx \le \int_{\mathbb{R}^N} \frac{u}{|x|^N} (|x|^2 \Delta\varphi + 2(2-N)(x, \nabla\varphi)) dx. \tag{1.66}$$

Introduce the notation

$$f \equiv |x|^2 \Delta\varphi + 2(2-N)(x, \nabla\varphi).$$

Using the Hölder inequality, we arrive at the inequality

$$\int_{\mathbb{R}^N} \frac{|u|}{|x|^N}|f|\,dx = \int_{\mathbb{R}^N} \frac{|u|\varphi^{1/q}}{|x|^{N/q}} \frac{|f|}{\varphi^{1/q}|x|^{N-N/q}}\,dx$$
$$\leq \Big(\int_{\mathbb{R}^N} \frac{|u|^q}{|x|^N}\varphi\,dx\Big)^{1/q}\Big(\int_{\mathbb{R}^N} \frac{|f|^{q'}}{\varphi^{q'-1}|x|^N}\,dx\Big)^{1/q'}.$$

Due to (1.66), we obtain the inequality

$$\int_{\mathbb{R}^N} \frac{|u|^q}{|x|^N}\varphi\,dx \leq \int_{\mathbb{R}^N} \frac{||x|^2\Delta\varphi + 2(2-N)(x,\nabla\varphi)|^{q'}}{|x|^N\varphi^{q'-1}}\,dx, \quad \frac{1}{q}+\frac{1}{q'} = 1. \tag{1.67}$$

Now we assume that the right-hand side of this inequality is finite; this can be done by an appropriate choice of the function φ. Take a radial function $\varphi \in \mathbb{C}_0^\infty(\mathbb{R}^N\backslash\{0\})$. Then (1.67) implies

$$\int_{\mathbb{R}^N} \frac{|u|^q}{|x|^N}\varphi(|x|)\,dx \leq \int_{\mathbb{R}^N} \frac{|r^2\varphi_{rr} + 2(2-N)r\varphi_r|^{q'}}{r^N\varphi^{q'-1}}\,dx, \tag{1.68}$$

where $|x| = r$. Using the change of variables $s = \ln r$, $-\infty < s < +\infty$, and setting $\psi(s) = \varphi(e^s)$, we see that the right-hand side of (1.68) takes the form

$$\int_{\mathbb{R}^N} \frac{|u|^q}{|x|^N}\psi(|x|)\,dx \leq |S^{N-1}| \int_{\mathbb{R}^1} \frac{|\psi_{ss} + 2(2-N)\psi_s|^{q'}}{\psi^{q'-1}}\,ds, \tag{1.69}$$

where $|S^{N-1}|$ is the measure on the unit sphere in $\mathbb{R}^N$. Then we choose $\varphi_0 \in \mathbb{C}_0^\infty(\mathbb{R}^N)$ such that

$$\varphi_0(t) = \begin{cases} 1 & \text{if } |t| \leq 1, \\ 0 & \text{if } |t| \geq 2, \end{cases}$$

and set

$$\psi(s) = \varphi_0\Big(\frac{s}{R}\Big), \quad R > 0.$$

After the choice of variables $s = R\tau$, we see that (1.69) for $R \geq 1$ takes the form

$$\int_{\mathbb{R}^N} \frac{|u|^q}{|x|^N}\psi(|x|)\,dx \leq R^{1-q'}|S^{N-1}| \int_{\mathbb{R}^1} \frac{|\varphi_0''(\tau) + 2(2-N)\varphi_0'(\tau)|^{q'}}{\varphi_0^{q'-1}(\tau)}\,d\tau.$$

Then, due to the choice of φ_0, we have

$$\int_{-R}^{R}\int_{S^{N-1}} |u|^q\,ds\,d\omega \leq R^{1-q'}A_{\varphi_0}, \tag{1.70}$$

where

$$A_{\varphi_0} = |S^{N-1}| \int\limits_{1\le|\tau|\le 2} \frac{|\varphi_0'' + 2(2-N)\varphi_0'|^{q'}}{\varphi_0^{q'-1}}\,d\tau < +\infty.$$

Finally, tending $R \to +\infty$ in (1.70) and using the inequality $q' > 1$ (hence the right-hand side of (1.70) tends to zero), we conclude that $u \equiv 0$. The proof is complete. □

1.4 Parabolic differential inequalities

1.4.1 Example

Consider the following classical problem:

$$\frac{\partial u}{\partial t} - \Delta u \ge u^q, \quad q > 1,\ u \ge 0,\ x \in \mathbb{R}^N,\ t > 0, \tag{1.71}$$

$$u(x,0) = u_0(x) \ge 0. \tag{1.72}$$

First, we state the definition of a weak solution for this problem.

Definition 1.5. A *weak generalized solution* of problem (1.71) is a function

$$u(x,t) \in \mathbb{L}^q_{\mathrm{loc}}(\mathbb{R}^N \times \mathbb{R}_+), \quad u_0(x) \in \mathbb{L}^1(\mathbb{R}^N), \tag{1.73}$$

satisfying the inequality

$$\begin{aligned} &-\int\limits_{\mathbb{R}_+}\int\limits_{\mathbb{R}^N} \Big(\frac{\partial\varphi}{\partial t} + \Delta\varphi(x,t)\Big) u\,dx\,dt \\ &\quad - \int\limits_{\mathbb{R}^N} \varphi(x,0)u_0(x)\,dx \ge \int\limits_{\mathbb{R}_+}\int\limits_{\mathbb{R}^N} \varphi(x,t)|u|^q\,dx\,dt \end{aligned} \tag{1.74}$$

for all $\varphi(x,t) \in \mathbb{C}_0^\infty(\mathbb{R}^N \times \mathbb{R}_+)$, $\varphi(x,t) \ge 0$.

We prove that, for parabolic inequalities, the method of nonlinear capacity allows us to deduce conditions of the nonexistence of nontrivial weak generalized solutions. We have the following inequalities:

$$\begin{aligned} &\Big|\int\limits_{\mathbb{R}_+}\int\limits_{\mathbb{R}^N} \Big(\frac{\partial\varphi}{\partial t} + \Delta\varphi(x,t)\Big) u\,dx\,dt\Big| \\ &\quad \le \int\limits_{\mathbb{R}_+}\int\limits_{\mathbb{R}^N} \Big|\frac{\partial\varphi}{\partial t} + \Delta\varphi(x,t)\Big| |u|\,dx\,dt \end{aligned}$$

$$= \int_{\mathbb{R}_+}\int_{\mathbb{R}^N} \frac{|\frac{\partial\varphi}{\partial t} + \Delta\varphi(x,t)|}{\varphi^{1/q}}\varphi^{1/q}|u|\,dx\,dt$$

$$\le \left(\int_{\mathbb{R}_+}\int_{\mathbb{R}^N} \frac{|A(\varphi)|^{q'}}{\varphi^{q'-1}}\,dx\,dt\right)^{1/q'}\left(\int_{\mathbb{R}_+}\int_{\mathbb{R}^N} \varphi(x,t)|u|^q\,dx\,dt\right)^{1/q}$$

$$\le \frac{1}{q'}\int_{\mathbb{R}_+}\int_{\mathbb{R}^N} \frac{|A(\varphi)|^{q'}}{\varphi^{q'-1}}\,dx\,dt + \frac{1}{q}\int_{\mathbb{R}_+}\int_{\mathbb{R}^N} \varphi(x,t)|u|^q\,dx\,dt, \tag{1.75}$$

where

$$A(\varphi) = \frac{\partial\varphi}{\partial t} + \Delta\varphi(x,t).$$

Then (1.74) and (1.75) imply the inequality

$$\int_{\mathbb{R}_+}\int_{\mathbb{R}^N} \frac{|A(\varphi)|^{q'}}{\varphi^{q'-1}}\,dx\,dt - q'\int_{\mathbb{R}^N} \varphi(x,0)u_0(x)\,dx \ge \int_{\mathbb{R}_+}\int_{\mathbb{R}^N} \varphi(x,t)|u|^q\,dx\,dt. \tag{1.76}$$

Now we choose a test function $\varphi(x,t)$ of the form

$$\varphi(x) = \varphi_0\left(\frac{t}{R^2} + \frac{|x|^2}{R^2}\right), \quad \varphi_0(s) = \begin{cases} 1 & \text{if } 0 \le s \le 1, \\ 0 & \text{if } s \ge 2, \end{cases} \tag{1.77}$$

and consider the nonlinear capacity obtained:

$$\int_{\mathbb{R}_+}\int_{\mathbb{R}^N} \frac{|A(\varphi)|^{q'}}{\varphi^{q'-1}}\,dx\,dt. \tag{1.78}$$

After the change of variables $t = R^2\tau$, $x = R\xi$, we obtain the expression

$$\int_{\mathbb{R}_+}\int_{\mathbb{R}^N} \frac{|A(\varphi)|^{q'}}{\varphi^{q'-1}}\,dx\,dt = c_0 R^{N+2-2q'}, \tag{1.79}$$

where

$$c_0 = \int_{1\le\tau+|\xi|\le\sqrt{2}} \frac{|A(\varphi_0)|^{q'}}{\varphi_0^{q'-1}}\,d\tau\,d\xi$$

is the integral of the nonlinear capacity. As before, we can choose a test function $\varphi_0(s)$ such that the nonlinear capacity $c_0 > 0$ is finite. Therefore, inequality (1.76) implies the estimate

$$c_0 R^{N+2-2q'} - q'\int_{\mathbb{R}^N} \varphi(x,0)u_0(x)\,dx \ge \int_{\mathbb{R}_+}\int_{\mathbb{R}^N} \varphi(x,t)|u|^q\,dx\,dt. \tag{1.80}$$

Therefore, under the condition $N + 2 < 2q'$, passing in inequality (1.80) to the limit as $R \to +\infty$, we get the inequality

$$-q' \int_{\mathbb{R}^N} u_0(x)\,dx \geq \int_{\mathbb{R}_+} \int_{\mathbb{R}^N} \varphi(x,t)|u|^q\,dx\,dt, \tag{1.81}$$

which implies that, under the additional condition on the initial function

$$\int_{\mathbb{R}^N} u_0(x)\,dx \geq 0,$$

which obviously holds in the cases considered, there are no nontrivial global solutions of problem (1.71)–(1.72).

Now we discuss the problem on the nonexistence of global solutions of problem (1.71)–(1.72) for the critical parameter

$$N + 2 = 2q'.$$

To this end, as before, we note that the support of the function $A(\varphi)(x,t)$ is contained in the segment

$$R^2 \leq t + |x|^2 \leq 2R^2.$$

Therefore, inequalities (1.75) imply that

$$\begin{aligned}
&\left| \int_{\mathbb{R}_+} \int_{\mathbb{R}^N} \left(\frac{\partial\varphi}{\partial t} + \Delta\varphi(x,t) \right) u\,dx\,dt \right| \\
&\quad \leq \left(\int_{R^2 \leq t+|x|^2 \leq 2R^2} \frac{|A(\varphi)|^{q'}}{\varphi^{q'-1}}\,dx\,dt \right)^{1/q'} \\
&\qquad \times \left(\int_{R^2 \leq t+|x|^2 \leq 2R^2} \varphi(x,t)|u|^q\,dx\,dt \right)^{1/q} - \int_{\mathbb{R}^N} \varphi(x,0)u_0(x)\,dx = \mathrm{I},
\end{aligned} \tag{1.82}$$

where

$$\mathrm{I} \geq \int_{\mathbb{R}_+} \int_{\mathbb{R}^N} \varphi(x,t)|u|^q\,dx\,dt.$$

Due to inequality (1.80), we conclude that the integral in the right-hand side

$$\int_{\mathbb{R}_+} \int_{\mathbb{R}^N} \varphi(x,t)|u|^q\,dx\,dt \tag{1.83}$$

converges. Hence the following limit relation is valid:

$$\lim_{R\to+\infty} \int\limits_{R^2\le t+|x|^2\le 2R^2} \varphi(x,t)|u|^q\,dx\,dt = 0. \tag{1.84}$$

Thus, passing to the limit as $R \to +\infty$, we arrive at the inequality

$$-\int\limits_{\mathbb{R}^N} u_0(x)\,dx \ge \int\limits_{\mathbb{R}_+}\int\limits_{\mathbb{R}^N} |u|^q\,dx\,dt. \tag{1.85}$$

This is impossible under the condition

$$\int\limits_{\mathbb{R}^N} u_0(x)\,dx \ge 0,$$

which obviously holds in the case considered.

1.4.2 Singular Emden–Fowler-type differential inequality

Now we consider a singular parabolic problem. Let $N \ge 2$, $q > 1$, and

$$\begin{aligned} &\frac{\partial u}{\partial t} - |x|^2\Delta u \ge |u|^q \quad \text{in } \mathbb{R}^N\backslash\{0\}\times\mathbb{R}_+, \\ &u(x,0) = u_0(x) \in \mathbb{L}^1_{\mathrm{loc}}(\mathbb{R}^N\backslash\{0\}) \quad \text{in } \mathbb{R}^N\backslash\{0\}. \end{aligned} \tag{1.86}$$

We follow the way described in [74].

Definition 1.6. A function u is called a *weak generalized solution* of problem (1.86) if

$$u \in \mathbb{L}^q_{\mathrm{loc}}(\mathbb{R}^N\backslash\{0\}\times\mathbb{R}_+)$$

and

$$\begin{aligned} \int\limits_0^{+\infty}\int\limits_{\mathbb{R}^N} \frac{|u|^q}{|x|^N}\varphi\,dx\,dt \le &-\int\limits_0^{+\infty}\int\limits_{\mathbb{R}^N} u\Delta(|x|^{2-N}\varphi)\,dx\,dt \\ &-\int\limits_0^{+\infty}\int\limits_{\mathbb{R}^N} \frac{u}{|x|^N}\frac{\partial\varphi}{\partial t}\,dx\,dt - \int\limits_{\mathbb{R}^N} \frac{u_0(x)}{|x|^N}\varphi(x,0)\,dx \end{aligned} \tag{1.87}$$

for any nonnegative function $\varphi \in \mathbb{C}_0^\infty(\mathbb{R}^N\backslash\{0\}\times\mathbb{R}_+)$.

Theorem 1.3. *Let the following conditions hold:*

(i) $u_0(x) \in \mathbb{L}^1_{\mathrm{loc}}(\mathbb{R}^N\backslash\{0\})$,

(ii) $\liminf\limits_{R\to+\infty} \int\limits_{e^{-R}\le|x|\le e^R} \frac{u_0(x)}{|x|^N}\,dx \ge 0$ *(perhaps, infinite).*

If $1 < q < 3$, *then problem* (1.86) *has no weak generalized solutions.*

Proof. First, we note that

$$\Delta(|x|^{2-N}\varphi) = \frac{1}{|x|^N}(|x|^2\Delta\varphi + 2(2-N)(x,\nabla\varphi)).$$

Let

$$f \equiv |x|^2\Delta\varphi + 2(2-N)(x,\nabla\varphi) + \varphi'.$$

By the Hölder and Young inequalities we obtain the following inequalities:

$$\begin{aligned}
\int_0^{+\infty}\int_{\mathbb{R}^N} \frac{uf}{|x|^N}\,dx\,dt &= \int_0^{+\infty}\int_{\mathbb{R}^N} \frac{|u|}{|x|^{N/q}}\varphi^{1/q}\frac{f}{|x|^{N(1-1/q)}}\frac{1}{\varphi^{1/q}}\,dx\,dt \\
&\le \Bigg(\int_0^{+\infty}\int_{\mathbb{R}^N} \frac{|u|^q\varphi}{|x|^N}\,dx\,dt\Bigg)^{1/q}\Bigg(\int_0^{+\infty}\int_{\mathbb{R}^N} \frac{|f|^{q'}}{|x|^N\varphi^{q'-1}}\,dx\,dt\Bigg)^{1/q'} \\
&\le \frac{1}{q}\int_0^{+\infty}\int_{\mathbb{R}^N} \frac{|u|^q\varphi}{|x|^N}\,dx\,dt + \frac{1}{q'}\int_0^{+\infty}\int_{\mathbb{R}^N} \frac{|f|^{q'}}{|x|^N\varphi^{q'-1}}\,dx\,dt.
\end{aligned}$$

Under condition (1.87), this implies

$$\begin{aligned}
\int_0^{+\infty}\int_{\mathbb{R}^N} \frac{|u|^q\varphi}{|x|^N}\,dx\,dt \le{}& \frac{1}{q}\int_0^{+\infty}\int_{\mathbb{R}^N} \frac{|u|^q\varphi}{|x|^N}\,dx\,dt + \frac{1}{q'}\int_0^{+\infty}\int_{\mathbb{R}^N} \frac{|f|^{q'}}{|x|^N\varphi^{q'-1}}\,dx\,dt \\
&- \int_{\mathbb{R}^N} \frac{u_0(x)}{|x|^N}\varphi(x,0)\,dx.
\end{aligned}$$

Then we obtain the estimate

$$\begin{aligned}
\int_0^{+\infty}\int_{\mathbb{R}^N} \frac{|u|^q\varphi}{|x|^N}\,dx\,dt \le{}& \int_0^{+\infty}\int_{\mathbb{R}^N} \frac{||x|^2\Delta\varphi + 2(2-N)(x,\nabla\varphi) + \varphi'|^{q'}}{|x|^N\varphi^{q'-1}}\,dx\,dt \\
&- q'\int_{\mathbb{R}^N} \frac{u_0(x)}{|x|^N}\varphi(x,0)\,dx,
\end{aligned} \tag{1.88}$$

where

$$\frac{1}{q} + \frac{1}{q'} = 1 \quad\text{and}\quad 0 \le \varphi \in C_0^\infty(\mathbb{R}^N\backslash\{0\} \times \mathbb{R}_+).$$

Choosing $\varphi(x,t) = \psi(|x|,t)$ and performing the change of variables $s = \ln r$, $r = |x|$, $s \in \mathbb{R}^1$, from (1.88) we obtain

$$\int_0^{+\infty}\int_{\mathbb{R}^N} \frac{|u|^q\varphi(x,t)}{|x|^N}\,dx\,dt \le |S^{N-1}|\int_0^{+\infty}\int_{\mathbb{R}^1} \frac{|\psi_{ss} + 2(2-N)\psi_s + \psi'|^{q'}}{|x|^N\psi^{q'-1}}\,ds\,dt$$

$$-q'\int_{\mathbb{R}^1}\int_{S^{N-1}} u_0(e^s,\omega)\psi(e^s,0)\,ds\,d\omega, \tag{1.89}$$

where

$$\omega = \frac{x}{|x|}, \quad x \neq 0.$$

Now we specify the choice of the function φ:

$$\psi(e^s,t) = \psi_0\left(\frac{t}{R^2}\right)\psi_1\left(\frac{a(s,t)}{R}\right),$$

where ψ_0, ψ_1, and $a(s,t)$ are sufficiently smooth functions defined for $R > 0$.

Introduce the operator

$$\Gamma = \frac{\partial^2}{\partial s^2} + 2(2-N)\frac{\partial}{\partial s} + \frac{\partial}{\partial t}$$

acting on functions of the class $\mathbb{C}_0^\infty(\mathbb{R}^2)$. We have

$$\begin{aligned}\Gamma(\varphi(s,t)) = {}& \frac{1}{R^2}\left[\psi_0\left(\frac{t}{R^2}\right)\psi_1''\left(\frac{a(s,t)}{R}\right)a_s^2(s,t) + \psi_0'\left(\frac{t}{R^2}\right)\psi_1\left(\frac{a(s,t)}{R}\right)\right]\\ &+ \frac{1}{R}[a_{ss}(s,t) + 2(2-N)a_s(s,t) + a_t(s,t)]\psi_0\left(\frac{t}{R^2}\right)\psi_1'\left(\frac{a(s,t)}{R}\right).\end{aligned}$$

Take a function $a(s,t)$ such that

$$a_{ss}(s,t) + 2(2-N)a_s(s,t) + a_t(s,t) = 0;$$

for example, let $a(s,t) = s + 2(N-2)t$. Next, we take $\psi_0, \psi_1 \in \mathbb{C}_0^\infty(\mathbb{R})$ such that

$$\psi_0(y) = \begin{cases} 1 & \text{if } 0 \le y \le 1,\\ 0 & \text{if } y \ge 2,\end{cases} \tag{1.90}$$

$$\psi_1(y) = \begin{cases} 1 & \text{if } |y| \le 1,\\ 0 & \text{if } |y| \ge 2,\end{cases} \tag{1.91}$$

and set

$$\psi(e^s,t) = \psi_0\left(\frac{t}{R^2}\right)\psi_1\left(\frac{s+2(N-2)t}{R}\right).$$

After the change of variables $t = R^2\tau$, $s = R\xi$, we obtain

$$\psi(e^s,t) = \psi_0(\tau)\psi_1(\xi + 2(N-2)R\tau)$$

and, moreover,

$$\Gamma_{\xi,\tau}(\psi) = \frac{1}{R^2}[\psi_0(\tau)\psi_1''(\xi + 2(N-2)\tau) + \psi_0'(\tau)\psi_1(\xi + 2(N-2)R\tau)],$$

$$\int\limits_{A_{R,t}} \frac{|u|^q}{|x|^N}\,dx\,dt \le C' R^{3-2q'} - q' \int\limits_{-\infty}^{+\infty} \int\limits_{S^{N-1}} u_0(e^s,\omega)\psi_1\left(\frac{s}{R}\right) ds\,d\omega, \tag{1.92}$$

where

$$\begin{aligned}
A_{R,t} &\equiv \{(x,t) \in \mathbb{R}^N \times \mathbb{R}_+ : \big|\ln|x| + 2(2-N)t\big| \le R\},\\
C' &\equiv |S^{N-1}| \iint\limits_{B_{\xi,\tau}} |\psi_0(\tau)\psi_1''(\xi + 2(N-2)\tau)\\
&\qquad + \psi_0'(\tau)\psi_1(\xi + 2(N-2)R\tau)|^{q'} \psi^{1-q'}\,d\tau\,d\xi,\\
B_{\xi,\tau} &\equiv \{(\xi,\tau) \in \mathbb{R} \times \mathbb{R}_+ : |\xi + 2(2-N)R\tau| \le 2\}.
\end{aligned}$$

In a standard way, we obtain $C' < +\infty$. Therefore

$$\int\limits_0^{+\infty} \int\limits_{\mathbb{R}^N} \frac{|u|^q}{|x|^N}\,dx\,dt \le -q' \liminf_{R\to+\infty} \int\limits_{e^{-R}\le|x|\le e^R} \frac{u_0(x)}{|x|^N}\,dx \le 0. \tag{1.93}$$

We see that in the case $1 < q < 3$, inequality (1.93) yields the required assertion. □

1.5 Hyperbolic differential inequalities

1.5.1 Example

Consider the following problem:

$$\frac{\partial^2 u}{\partial t^2} - \Delta u \ge |u|^q, \quad q > 1,\ x \in \mathbb{R}^N,\ t > 0, \tag{1.94}$$

$$u(x,0) = u_0(x), \quad u'(x,0) = u_1(x). \tag{1.95}$$

We state the definition of a weak generalized solution.

Definition 1.7. A *weak generalized solution* of problem (1.94)–(1.95) is a function $u(x,t)$ of the class

$$u(x,t) \in \mathbb{L}^q_{\mathrm{loc}}(\mathbb{R}^N \times \mathbb{R}_+), \quad u_0(x), u_1(x) \in \mathbb{L}^1_{\mathrm{loc}}(\mathbb{R}^N),$$

satisfying the inequality

$$\int\limits_{\mathbb{R}^N} (u_0(x)\varphi'(0,x) - u_1(x)\varphi(0,x))\,dx$$

$$+\int_{\mathbb{R}_+}\int_{\mathbb{R}^N} \varphi'' u\,dx\,dt - \int_{\mathbb{R}_+}\int_{\mathbb{R}^N} \Delta\varphi\, u\,dx\,dt$$
$$\geq \int_{\mathbb{R}_+}\int_{\mathbb{R}^N} \varphi(x,t)|u|^q\,dx\,dt \tag{1.96}$$

for all $\varphi(x,t) \in \mathbb{C}_0^\infty(\mathbb{R}^N \times \mathbb{R}_+)$.

Recall the Young inequality with a parameter:

$$ab \leq \frac{\varepsilon}{q}a^q + \frac{1}{q'}\frac{1}{\varepsilon^{q'-1}}b^{q'}, \quad \frac{1}{q}+\frac{1}{q'} = 1. \tag{1.97}$$

First, we consider the second integral in (1.96),

$$\int_{\mathbb{R}_+}\int_{\mathbb{R}^N} \varphi'' u\,dx\,dt. \tag{1.98}$$

We have

$$\left|\int_{\mathbb{R}_+}\int_{\mathbb{R}^N} \varphi'' u\,dx\,dt\right| \leq \int_{\mathbb{R}_+}\int_{\mathbb{R}^N} |\varphi''||u|\,dx\,dt = \int_{\mathbb{R}_+}\int_{\mathbb{R}^N} \frac{|\varphi''|}{\varphi^{1/q}}|u|\varphi^{1/q}\,dx\,dt$$
$$\leq \left(\int_{\mathbb{R}_+}\int_{\mathbb{R}^N} \frac{|\varphi''|^{q'}}{\varphi^{q'-1}}\,dx\,dt\right)^{1/q}\left(\int_{\mathbb{R}_+}\int_{\mathbb{R}^N} |u|^q\varphi\,dx\,dt\right)^{1/q'}$$
$$\leq \frac{\varepsilon}{q}\int_{\mathbb{R}_+}\int_{\mathbb{R}^N} |u|^q\varphi\,dx\,dt + \frac{1}{q'}\frac{1}{\varepsilon^{q'-1}}\int_{\mathbb{R}_+}\int_{\mathbb{R}^N} \frac{|\varphi''|^{q'}}{\varphi^{q'-1}}\,dx\,dt. \tag{1.99}$$

Now we consider the third integral in (1.96),

$$\int_{\mathbb{R}_+}\int_{\mathbb{R}^N} \Delta\varphi u\,dx\,dt. \tag{1.100}$$

We have the following inequalities:

$$\left|\int_{\mathbb{R}_+}\int_{\mathbb{R}^N} \Delta\varphi u\,dx\,dt\right| \leq \int_{\mathbb{R}_+}\int_{\mathbb{R}^N} |\Delta\varphi||u|\,dx\,dt = \int_{\mathbb{R}_+}\int_{\mathbb{R}^N} \frac{|\Delta\varphi|}{\varphi^{1/q}}\varphi^{1/q}|u|\,dx\,dt$$
$$\leq \left(\int_{\mathbb{R}_+}\int_{\mathbb{R}^N} \varphi|u|^q\,dx\,dt\right)^{1/q}\left(\int_{\mathbb{R}_+}\int_{\mathbb{R}^N} \frac{|\Delta\varphi|^{q'}}{\varphi^{q'-1}}\,dx\,dt\right)^{1/q'}$$
$$\leq \frac{\varepsilon}{q}\int_{\mathbb{R}_+}\int_{\mathbb{R}^N} \varphi|u|^q\,dx\,dt + \frac{1}{q'}\frac{1}{\varepsilon^{q'-1}}\int_{\mathbb{R}_+}\int_{\mathbb{R}^N} \frac{|\Delta\varphi|^{q'}}{\varphi^{q'-1}}\,dx\,dt. \tag{1.101}$$

From inequalities (1.99), (1.101), and (1.96) we obtain the estimate

$$\int_{\mathbb{R}^N} (u_0(x)\varphi'(0,x) - u_1(x)\varphi(0,x))dx$$
$$+ \frac{1}{q'}\frac{1}{\varepsilon^{q'-1}} \int_{\mathbb{R}_+}\int_{\mathbb{R}^N} \frac{|\varphi''|^{q'}}{\varphi^{q'-1}}\,dx\,dt + \frac{1}{q'}\frac{1}{\varepsilon^{q'-1}} \int_{\mathbb{R}_+}\int_{\mathbb{R}^N} \frac{|\Delta\varphi|^{q'}}{\varphi^{q'-1}}\,dx\,dt$$
$$\geq \left(1 - \frac{2\varepsilon}{q}\right) \int_{\mathbb{R}_+}\int_{\mathbb{R}^N} \varphi(x,t)|u|^q\,dx\,dt. \tag{1.102}$$

Now we take $\varepsilon = q/4$. Moreover, we choose a test function $\varphi(x,t)$ so that $\varphi'(x,0) = 0$. Then from (1.102) we obtain the inequality

$$-\int_{\mathbb{R}^N} u_1(x)\varphi(0,x)\,dx$$
$$+ \frac{1}{q'}\frac{1}{\varepsilon^{q'-1}} \int_{\mathbb{R}_+}\int_{\mathbb{R}^N} \frac{|\varphi''|^{q'}}{\varphi^{q'-1}}\,dx\,dt + \frac{1}{q'}\frac{1}{\varepsilon^{q'-1}} \int_{\mathbb{R}_+}\int_{\mathbb{R}^N} \frac{|\Delta\varphi|^{q'}}{\varphi^{q'-1}}\,dx\,dt$$
$$\geq \frac{1}{2} \int_{\mathbb{R}_+}\int_{\mathbb{R}^N} \varphi(x,t)|u|^q\,dx\,dt. \tag{1.103}$$

Introduce the shear function

$$\varphi(t,x) \equiv \varphi_0\left(\frac{t^2 + |x|^2}{R^2}\right), \tag{1.104}$$

where $\varphi_0 : \mathbb{R}_+ \to \mathbb{R}_+$ is a smooth nonnegative function such that

$$\varphi_0(\xi) = \begin{cases} 1 & \text{if } 0 \leq \xi \leq 1, \\ 0 & \text{if } \xi \geq 2. \end{cases} \tag{1.105}$$

Taking into account (1.104), we perform the change of variables $(t,x) \to (\tau,\eta)$ by the formulas

$$t = R\tau, \quad x = R\eta. \tag{1.106}$$

Substituting the function φ_0 into (1.103) and using the new variables (1.106), we obtain

$$\frac{1}{2} \int_{\mathbb{R}_+}\int_{\mathbb{R}^N} |u|^q\varphi\,dx\,dt \leq c_1R^\theta + c_2R^\theta - \int_{\mathbb{R}^N} u_1(x)\varphi(0,x)\,dx, \tag{1.107}$$

where $\theta = N + 1 - 2q'$,

$$c_1 = \frac{1}{q'}\frac{1}{\varepsilon^{q'-1}} \int_{1\leq\tau+|\eta|\leq\sqrt{2}} \frac{|\varphi_0''|^{q'}}{\varphi_0^{q'-1}}\,d\eta\,d\tau,$$

$$c_2 = \frac{1}{q'} \frac{1}{\varepsilon^{q'-1}} \int\limits_{1 \le \tau + |\eta| \le \sqrt{2}} \frac{|\Delta \varphi_0|^{q'}}{\varphi_0^{q'-1}} \, d\eta \, d\tau.$$

Now we assume that

$$N + 1 < 2q'. \tag{1.108}$$

Passing to the limit as $R \to +\infty$ in inequality (1.107), we obtain the inequality

$$\frac{1}{2} \int\limits_{\mathbb{R}_+} \int\limits_{\mathbb{R}^N} |u|^q \, dx \, dt \le - \int\limits_{\mathbb{R}^N} u_1(x) \, dx. \tag{1.109}$$

Under the additional condition

$$\int\limits_{\mathbb{R}^N} u_1(x) \, dx \ge 0,$$

inequality (1.109) implies that there are no global nontrivial solutions of problem (1.94)–(1.95). In a standard way we can easily show that, under the condition

$$N + 1 = 2q', \tag{1.110}$$

nontrivial global solutions of the problem do not exist.

1.5.2 Problem with nonlocal nonlinearity

Consider the Cauchy problem for the following equation with nonlocal nonlinearity:

$$\begin{gathered} \frac{\partial^2 u}{\partial t^2} - a\Big(\int\limits_{\mathbb{R}^N} |\nabla u|^2 \, dx \Big) \Delta u \ge |u|^q \quad \text{in } \mathbb{R}_+^{N+1}, \\ u(0,x) = u_0(x), \quad \frac{\partial u}{\partial t}(0,x) = u_1(x), \quad x \in \mathbb{R}^N, \ q > 1, \end{gathered} \tag{1.111}$$

where $a : \mathbb{R}_+ \to \mathbb{R}$ is a continuous bounded function such that

$$|a(s)| \le c_0 \quad \forall s \in \mathbb{R}_+. \tag{1.112}$$

As usual, we assume that the initial data satisfy the following conditions:

$$u_0, u_1 \in \mathbb{L}^1_{\text{loc}}(\mathbb{R}^N), \quad \int\limits_{\mathbb{R}^N} u_1 \, dx \ge 0. \tag{1.113}$$

We follow the paper [74]. First, we state the definition of a weak solution of problem (1.111).

Definition 1.8. A *weak generalized solution* of problem (1.111) is a function

$$u \in \mathbb{L}^q_{\mathrm{loc}}(\mathbb{R}^{N+1}_+), \quad \nabla u \in \mathbb{L}^2_{\mathrm{loc}}(\mathbb{R}^{N+1}_+)$$

satisfying the condition

$$\iint\limits_{\mathbb{R}^{N+1}_+} |u|^q \varphi \, dt\, dx \le \iint\limits_{\mathbb{R}^{N+1}_+} u\varphi'' \, dt\, dx - \iint\limits_{\mathbb{R}^{N+1}_+} a\Big(\int\limits_{\mathbb{R}^N} |\nabla u|^2 \, dx \Big) u \Delta\varphi \, dt\, dx + \int\limits_{\mathbb{R}^N} (u_1(x)\varphi(0,x) - u_0(x)\varphi'(0,x)) \, dx \tag{1.114}$$

for any nonnegative shear function $\varphi \in \mathbb{C}^{2,2}_0(\mathbb{R}^{N+1}_+)$.

Note that, due to (1.112), the coefficient $a(\cdot)$ has sense in the case where

$$\int\limits_{\mathbb{R}^N} |\nabla u|^2 dx = +\infty.$$

Theorem 1.4. *Let conditions* (1.112) *and* (1.113) *hold. Then problem* (1.111) *has no global nontrivial generalized solutions of the class specified for any* $q > 1$ *in the case* $N = 1$ *or for*

$$1 < q \le \frac{N+1}{N-1}$$

in the case $N > 1$.

Proof. By inequality (1.114) and condition (1.112) we have

$$\iint\limits_{\mathbb{R}^{N+1}_+} |u|^q \varphi \, dt\, dx \le \iint\limits_{\mathbb{R}^{N+1}_+} u\varphi'' \, dt\, dx + c_0 \iint\limits_{\mathbb{R}^{N+1}_+} |u||\Delta\varphi| \, dt\, dx + \int\limits_{\mathbb{R}^N} (u_1(x)\varphi(0,x) - u_0(x)\varphi'(0,x)) \, dx. \tag{1.115}$$

Denote

$$f = |\varphi''| + c_0|\Delta\varphi|.$$

Then

$$\iint\limits_{\mathbb{R}^{N+1}_+} |u| f \, dt\, dx = \iint\limits_{\mathbb{R}^{N+1}_+} |u|\varphi^{1/q} \frac{f}{\varphi^{1/q}} \, dt\, dx \le \Big(\iint\limits_{\mathbb{R}^{N+1}_+} |u|^q \varphi \, dt\, dx \Big)^{1/q} \Big(\iint\limits_{\mathbb{R}^{N+1}_+} \frac{f^{q'}}{\varphi^{q'-1}} \, dt\, dx \Big)^{1/q'}.$$

Due to (1.115), we obtain the inequality

$$\iint\limits_{\mathbb{R}_+^{N+1}} |u|^q \varphi \, dt \, dx \le \iint\limits_{\mathbb{R}_+^{N+1}} \frac{(|\varphi''| + c_0|\Delta\varphi|)^{q'}}{\varphi^{q'-1}} \, dt \, dx + q' \int\limits_{\mathbb{R}^N} (u_1(x)\varphi(0,x) - u_0(x)\varphi'(0,x)) \, dx. \tag{1.116}$$

We take the shear function of the form

$$\varphi(t,x) \equiv \varphi_0\left(\frac{t^2 + |x|^2}{R^2}\right), \tag{1.117}$$

where $\varphi_0 \in \mathbb{C}_0^{(2)}(\mathbb{R}_+)$ is such that

$$\varphi_0(\xi) = \begin{cases} 1 & \text{if } 0 \le \xi \le 1, \\ 0 & \text{if } \xi \ge 2, \end{cases} \tag{1.118}$$

and $\varphi_0 \ge 0$. Note that $\varphi'(0,x) = 0$.

Substituting this function φ into (1.116) and performing the change of variables $(t,x) \to (\tau,\eta)$, where $t = R\tau$ and $x = R\eta$, we arrive at the inequality

$$\iint\limits_{\mathbb{R}_+^{N+1}} |u|^q \varphi \, dt \, dx \le c_1 R^{N+1-2q'} - q' \int\limits_{\mathbb{R}^N} (u_1(x)\varphi(0,x) - u_0(x)\varphi'(0,x)) \, dx, \tag{1.119}$$

where

$$c_1 = \iint\limits_{1 \le \tau^2 + |\eta|^2 \le 2} \frac{||\varphi_0''| + c_0|\Delta_\eta \varphi_0||^{q'}}{\varphi_0^{q'-1}} \, d\tau \, d\eta.$$

Clearly, there exists a function φ_0 of the required form such that $c_1 < +\infty$.

Analyzing inequality (1.119) in a standard way under the condition $N+1 < 2q'$ and applying the usual arguments in the limit case $N + 1 = 2q'$, we complete the proof of Theorem 1.4. □

1.6 Bibliographical notes

In this chapter, we have presented the results obtained in [74, 35]. Note that, apparently, the method of test functions if the basic method for the study of the blow-up of solutions to nonlinear partial differential equations. However, an essential applicability condition is the existence of a lower estimate of the "source" in the equation:

$$f(x,u) \ge c|u|^q, \quad q > 1.$$

However, S. I. Pokhozhaev in his last works applied the method of test functions to sign-alternating solutions of the Cauchy problem for the semilinear parabolic equation

$$\frac{\partial u}{\partial t} = \Delta u + |u|^{q-1}u, \quad q > 1,$$

and obtained the values of critical exponents.

2 Method of self-similar solutions of V. A. Galaktionov

The method of self-similar solutions proposed by V. A. Galaktionov is designed for the study of the blow-up of solutions to quasilinear parabolic equations. Equations of such type are used in various models of mechanics, physics, biology, and ecology; for example, they describe processes of electron and ion heat conduction in plasma, adiabatic filtration of gases in porous media, processes of chemical kinetics, migration, population growth, etc.

This method is based on the analysis of self-similar solutions and comparison theorems. Self-similar solutions can be considered as a certain "basis" of a wide class of arbitrary solutions; comparison theorems characterize a peculiar "monotonicity" of solutions with respect to initial and boundary functions. Unfortunately, being restricted by the theme of the global insolvability, we have to omit the presentation of a significant part of possible applications of the method of self-similar solutions (see [31]). Moreover, our presentation of comparison theorems and maximum principles for linear and quasilinear parabolic equations is rather laconic; the reader can find more details in [31, 90, 21].

In this chapter, we examine the space-time structure and obtain conditions for appearance of unbounded solutions to the Cauchy problem for the following quasilinear equation with power nonlinearities:

$$u_t = \nabla(u^\sigma \nabla u) + u^\beta, \quad t > 0,\ x \in \mathbb{R}^N, \tag{2.1}$$

$$u(0,x) = u_0(x) \geq 0, \quad u_0^{\sigma+1} \in \mathbb{C}^{(1)}(\mathbb{R}^N), \quad \sigma > 0,\ \beta > 0. \tag{2.2}$$

By this example we demonstrate one of the most important advantages of the method of self-similar solutions, namely the possibility of the study of localization of unbounded solutions. Localization of a solution of problem (2.1)–(2.2) means the boundedness of the domain

$$\Omega_L = \Big\{x \in \mathbb{R}^N \;\Big|\; u(T_0^-, x) \equiv \lim_{t\to T_0^-} u(t,x) > 0\Big\}.$$

Certainly, a localized collapsing solution can unboundedly increase as $t \to T_0^-$ in a domain ω_L of finite dimension, which differs from the domain Ω_L.

Traditionally, solutions are divided into three classes: localized solutions of the S-mode for which $0 < \operatorname{meas}\omega_L < \infty$; solutions of the LS-mode that infinitely increase at a single point and are bounded from above uniformly with respect to $t \in (0, T_0)$ at other points, and nonlocalized unbounded solutions of the HS-mode. Using the method of self-similar solutions, we can prove that in problem (2.1)–(2.2) the combustion S-mode corresponds to $\beta = \sigma + 1$, the LS-mode to $\beta > \sigma + 1$, and the HS-mode (where localization does not occur) to $1 < \beta < \sigma + 1$.

https://doi.org/10.1515/9783110602074-002

2.1 Basic idea of the method

We demonstrate the basic idea of the method by an example of a simple linear boundary-value problem in a half-space. We further show that this study allows us to discover important properties of a sufficiently wide family of nonlinear problems. We discuss the notion of the asymptotic stability of self-similar solutions and prove that they are not only particular solutions that appear thanks to a happy coincidence, but also "centers of attraction" of a wide variety of solutions of the given equation and, moreover, other parabolic equations obtained from it by nonlinear perturbations. We follow the paper [31].

For the linear heat equation

$$u_t = u_{xx}, \quad t > 0,\ x > 0, \tag{2.3}$$

we consider the following problem:

$$u(0,x) = u_0(x) \geq 0, \quad u(t,0) = u_1(t) > 0. \tag{2.4}$$

Assume that the function $u_0(x)$ is Lipschitz-continuous for $x \geq 0$ and $\sup u_0 < \infty$.

Problem (2.3)–(2.4) describes the process of heat conduction with constant coefficient of thermal conductivity. We describe the process of heat conduction, obtain the law of motion of the heat wave, and find the dependence of the depth of penetration of heat on time. (By the depth of penetration we mean the value $x_{ef}(t)$ such that $u(t, x_{ef}(t)) = u(t,0)/2$.) The solution of problem (2.3)–(2.4) can be expressed through the heat potentials, which, however, do not allow us to obtain required characteristics of the process (see [31, 55]).

We proceed as follows. Consider the boundary regime of the following special form:

$$u_1(t) = (1+t)^m, \quad m > 0,\ t > 0. \tag{2.5}$$

For this boundary condition, problem (2.3)–(2.4) has the self-similar solution

$$u_A(t,x) = (1+t)^m \theta_A(\xi), \quad \xi = \frac{x}{\sqrt{1+t}}, \tag{2.6}$$

and the function $\theta_A(\xi)$ is a solution of the equation

$$\theta_A'' + \frac{1}{2}\theta_A'\xi - m\theta_A = 0, \quad \theta_A(0) = 1, \quad \theta_A(\infty) = 0. \tag{2.7}$$

Thus, the construction of a self-similar solution is reduced to the ordinary differential equation (2.7), whose solution exists, is unique, monotone, and strictly positive:

$$\theta_A(\xi) = 2^{2m+1}\frac{\Gamma(1+m)}{\pi^{1/2}} \exp\left\{-\frac{\xi^2}{4}\right\} H_{-(2m+1)}\left(\frac{\xi}{2}\right), \tag{2.8}$$

where $H_\nu(z)$ is the Hermite function (see [78]). The self-similar solution constructed has a simple space-time structure; it allows us to find the dependence on time of the depth of penetration of a heat wave:

$$x_{ef}^A(t) = \xi_{ef}\sqrt{1+t}, \quad \text{where } \theta_A(\xi_{ef}) = \frac{\theta_A(0)}{2} = \frac{1}{2}.$$

We further need comparison theorems; for brevity, we present only their statements (the interested reader can find proofs in [21, 90]). We state these theorems for boundary-value problems, but they can also be easily reformulated for Cauchy problems.

Consider the following problem for a quasilinear parabolic equation:

$$u_t = B(u) = \nabla(k(u)\nabla u) + Q(u), \tag{2.9}$$

$$u(0,x) = u_0(x) \geq 0, \quad x \in \Omega, \tag{2.10}$$

$$u(t,x) = u_1(t,x) \geq 0, \quad t \in (0,T),\ x \in \partial\Omega, \tag{2.11}$$

where

$$k(u) \in \mathbb{C}^{(2)}((0,\infty)) \cap \mathbb{C}([0,\infty)), \quad Q(u) \in \mathbb{C}^{(1)}([0,\infty)),$$
$$u_0 \in \mathbb{C}(\Omega), \quad \sup u_0 < \infty, \quad u_1 \in \mathbb{C}([0,T) \times \partial\Omega), \quad \sup u_1 < \infty.$$

Proposition 2.1 (first comparison theorem). *Let $u^{(1)}$ and $u^{(2)}$ be nonnegative classical solutions of equation* (2.9) *in $(0,T) \times \Omega$ such that*

$$u^{(2)}(0,x) \geq u^{(1)}(0,x) \quad \text{for } x \in \Omega,$$
$$u^{(2)}(t,x) \geq u^{(1)}(t,x) \quad \text{for } t \in [0,T),\ x \in \partial\Omega.$$

Then

$$u^{(2)}(t,x) \geq u^{(1)}(t,x) \quad \text{in } [0,T) \times \Omega.$$

Proposition 2.2 (second comparison theorem). *Let $u(t,x) \geq 0$ be a classical solution of problem* (2.9)–(2.11) *in $[0,T) \times \overline{\Omega}$, and let functions $u_\pm(t,x) \in \mathbb{C}_{t,x}^{(1,2)}((0,T) \times \Omega) \cap \mathbb{C}([0,T) \times \overline{\Omega})$ (called the upper and lower solutions) satisfy the inequalities*

$$\frac{\partial u_+}{\partial t} \geq B(u_+), \quad \frac{\partial u_-}{\partial t} \leq B(u_-) \quad \text{in } (0,T) \times \Omega,$$
$$u_-(0,x) \leq u_0(x) \leq u_+(0,x), \quad x \in \Omega,$$
$$u_-(t,x) \leq u(t,x) \leq u_+(t,x), \quad t \in [0,T),\ x \in \partial\Omega.$$

Then

$$u_-(t,x) \leq u(t,x) \leq u_+(t,x) \quad \text{in } [0,T) \times \Omega.$$

For the case of a Cauchy problem, we must replace the domain Ω by $\mathbb{R}^N$ and remove the conditions for boundary data. If an equation does not satisfy the uniform parabolicity conditions (for example, it is degenerate, i. e., $k(0) = 0$), then we can state analogs of these theorems for generalized solution; however, in this case, we must assume that $u_\pm(t,x) \in \mathbb{C}([0,T)\times\overline{\Omega})$ and $u_\pm(t,x) \in \mathbb{C}^{(1,2)}_{t,x}$ everywhere in $((0,T)\times\overline{\Omega})$ outside a finite number of smooth nonintersecting surfaces $(0,T)\times S_i(t)$ on which the function $k(u)\nabla u$ is continuous.

Now we turn to problem (2.3)–(2.4). By the first comparison theorem the self-similar solution u_A majorizes a large variety of solutions. For example, we have the following:

Assertion 2.1. *Let*

$$u_1(t) \le (1+t)^m, \quad t>0, \quad \textit{and} \quad u_0(x) \le \theta_A(x), \quad x>0.$$

Then the solution of problem (2.3)–(2.4) *satisfies the inequality*

$$u(t,x) \le (1+t)^m \theta_A\left(\frac{x}{\sqrt{1+t}}\right), \quad t>0,\ x>0.$$

Thus the self-similar solution provides an upper estimate, which allows us to determine the character of the heat transfer process and to obtain certain information on the spatial profile of the heat wave (not necessarily self-similar).

Consider another aspect of the problem. Assume that the restrictions for the initial function $u_0(x)$ do not hold, for example, $u_0(x) \equiv 1$ for $x \ge 0$; then the inequality $u_0(x) \le \theta_A(x)$ is invalid for sufficiently large $x > 0$ since $\theta_A(x) \to 0$ as $x \to \infty$. In this case, self-similar solutions also allow us to obtain exact estimates of the space-time structure of heat waves for sufficiently large t.

Assume that the boundary regime is self-similar, that is, it is of the form (2.5). Introduce the self-similar transformation of a solution of problem (2.3)–(2.4) by the formula

$$\theta(t,\xi) = \frac{1}{(1+t)^m} u(t, \xi\sqrt{1+t}), \quad t>0,\ \xi>0. \tag{2.12}$$

The self-similar transformation of the solution $u_A(t,x)$ yields exactly the function $\theta_A(\xi)$. We prove the following assertion on the asymptotic stability of solution (2.6) with respect to perturbations of the initial function.

Assertion 2.2. *Let $u_1(t) = (1+t)^m$ for $t > 0$. The self-similar solution* (2.6) *is asymptotically stable with respect to arbitrary (bounded) perturbations of the initial function. In other words, for any $u_0(x)$, we have*

$$\begin{aligned}\|\theta(t,\cdot) - \theta_A(\cdot)\|_{\mathbb{C}(\mathbb{R}_+)} &\equiv \sup_{\xi>0} |\theta(t,\xi) - \theta_A(\xi)| \\ &= O((1+t)^{-m}) \to 0 \quad \textit{as}\ t \to \infty.\end{aligned} \tag{2.13}$$

Indeed, set $z(t,x) = u(t,x) - u_A(t,x)$. Then z satisfies the heat equation

$$z_t = z_{xx}, \quad t > 0,\ x > 0,$$

where $z(t,0) = 0$ for $t > 0$, and $\sup_{x>0} |z(0,x)| < \infty$. The first comparison theorem yields

$$|z(t,x)| \le M = \sup_{x>0} |z(0,x)|. \tag{2.14}$$

This immediately implies that

$$|\theta(t,\xi) - \theta_A(\xi)| \le M(1+t)^{-m} \to 0 \quad \text{as } t \to \infty \tag{2.15}$$

for all $\xi \ge 0$. Thus, for any initial function, the solution of the problem with power boundary regime approaches the self-similar solution after a certain time.

This does not exhaust the properties of the self-similar solution. Similarly to the above, we can prove that, in addition, the self-similar solution is stable with respect to small perturbations of the boundary regime. The general result that gives a clear idea of the nature of the heat transfer for arbitrary initial perturbations and boundary regimes can be stated as follows.

Assertion 2.3. *Let $u_1(t)/(1+t)^m \to 1$ as $t \to \infty$. Then*

$$\|\theta(t,\cdot) - \theta_A(\cdot)\|_{C(\mathbb{R}_+)} = O\big[\max\{t^{-m},\ |1 - u_1(t)/t^m|\}\big] \to 0 \quad \text{as } t \to \infty. \tag{2.16}$$

Now we show that the space-time structure of the self-similar solution is also preserved for large time in the case of "small nonlinear perturbation" of the initial parabolic heat equation.

Assume that the sufficiently smooth coefficient of thermal conductivity is not constant but is close to a constant for large temperatures:

$$k(u) \to 1, \quad u \to \infty. \tag{2.17}$$

Consider the heat process (2.4) in a nonlinear heat-conducting medium:

$$u_t = (k(u)u_x)_x, \quad t > 0,\ x > 0.$$

For all $u \ge 0$, introduce the function

$$G_k(u) = \int_0^u |1 - k^{1/2}(\eta)|^2\, d\eta \tag{2.18}$$

and prove the asymptotic stability under "small nonlinear perturbations."

Lemma 2.1. *Let $u_1(t) = (1+t)^m$, let the initial function*

$$u_0(x) \in \mathbb{L}^2(\mathbb{R}_+)$$

be nonincreasing in x, and let condition (2.17) be fulfilled. Then the self-similar solution (2.6) is stable with respect to perturbations of the coefficient of thermal conductivity specified, and we have the following estimate:

$$\|\theta(t,\cdot) - \theta_A(\cdot)\|^2_{\mathbb{L}^2(\mathbb{R}_+)} \equiv \int_0^\infty [\theta(t,\xi) - \theta_A(\xi)]^2 \, d\xi$$

$$= O\left[\frac{1}{(1+t)^{2m+1/2}} \max\left\{1, \int_0^t (1+\tau)^{m-1/2} G_k[(1+\tau)^m] \, d\tau\right\}\right] \to 0 \tag{2.19}$$

as $t \to \infty$.

Proof. First, we note that the convergence $\theta(t,\cdot) \to \theta_A(\cdot)$ as $t \to \infty$ with respect to the norm of $\mathbb{L}^2$ means, in particular, the pointwise convergence almost everywhere. Indeed, under condition (2.17), the right-hand side of the estimate (2.19) tends to zero as $t \to \infty$:

$$\lim_{t\to\infty} \frac{1}{(1+t)^{2m+1/2}} \int_0^t (1+\tau)^{m-1/2} G_k[(1+\tau)^m] \, d\tau$$

$$= \frac{1}{2m+1/2} \lim_{s\to\infty} \frac{G_k(s)}{s} = \frac{1}{2m+1/2} \lim_{s\to\infty} (1 - k^{1/2}(s))^2 = 0. \tag{2.20}$$

Consider the function $w = u - u_A$ satisfying, by (2.9) and (2.17), the equation

$$w_t = (k(u)u_x - (u_A)_x)_x, \tag{2.21}$$

where the boundary condition is of the form $w(t,0) \equiv 0$. By the condition of the lemma, $u_0(x) \in \mathbb{L}^2(\mathbb{R}_+)$, and (2.8) and the asymptotics of the Hermite function as $x \to \infty$ imply

$$u_A(t,\cdot) \in \mathbb{L}^2(\mathbb{R}_+).$$

Therefore, we can conclude that, at the initial time moment,

$$w(0,x) \in \mathbb{L}^2(\mathbb{R}_+).$$

Thus, we may multiply the function w_t by w in $\mathbb{L}^2(\mathbb{R}_+)$ in the sense of the inner product and integrate by parts (which is possible due to the uniform boundedness of the derivatives u_x and $(u_A)_x$ in $\mathbb{R}_+ \times \mathbb{R}_+$ and the condition $w(t,x) \to 0$ as $x \to \infty$). As a result, we have

$$\frac{1}{2}\frac{d}{dt}\|w\|^2_{\mathbb{L}^2} = -(w_x, k(u)u_x - (u_A)_x). \tag{2.22}$$

The identity

$$
\begin{aligned}
&-(u_x-(u_A)_x)(k(u)u_x-(u_A)_x)\\
&\quad=-(k^{1/2}(u)u_x-(u_A)_x)^2+(1-k^{1/2}(u))^2u_x(u_A)_x\\
&\quad\equiv-(k^{1/2}(u)u_x-(u_A)_x)^2+\frac{\partial}{\partial x}G_k(u)(u_A)_x,
\end{aligned}
\tag{2.23}
$$

which can be easily verified, allows us to transform equation (2.22) to the form

$$
\frac{1}{2}\frac{d}{dt}\|w\|^2_{\mathbb{L}^2}=-\|k^{1/2}(u)u_x-(u_A)_x\|^2_{\mathbb{L}^2}+\left(\frac{\partial}{\partial x}G_k(u),(u_A)_x\right).
\tag{2.24}
$$

Doe to the maximum principle and the monotonicity of $u_0(x)$ ($[u_0(x)]_x\le 0$), we have $u_x(t,x)\le 0$ for all $t>0$ in $\mathbb{R}_+$. Since $(u_A)_x<0$, from (2.24) we obtain the estimate

$$
\begin{aligned}
\frac{1}{2}\frac{d}{dt}\|w\|^2_{\mathbb{L}^2}&\le\left(\frac{\partial}{\partial x}G_k(u),(u_A)_x\right)\\
&\le-\sup_x|u_A(t,x)_x|\int_0^\infty\frac{\partial}{\partial x}G_k(u(t,x))\,dx.
\end{aligned}
\tag{2.25}
$$

It is easy to prove that

$$
\left|[u_A(t,x)]_x\right|\equiv(1+t)^{m-1/2}|\theta_A'(\xi)|,\quad \sup|\theta_A'(\xi)|=q_A<\infty.
\tag{2.26}
$$

Then estimate (2.25) implies

$$
\frac{1}{2}\frac{d}{dt}\|w\|^2_{\mathbb{L}^2}\le q_A(1+t)^{m-1/2}G_k[(1+t)^m].
\tag{2.27}
$$

From the form of the self-similar transformation we obtain

$$
\|w(t,\cdot)\|^2_{\mathbb{L}^2}\equiv(1+t)^{2m+1/2}\|\theta(t,\cdot)-\theta_A(\cdot)\|^2_{\mathbb{L}^2};
\tag{2.28}
$$

therefore (2.27) implies the inequality

$$
\begin{aligned}
\|\theta(t,\cdot)-\theta_A(\cdot)\|^2_{\mathbb{L}^2}&\le\|u_0(\cdot)-\theta_A(\cdot)\|^2_{\mathbb{L}^2}(1+t)^{-2m-1/2}\\
&\quad+2q_A(1+t)^{-2m-1/2}\int_0^t(1+\tau)^{m-1/2}G_k[(1+\tau)^m]\,d\tau,
\end{aligned}
\tag{2.29}
$$

which coincides with (2.29). □

Thus, for large time, the self-similar solution adequately describes properties of solutions of a wide class of quasilinear parabolic equations. In addition, the function $\theta_A(\xi)$ determines the spatial shape as $t\to\infty$. Thus, using only the self-similar solution, we can describe asymptotic properties of solutions for problems with various initial conditions and various parabolic equations.

Now, having got acquainted with the concept of asymptotic stability, we pass to collapsing solutions and the notion of localization.

2.2 Three types of self-similar solutions

We turn to the study of a rather complicated nonlinear problem for a quasilinear equation with power nonlinearities,

$$u_t = \nabla(u^\sigma \nabla u) + u^\beta, \quad t > 0,\ x \in \mathbb{R}^N, \tag{2.30}$$

$$u(0,x) = u_0(x) \geq 0, \quad u_0^{\sigma+1} \in \mathbb{C}^{(1)}(\mathbb{R}^N),\ \sigma > 0,\ \beta > 1. \tag{2.31}$$

We start by analysis of particular self-similar solutions of equation (2.30). We construct unbounded solutions whose space-time structures in the *HS*-mode, *S*-mode, and *LS*-mode are substantially different. Although these particular solutions are realized for certain specific choice of the initial function, the analysis of their structures allows us to draw informed conclusions about the combustion process with hyperbolic growth. Moreover, these solutions are useful in the study of conditions of global insolvability of the problem. The space-time structure of unbounded self-similar solution contains important and exhaustive information on general properties of evolution of unbounded solutions. Therefore, it is not an exaggeration to say that the particular solutions mentioned form a "basis" of the nonlinear problem.

We make two remarks. First, from the physical standpoint, equation (2.30) describes processes with finite perturbation propagation speed (see [31]). Therefore, if u_0 is a finite function, then $u(t,x)$ also is a finite function at all subsequent time moments. Second, we will not focus on the asymptotic stability of self-similar collapsing solutions, which can be proved in analogy with the stability for the heat equation considered in the previous section (the proof can be found in [31]). We also omit the discussion of the global solvability and effective localization, which are not directly related to the theory of blow-up. Thus, for any $\sigma > 0$ and $\beta > 1$, equation (2.30) has an unbounded self-similar solution

$$u_A(t,x) = \frac{\theta_A(\xi)}{(T_0 - t)^{1/(\beta-1)}}, \tag{2.32}$$

where

$$\xi = \frac{x}{(T_0 - t)^m}, \quad m = \frac{\beta - (\sigma+1)}{2(\beta-1)},$$

$T_0 > 0$ is the lifetime of the solution u_A, and the function $\theta_A(\xi) \geq 0$ satisfies the following elliptic equation:

$$\nabla_\xi(\theta_A^\sigma \nabla_\xi \theta_A) - m(\nabla_\xi \theta_A, \xi) - \frac{1}{\beta-1}\theta_A + \theta_A^\beta = 0, \quad \xi \in \mathbb{R}^N. \tag{2.33}$$

It is easy to verify that equation (2.33) has the trivial and spatially homogeneous solutions

$$\theta_A \equiv 0, \quad \theta_A = \theta_H \equiv (\beta-1)^{-1/(\beta-1)},$$

corresponding to the spatially homogeneous combustion with blow-up (homothermic combustion). We confine ourselves to the analysis of radially symmetric self-similar solutions

$$\xi = \frac{r}{(T_0 - t)^m}, \quad r = |x|. \tag{2.34}$$

In this case, equation (2.33) becomes the ordinary differential equation

$$\frac{1}{\xi^{N-1}}(\xi^{N-1}\theta_A^\sigma\theta_A')' - m\theta_A'\xi - \frac{1}{\beta - 1}\theta_A + \theta_A^\beta = 0, \quad \xi > 0. \tag{2.35}$$

The symmetric solution (2.35) is defined in $\mathbb{R}^N$ under the condition $\theta_A'(0) = 0$. Moreover, we assume that the solution vanishes at infinity: $\theta_A(\infty) = 0$.

Note that equation (2.35) degenerates at $\theta_A = 0$; therefore, in general, it admits only generalized solutions that have no necessary smoothness at degeneration points. However, in all cases, the self-similar heat flow $\xi^{N-1}\theta_A^\sigma\theta_A'$ is continuous, which means that $\theta_A^\sigma\theta_A' = 0$ at all points where $\theta_A = 0$, and therefore we can apply the second comparison theorem in the generalized form.

2.3 Self-similar *S*-mode, $\beta = \sigma + 1$

In this case, the radially symmetric problem for equation (2.35) is of the form

$$\frac{1}{\xi^{N-1}}(\xi^{N-1}\theta_A^\sigma\theta_A')' - \frac{1}{\sigma}\theta_A + \theta_A^{\sigma+1} = 0, \quad \xi > 0, \tag{2.36}$$

$$\theta_A'(0) = 0 \ (\theta_A(0) > 0), \quad \theta_A(\infty) = 0. \tag{2.37}$$

In the one-dimensional case ($N = 1$), equation (2.36) becomes autonomous and can be integrated explicitly. In particular, it is easy to obtain the following solution:

$$\theta_A(\xi) = \left(\frac{2(\sigma+1)}{\sigma(\sigma+2)}\cos^2\left(\frac{\pi\xi}{L_S}\right)\right)^{1/\sigma}, \quad L_S = \frac{2\pi}{\sigma}\sqrt{\sigma+1}. \tag{2.38}$$

Formula (2.32) implies that, for $\beta = \sigma + 1$, the self-similar solution can be written in separated variables:

$$u_A(t,x) = \frac{1}{(T_0 - t)^{1/\sigma}}\theta_A(x), \quad 0 < t < T_0, \ x \in \mathbb{R}. \tag{2.39}$$

This solution has a rather unusual form from the standpoint of traditional ideas about the nature of heat propagation in media with diffusion. Indeed, the periodic function $\theta_A(x)$ vanishes at the points $x_k = (1/2 \pm k)L_S$, $k = 0, 1, \ldots$. Since the heat flow is continuous as $x \to x_k$ and tends to zero, the generalized solution of the problem can contain only one "wave"; in particular, it is of the form

$$u_A(t,x) = \begin{cases} \frac{1}{(T_0-t)^{1/\sigma}}\left(\frac{2(\sigma+1)}{\sigma(\sigma+2)}\cos^2\left(\frac{\pi x}{L_S}\right)\right)^{1/\sigma} & \text{for } |x| < \frac{L_S}{2}, \\ 0 & \text{for } |x| \geq \frac{L_S}{2}. \end{cases} \tag{2.40}$$

The elementary structure of the self-similar collapsing solution is localized during its lifetime. In spite of the unbounded increase as $t \to T_0$, heat perturbations do not penetrate into the surrounding cold space.

Numerical simulation shows that in the case $\beta = \sigma + 1$, practically for arbitrary nonmonotonic initial perturbations on a segment of small length, the solution blurs to the resonance length, localizes at a segment of length L_S, and blows up (see [31]).

Self-similar solutions of the S-mode also exist in spaces of arbitrary dimension $N > 1$.

Theorem 2.1. *For any $N > 1$, there exists a finite solution $\theta_A(\xi)$ of problem (2.36)–(2.37). The function θ_A decreases at all points where it is positive. The problem has no nonmonotonic solutions.*

Proof. First, we note that the finiteness of any possible solution θ_A follows from the analysis of the equation for small $\theta_A > 0$, reduced to the equivalent integral equation, by the fixed-point theorem for continuous mappings. This simple analysis also yields a unique possible asymptotics of the function θ_A: it is finite, and if $\operatorname{meas}\operatorname{supp}\theta_A = \xi_0 > 0$, then

$$\theta_A(\xi) = (1 + \varepsilon(\xi))\left[\frac{\sigma}{2(\sigma+2)}(\xi_0 - \xi)^2\right]^{1/\sigma}, \tag{2.41}$$

where

$$\varepsilon(\xi) \to 0 \quad \text{as } \xi \to \xi_0^-,$$

and $\theta_A(\xi) \equiv 0$ for all $\xi \geq \xi_0$.

To prove the existence of such a solution, consider the family of Cauchy problems for equation (2.36):

$$\frac{1}{\xi^{N-1}}(\xi^{N-1}|\theta|^\sigma\theta')' - \frac{1}{\sigma}\theta + |\theta|^\sigma\theta = 0, \quad \xi > 0, \tag{2.42}$$

$$\theta(0) = \mu, \quad \theta'(0) = 0. \tag{2.43}$$

At points where $\theta \geq 0$, equation (2.42) coincides with (2.36). Thus, we must choose the parameter $\mu = \theta_0 > 0$ such that the solution $\theta(\xi, \mu)$ of the family of the Cauchy problem is nonnegative for all $\xi \geq 0$ and satisfies the condition $\theta(\infty; \mu) = 0$. The local existence and uniqueness of the solution of problem (2.42)–(2.43) can be proved by using the Banach contraction mapping theorem.

We need the following three lemmas.

Lemma 2.2. *Let*

$$0 < \mu < \mu^* = \left[\frac{2(\sigma+1)}{\sigma(\sigma+2)}\right]^{1/\sigma}.$$

Then $\theta(\xi;\mu) > 0$ for all $\xi > 0$. Moreover, for any $\mu > \theta_H = \sigma^{-1/\sigma}$, the solution is bounded in $\mathbb{R}_+$:

$$|\theta(\xi;\mu)| < \mu \quad \textit{for } \xi > 0.$$

Proof. Multiplying (2.42) by $|\theta|^\sigma\theta'$, integrating the equality obtained over the interval $(0,\xi)$, and taking into account conditions (2.43), we obtain the equation

$$\frac{1}{2}(|\theta|^\sigma\theta')^2(\xi) + (N-1)\int_0^\xi (|\theta|^\sigma\theta')^2(\eta)\frac{d\eta}{\eta} + \Phi(|\theta(\xi)|) = \Phi(\mu), \tag{2.44}$$

where

$$\Phi(\mu) = \frac{\mu^{2\sigma+2}}{2(\sigma+1)} - \frac{\mu^{\sigma+2}}{\sigma(\sigma+2)}, \quad \mu \geq 0. \tag{2.45}$$

The graph of the function $\Phi(\mu)$ is a curve decreasing from $\mu = 0$ ($\Phi(0) = 0$) to $\mu = \theta_H$ and then monotonically increasing and vanishing at the point $\mu = \mu^*$ ($\Phi(\mu^*) = 0$).

Further, from (2.44) we conclude that $\Phi(|\theta(\xi)|) \leq \Phi(\mu)$ for all $\xi \geq 0$, and the equality is achieved only for the stationary solution $\theta \equiv \mu = \theta_H$. This immediately implies the boundedness of the solution for $\mu > \theta_H$. Moreover, if $\mu < \mu^*$, then we have the following estimate:

$$\mu_1 < \theta(\xi;\mu) < \mu, \quad \xi > 0,\ \mu \neq \theta_H, \tag{2.46}$$

where μ_1 is the root of the equation $\Phi(\mu_1) = \Phi(\mu)$ distinct from μ. The lemma is proved. □

Now we show that, for a sufficiently large μ, the solution θ is not strictly positive.

Lemma 2.3. *There exists $\mu = \mu^* > \theta_H$ such that the solution of problem* (2.42)–(2.43) *vanishes.*

Proof. We prove the lemma by contradiction. Assume that $\theta(\xi;\mu) > 0$ for all $\mu > \theta_H$. Problem (2.42)–(2.43) is equivalent to the integral equation

$$\varphi'(\xi) = (\sigma+1)\xi^{1-N}\int_0^\xi \eta^{N-1}\varphi(\eta)\left[\frac{|\varphi|^{-\sigma/(\sigma+1)}}{\sigma} - 1\right]d\eta, \quad \xi > 0, \tag{2.47}$$

for the function

$$\varphi(\xi) = |\theta|^\sigma\theta(\xi;\mu), \quad \varphi(0) = \mu^{\sigma+1}.$$

We set $\psi_\mu(\xi) = \varphi(\xi)/\varphi(0)$. Then equation (2.47) can be rewritten in the form

$$\psi_\mu'(\xi) = (\sigma+1)\xi^{1-N}\int_0^\xi \eta^{N-1}\psi_\mu(\eta)\left[\frac{\mu^{-\sigma}|\psi_\mu|^{-\sigma/(\sigma+1)}}{\sigma} - 1\right]d\eta; \tag{2.48}$$

moreover, due to the boundedness of the solution, Lemma 2.2 implies that

$$|\psi_\mu(\xi)| \le 1, \quad \xi \ge 0,\ \mu > \theta_H.$$

Thus, from (2.48) we obtain the estimate

$$|\psi'_\mu(\xi)| \le \frac{\sigma+1}{\xi^{N-1}} \int_0^\xi \eta^{N-1}\left[\frac{\mu^{-\sigma}}{\sigma}+1\right] d\eta = \frac{\sigma+1}{N}\xi\left[\frac{\mu^{-\sigma}}{\sigma}+1\right]. \tag{2.49}$$

This implies that, for any $\mu > \theta_H$, the functions ψ_μ and ψ'_μ are uniformly bounded on any compact set $[0,\xi_m]$. Then the Arzelà–Ascoli theorem implies the existence of a sequence

$$\mu_k \to \infty \quad \text{as } k \to \infty$$

such that the corresponding sequence $\psi_{\mu_k}(\xi)$ converges uniformly on $[0,\xi_m]$ to a certain function $w(\xi)$. The equation for w is obtained from (2.48) by passing to the limit as $\mu = \mu_k \to \infty$:

$$\begin{gathered} w'(\xi) = -(\sigma+1)\xi^{1-N}\int_0^\xi \eta^{N-1} w(\eta)\, d\eta, \quad w(0)=1, \\ w \in \mathbb{C}([0,\infty)) \cap \mathbb{C}^{(1)}(\mathbb{R}_+). \end{gathered} \tag{2.50}$$

By the assumption $\psi_\mu > 0$ for any $\mu > \theta_H$ we have the inequality $w \ge 0$ in $\mathbb{R}_+$. However, due to (2.50), the function $w(\xi)$ strictly decreases, and hence $w > 0$ on a finite interval since in the opposite case this function becomes negative in finite time. This leads to a contradiction since problem (2.50) is equivalent to the boundary-value problem

$$w'' + \frac{N-1}{\xi}w' + (\sigma+1)w = 0, \quad w'(0) = 0, \quad w(0) = 1. \tag{2.51}$$

Its solution can be expressed in terms of a Bessel function as

$$w = C_1\xi^{(2-N)/2} J_{(N-2)/2}(\xi(\sigma+1)^{1/2})$$

and therefore vanishes at the point

$$\xi = \xi_1 \equiv \frac{z_N^{(1)}}{(\sigma+1)^{1/2}},$$

where $z_N^{(1)} > 0$ is the first root of the Bessel function of order $(N-2)/2$. The lemma is proved. □

Lemma 2.2 implies that the values $\mu_1 > \theta_H$ correspond to strictly positive solutions $\theta(\xi, \mu)$, and Lemma 2.3 implies that the values $\mu_2 > \mu_1$ correspond to vanishing solutions. Therefore, there can exist $\mu = \theta_0 \in (\mu_1, \mu_2]$ such that the graph of the function $\theta^{\sigma+1}(\xi; \theta_0)$ touches the axis, which allows the zero extension of $\theta(\xi; \theta_0)$ in the domain $\xi > \xi_0$.

As a result, we obtain a localized generalized solution of the original problem with continuous heat flow $-\xi^{N-1}|\theta_A|^\sigma \theta_A'$. To use the properties of the solution, we also need the condition of continuous dependence of $\theta(\xi; \mu)$ on the parameter μ.

Lemma 2.4. *Let a solution $\theta = \theta(\xi, \mu_1)$, $\mu_1 > 0$, be such that the compact set $K = [0, \xi_m]$ contains no points at which $|\theta|^\sigma \theta = (|\theta|^\sigma \theta)' = 0$. Then $\theta(\xi; \mu)$ and $|\theta|^\sigma \theta'(\xi; \mu)$ continuously depend on the parameter μ in a neighborhood of $\mu = \mu_1$ in K.*

Proof. Consider equation (2.47), which is equivalent to problem (2.42)–(2.43) for $\mu = \mu_1$. The integrand contains the function $|\varphi|^{-\sigma/(\sigma+1)}\varphi$, which is not differentiable at $\varphi = 0$. Naturally, if $\theta(\xi; \mu_1) > 0$ on K, then we have the continuous dependence. Let $\xi = \xi_1$ be the first point at which $\theta(\xi; \mu_1)$ vanishes. By the condition

$$(|\theta|^\sigma \theta)'(\xi_1; \mu_1) \neq 0$$

we have the continuous dependence on the parameter μ on any segment $[0, \xi_1 - \varepsilon]$ for sufficiently small $\varepsilon > 0$. In a neighborhood of $\varphi = 0$, the operator in the right-hand side of (2.47) is not a contraction operator; however, the nondifferentiable term $|\varphi|^{-\sigma/(\sigma+1)}$ is small on the interval $(\xi_1 - \varepsilon, \xi_1 + \varepsilon)$. Therefore, for the extension of the solution $\theta(\xi; \mu)$ (where $|\mu - \mu_1|$ is small) in this domain, the derivatives $\varphi'(\xi)$ with respect to ξ and μ are continuous; hence the solution

$$\varphi(\xi) = (|\theta|^\sigma \theta)(\xi; \mu)$$

also is continuous, and the extension is unique. Similarly, we can construct an extension of $\theta(\xi; \mu)$ to the whole compact set K, and the continuous dependence of φ and φ' on the parameter μ in a neighborhood of $\mu = \mu_1$ is preserved in this extension. The lemma is proved. □

Now we complete the proof of Theorem 2.1. It is based on Lemmas 2.2–2.4. Introduce the set

$$M = \{\mu^0 > 0 \mid \theta(\xi; \mu) > 0,\ 0 < \mu < \mu^0\}.$$

Lemma 2.2 implies that $M \neq \{\mu^0 \leq \theta_H\}$, whereas Lemma 2.3 implies the boundedness of M from above. Therefore $\theta_0 := \sup M < \infty$. Lemma 2.4 implies that the solution of problem (2.42)–(2.43) for $\mu = \theta_0$ is a required self-similar function θ_A with asymptotics (2.41).

The monotonicity of any nonnegative solution of problem (2.36)–(2.37) immediately follows from Lemma 2.2. Indeed, due to (2.44), possible oscillations of θ near θ_H are damped, that is, if $\xi_1 < \xi_2$ are two maximum (minimum) points of the function $\theta \geq 0$, then $\theta(\xi_1; \mu) > \theta(\xi_2; \mu)$ (respectively, $\theta(\xi_1; \mu) < \theta(\xi_2; \mu)$). The theorem is proved. □

2.4 Self-similar *HS*-mode, $\beta < \sigma + 1$

In this section, we prove the theorem on the solvability of the self-similar problem (2.35) for $\beta \in (1, \sigma+1)$ by the same method as before. An immediate analysis of the equation shows that a solution $\theta_A(\xi)$ must be a finite function with the asymptotics in a neighborhood of the degeneration point $\xi_0 = \operatorname{meas}\operatorname{supp}\theta_A$ distinct from (2.41):

$$\theta_A(\xi) = \left|\frac{\sigma(\sigma+1-\beta)}{2(\beta-1)}\xi_0(\xi_0-\xi)\right|^{1/\sigma}(1+w(\xi)), \tag{2.52}$$

where

$$w(\xi) \to 0 \quad \text{as } \xi \to \xi_0^-.$$

The existence of such a solution is explained by the following theorem.

Theorem 2.2. *For any values of the parameters β and σ such that $1 < \beta < \sigma + 1$, there exists a finite solution $\theta_A(\xi)$ of problem (2.35). The function θ_A strictly decreases at all points where it is positive.*

Proof. The proof is similar to that of Theorem 2.1. We briefly discuss the key points. The identity similar to (2.44) is of the form

$$\frac{1}{2}(|\theta|^\sigma\theta')^2(\xi) + (N-1)\int_0^\xi (|\theta|^\sigma\theta')^2(\eta)\frac{d\eta}{\eta} - m\int_0^\xi \eta|\theta|^\sigma(\theta')^2(\eta)\,d\eta + \Phi(|\theta(\xi)|) = \Phi(\mu), \tag{2.53}$$

where

$$\Phi(\mu) = \frac{\mu^{\beta+\sigma+1}}{\beta+\sigma+1} - \frac{\mu^{\sigma+2}}{(\beta-1)(\sigma+2)}, \quad \mu \ge 0, \tag{2.54}$$

and

$$m = \frac{\beta-(\sigma+1)}{2(\beta-1)} < 0, \quad \mu \ge 0.$$

From (2.53) we obtain

$$\Phi(|\theta(\xi)|) \le \Phi(\mu) \quad \text{and} \quad \theta(\xi) > 0 \tag{2.55}$$

for all

$$0 < \mu < \mu_* = \left|\frac{\beta+\sigma+1}{(\beta-1)(\sigma+2)}\right|^{1/(\beta-1)}.$$

This implies an analog of Lemma 2.2 for the case $\beta < \sigma + 1$. For any $\mu > \theta_H = (\beta - 1)^{-1/(\beta-1)}$, the solution is uniformly bounded: $|\theta(\xi)| \leq \mu$ in $\mathbb{R}_+$.

To get an analog of Lemma 2.3, we extend equation (2.35) in the domain of negative values of θ, and the substitution $\varphi = |\theta|^\sigma \theta$ reduces the problem to the integral equation

$$\varphi'(\xi) = m(1+\sigma)\xi|\varphi|^{-\sigma/(\sigma+1)}\varphi + (\sigma+1)\xi^{1-N}\int_0^\xi \eta^{N-1}\left[\left(\frac{1}{\beta-1} - mN\right)|\varphi|^{-\sigma/(\sigma+1)} - |\varphi|^{(\beta-\sigma-1)/(\sigma+1)}\right]\varphi\, d\eta. \quad (2.56)$$

After the transformation

$$\psi_\mu(\xi) = \mu^{-(\sigma+1)}\varphi(\xi\mu^{((\sigma+1)-\beta)/2}), \quad (2.57)$$

equation (2.56) takes the form

$$\psi'_\mu(\xi) = m(\sigma+1)\mu^{1-\beta}\xi|\psi_\mu|^{-\sigma/(\sigma+1)}\psi_\mu(\sigma+1)\xi^{1-N} \times \int_0^\xi \eta^{N-1}\left[\left(\frac{1}{\beta-1} - mN\right)\mu^{1-\beta}|\psi_\mu|^{-\sigma/(\sigma+1)} - |\psi_\mu|^{(\beta-(\sigma+1))/(\sigma+1)}\right]\psi_\mu\, d\eta. \quad (2.58)$$

As in the proof of Lemma 2.3, we see that the assumption $\psi_\mu > 0$ for any $\mu > \theta_H$ leads, due to the compactness theorem, to the existence of a sequence $\{\mu_k\}$ such that $\psi_{\mu_k} \to w > 0$ as $\mu_k \to \infty$; moreover, the function $w(\xi)$ satisfies the problem

$$w' = -(\sigma+1)\xi^{1-N}\int_0^\xi \eta^{N-1}w^{\beta/(\sigma+1)}(\eta)\, d\eta, \quad w(0) = 1, \quad (2.59)$$

or, equivalently, the problem

$$w'' + \frac{N-1}{\xi}w' + (\sigma+1)w^{\beta/(\sigma+1)} = 0, \quad w'(0) = 0, \quad w(0) = 1, \quad (2.60)$$

whose solution vanishes. We complete the proof of Theorem 2.2 by using an assertion similar to Lemma 2.4 on the continuous dependence of $\theta(\xi;\mu)$ on the parameter μ, which can be proved similarly. The theorem is proved. □

We list the basic properties of the self-similar function $\theta_A(\xi)$ obtained before. For $1 < \beta < \sigma + 1$, the *HS*-mode of combustion with hyperbolic growth is realized: the unbounded solution is not localized. This directly follows from the law of the change in time of the support of the unbounded solution u_A. Indeed, from (2.34) we obtain the following expression for the radius of the spherical front of the propagating heat wave:

$$|x_{fr}(t)| = \xi_0(T_0 - t)^m = \xi_0(T_0 - t)^{(\beta-(\sigma+1))/(2\beta-2)}. \quad (2.61)$$

By the condition $\beta < \sigma + 1$ we see that $|x_{fr}| \to \infty$ as $t \to T_0^-$, that is, the heat wave fills the whole space in finite time. Moreover, in the *HS*-mode, we have

$$u_A(t,x) \to \infty \quad \text{as } t \to T_0^-. \tag{2.62}$$

A similar expression can be easily obtained for the half-width of the spatial profile:

$$|x_{ef}(t)| = \xi_*(T_0 - t)^{(\beta-\sigma-1)/(2\beta-2)}, \tag{2.63}$$

where

$$\theta_A(\xi_*) = \frac{\theta_A(0)}{2}.$$

Due to the monotonicity of θ_A, the value ξ_* is determined uniquely.

It turns out that self-similar patterns are preserved in the case of arbitrary initial perturbations $u_0(x)$. For $\beta < \sigma + 1$, all unbounded solutions $u(t,x)$ satisfy (2.62); they are not localized. The corresponding results on the asymptotic stability of self-similar solutions can be found in [31].

2.5 Self-similar *LS*-mode, $\beta > \sigma + 1$

Now we consider self-similar solutions that exhibit the localization property of the blow-up process more vividly than in the *S*-mode.

In the case where $\beta > \sigma + 1$, for the boundary-value problem (2.35), we prove the existence of a self-similar solution. We focus on the one-dimensional case.

Theorem 2.3. *Let $\beta > \sigma + 1$ and $N = 1$. Then problem* (2.35) *has a strictly monotonic positive solution*

Proof. As usual, we consider a family of Cauchy problems for the same equation

$$(|\theta|^\sigma\theta')' - m\theta'\xi - \frac{1}{\beta-1}\theta + |\theta|^{\beta-1}\theta = 0, \tag{2.64}$$

$$\theta(0;\mu) = \mu > 0, \quad \theta'_\xi(0;\mu) = 0, \quad m = \frac{\beta-(\sigma+1)}{2(\beta-1)} > 0. \tag{2.65}$$

We show that, for certain μ, the solution $\theta = \theta(\xi;\mu) \geq 0$ satisfies the condition at infinity $\theta(\infty;\mu) = 0$ and thus determines a required self-similar function θ_A.

We note one feature of problem (2.64)–(2.65). We proved for $\beta \leq \sigma + 1$ that, for various $\mu > 0$, there always exists a family of strictly positive functions θ that "oscillate" near the spatial homogeneous solution $\theta \equiv \theta_H$ and the amplitude of oscillations decreases as ξ grows. This provides the strict positiveness of these solutions. We will further see that in the case $\beta > \sigma + 1$ this property does not hold.

We get a fairly accurate picture of the nature of undamped oscillations for μ close to θ_H by examining solutions $v(\xi)$ of the problem obtained by the linearization of the original problem with respect to the homogeneous solution $\theta = \theta_H$.

For small values of the parameter $\varepsilon > 0$, we set

$$\theta(\xi;\mu) = \theta_H + \varepsilon v(\xi), \quad \theta_H = (\beta - 1)^{-1/(\beta-1)}. \tag{2.66}$$

Substituting (2.66) into (2.64), we obtain the problem

$$\theta_H^\sigma v'' - mv'\xi + v = \varepsilon\Phi_\varepsilon(v), \quad v(0) = v, \quad v'(0) = 0, \tag{2.67}$$

where $\Phi_\varepsilon : \mathbb{C}^2 \to \mathbb{C}$ is a bounded quasilinear second-order operator. The boundary conditions of problems (2.67) and (2.65) are related by the equality

$$\mu = \theta_H + \varepsilon v. \tag{2.68}$$

We conclude from (2.67) that, due to the continuous dependence of the solution on the parameter for sufficiently small $\varepsilon > 0$, the solution $v(\xi)$ of problem (2.67) is near to the solution of the corresponding linear problem

$$\theta_H^\sigma y'' - my'\xi + y = 0, \quad y(0) = v \neq 0, \quad y'(0) = 0. \tag{2.69}$$

The transformation of the independent variable

$$\xi = \sqrt{\frac{2\theta_H^\sigma \eta}{m}} \equiv \left|\frac{4(\beta-1)^{1-\sigma/(\beta-1)}}{\beta-(\sigma+1)}\right|^{1/2} \eta^{1/2} \tag{2.70}$$

reduces (2.69) to the confluent hypergeometric equation

$$\eta y''_{\eta\eta} + y'_\eta(c - \eta) - ay = 0, \quad \eta > 0, \quad y(0) = v, \tag{2.71}$$

where

$$c = \frac{1}{2}, \quad a = -\frac{1}{2m} = -\frac{\beta - 1}{\beta - \sigma - 1}. \tag{2.72}$$

The second boundary condition takes the form

$$\eta^{1/2} y'_\eta(\eta)\big|_{\eta=0} = 0.$$

Therefore an appropriate solution is a solution with bounded derivative at zero $y'_\eta(0)$. Such a solution of equation (2.71) exists; it is determined by the Kummer series (see [16])

$$y(\eta) = v\left(1 + \frac{a}{c}\frac{\eta}{1!} + \frac{a(a+1)}{c(c+1)}\frac{\eta^2}{2!} + \frac{a(a+1)(a+2)}{c(c+1)(c+2)}\frac{\eta^3}{3!} + \cdots\right), \tag{2.73}$$

which converges everywhere, or by the confluent hypergeometric function:

$$y(\eta) = vM(a, c, \eta) = ve^{\eta}\frac{\Gamma(c)}{\Gamma(c-a)\Gamma(a)}\int_0^1 e^{-\eta s}s^{c-a-1}(1-s)^{a-1}\,ds. \tag{2.74}$$

In general, this function is not monotonic. If

$$-a = \frac{\beta - 1}{\beta - (\sigma + 1)} = K, \tag{2.75}$$

where $K > 1$ is an integer, then the function $y(\eta)$ is a polynomial of degree K and has exactly K "zeros" for $\eta > 0$ (see [16]). Thus, the number of zeros of the solution of problem (2.69) is equal to

$$K = -[a], \quad a = -\frac{\beta - 1}{\beta - (\sigma + 1)} < 0,$$

and therefore, for any $\beta > \sigma + 1$, we have $K \geq 2$.

Returning to the original linearized problem (2.67), we obtain that, due to the continuous dependence of the solution v on the parameter ε in any compact set, there exists sufficient small $\varepsilon > 0$ such that, for all $|v| \leq 1$, the function $v(\xi)$ has at least K zeros for $\xi > 0$. For the original problem (2.64)–(2.65), this means that, for any

$$0 < \theta_H - \varepsilon < \mu < \theta_H + \varepsilon,$$

the solution $\theta(\xi, \mu)$ has at least K extrema.

Thus, we have established the behavior of $\theta(\xi; \mu)$ for all μ close to θ_H. Now we show that, for sufficiently large μ, the function $\theta(\xi; \mu)$ strictly decreases up to zero.

Lemma 2.5. *Let $N = 1$ and $\beta > \sigma + 1$. Then, for all*

$$\mu \geq \mu^* = \left|\frac{\beta + \sigma + 1}{(\beta - 1)(\sigma + 2)}\right|^{1/(\beta-1)},$$

the solution $\theta = \theta(\xi; \mu)$ of problem (2.64)–(2.65) *has zeros and strictly decreases for all ξ^* such that $\theta(\xi; \mu) > 0$ everywhere on $0 < \xi < \xi^*$.*

Proof. Consider identity (2.53), which in the one-dimensional case takes the form

$$\frac{1}{2}(|\theta|^{\sigma}\theta')^2(\xi) - m\int_0^{\xi} \eta|\theta|^{\sigma}(\theta')^2(\eta)\,d\eta + \Phi(|\theta(\xi)|) = \Phi(\mu), \tag{2.76}$$

where the function

$$\Phi(\mu) = \frac{\mu^{\beta+\sigma+1}}{\beta + \sigma + 1} - \frac{\mu^{\sigma+2}}{(\beta - 1)(\sigma + 2)}, \quad \mu \geq 0, \tag{2.77}$$

is of the same form as in the case $\beta \leq \sigma + 1$.

Assume the contrary. Under the condition $\mu \geq \mu^*$ of the lemma, let the solution $\theta(\xi, \mu)$ have a minimum at $\xi = \xi_* < \infty$ ($\theta'(\xi_*) = 0$); clearly, $\theta < \theta_H$ at this point (the solution oscillates near the stationary state θ_H). Setting $\xi = \xi_*$ in (2.76), from the condition $m > 0$ we obtain the inequality

$$\Phi(\theta(\xi_*)) > \Phi(\mu).$$

Thus we arrive at a contradiction, since the inequality can hold only if $\theta = \theta(\xi_*, \mu) > \mu^* > \theta_H$. Similarly, we can prove that, for $\mu \geq \mu^*$, the function $\theta(\xi; \mu)$ is not a positive solution in $\mathbb{R}_+$, and hence the case $\xi_* = \infty$ is impossible. □

Now we return to the proof of the solvability theorem. Denote by Q the set $\mu > \theta_H$, which contains a compact set $K = [0, \xi_K]$ such that $\theta(\xi; \mu) > 0$ on K and has at least one minimum. Analyzing the linearized equation, we see that this set in nonempty. Lemma 2.5 implies that Q in bounded from above. Therefore there exists

$$\sup Q = \theta_0 \in (\theta_H, \infty).$$

It is easy to see that, due to the choice of θ_0, the function $\theta(\xi; \theta_0)$, first, has no minima for $\xi > 0$ (continuous dependence on the parameter) and, second, cannot vanish (this follows from the lemma). Thus $\theta(\xi; \theta_0)$ is a positive strictly monotonic self-similar solution $\theta_A(\xi)$ of problem (2.35) for $\beta > \sigma + 1$. □

A multidimensional analog of this theorem is also valid.

Theorem 2.4. *Let $\sigma + 1 < \beta < \infty$ for $N = 1, 2$ and $\sigma + 1 < \beta < (\sigma + 1)\frac{N+2}{N-2}$ for $N \geq 3$. Then problem* (2.35) *has a strictly monotonic positive solution.*

The proof can be found in the monograph [31].

Now we verify that the blow-up is localized. As before, we write the formula expressing the dependence of the half-width of the self-similar thermal structure on time:

$$|x_{ef}(t)| = \xi_*(T_0 - t)^m = \xi_*(T_0 - t)^{(\beta-\sigma-1)/(2\beta-2)}, \quad 0 < t < T_0.$$

We see that in the *LS*-mode, the half-width decreases in time, and therefore intensive combustion occurs in a shrinking central domain of the structure. As a result, the hyperbolic growth develops at a single point; at other points of the space, the temperature is bounded from above uniformly with respect to t by the limit distribution $u_A(T_0^-, x)$.

Therefore, the self-similar solution is effectively localized. However, we cannot speak on the localization in the strict sense since $u_A(t, x)$ is strictly positive in $(0, T) \times \mathbb{R}^N$. Numerical simulations confirm the validity of self-similar estimates and, moreover, show the dependence of the localization domain on the initial functions (see [31]).

2.6 Blow-up of lower solutions

In this section, we present a simple but beautiful result obtained by using the comparison theorems and construction of self-similar lower solutions of the following Cauchy problem for the equation with power nonlinearities:

$$A(u) \equiv u_t - \nabla(u^\sigma \nabla u) - u^\beta = 0, \quad t > 0,\ x \in \mathbb{R}^N, \tag{2.78}$$

$$u(0,x) = u_0(x) \geq 0, \quad x \in \mathbb{R}^N,\ u_0 \in \mathbb{C}(\mathbb{R}^N),\ u_0^{\sigma+1} \in \mathbb{H}^1(\mathbb{R}^N). \tag{2.79}$$

Using the second comparison theorem (Proposition 2.1), we prove the following assertion.

Theorem 2.5. *Let $\beta \in (1, \sigma + 1 + 2/N)$ and $u_0(x) \not\equiv 0$. Then the solution of the Cauchy problem* (2.78)–(2.79) *blows up in finite time.*

Proof. First, we obtain a weaker result. Construct a collapsing lower solution of the form

$$u_-(t,x) = \frac{1}{(T-t)^{1/(\beta-1)}}\theta_-(\xi), \quad \theta_-(\xi) = A\left(1 - \frac{\xi^2}{a^2}\right)_+^{1/\sigma}, \tag{2.80}$$

$$\xi = \frac{|x|}{\zeta(t)}, \quad \zeta(t) = (T-t)^{\frac{\beta-(\sigma+1)}{2(\beta-1)}}, \quad 0 < t < T,\ \xi > 0, \tag{2.81}$$

where A, a, and T are positive constants. The function u_- blows up in accordance with self-similar laws. Let us find conditions under which this function is an unbounded lower solution. Assume that everywhere in $(0,T) \times \mathbb{R}_+$, except for the degeneration surface $(0,T) \times \{|x| = a\zeta(t)\}$, we have the following inequality:

$$A(u_-) \equiv (u_-)_t - \frac{1}{r^{N-1}}(r^{N-1}(u_-)^\sigma (u_-)'_r)'_r - (u_-)^\beta \leq 0. \tag{2.82}$$

This inequality can be reduced to

$$\frac{1}{\xi^{N-1}}(\xi^{N-1}\theta_-^\sigma \theta'_-)' - \frac{\beta - (\sigma+1)}{2(\beta-1)}\theta'_-\xi - \frac{1}{\beta-1}\theta_- + \theta_-^\beta \geq 0 \tag{2.83}$$

for $\xi \in (0,a)$. Substituting the function θ_- from (2.80) into (2.83), we obtain the following inequality, which is equivalent to (2.82):

$$\Phi_{\sigma\beta}(\Lambda) \equiv m - n\Lambda + A^{\beta-1}\Lambda^{(\beta-1)/(\sigma+1)} \geq 0, \tag{2.84}$$

where

$$\Lambda = \left(1 - \frac{\xi^2}{a^2}\right)_+,$$

$$m = \frac{4A^\sigma}{\sigma^2 a^2} + \frac{\beta - (\sigma+1)}{(\beta-1)\sigma}, \quad n = \frac{1}{\sigma}\left(1 + 2\frac{A^\sigma}{a^2}\left(N + \frac{2}{\sigma}\right)\right),$$

which must be valid for all $\Lambda \in (0,1]$. We find conditions for A and a.

First, from the inequality $\Phi_{\sigma\beta}(0) > 0$ we conclude that $m > 0$; therefore

$$\frac{4A^\sigma}{\sigma^2 a^2} > \frac{(\sigma+1)-\beta}{(\beta-1)\sigma}. \tag{2.85}$$

Second, it is easy to see that, for $m > 0$, inequality (2.84) is valid for all

$$\Lambda \in (0, \Lambda^*), \quad \Lambda^* = \frac{m}{n} \in (0,1).$$

This implies that (2.84) holds for all $\Lambda \in (0,1]$ if

$$m - n\Lambda + A^{\beta-1}(\Lambda^*)^{(\beta+\sigma-1)/\sigma} \geq 0, \quad \Lambda \in (\Lambda^*, 1), \tag{2.86}$$

which is equivalent to the condition

$$A^{\beta-1} \geq (n-m)(\Lambda^*)^{-(\beta+\sigma-1)/\sigma}.$$

Finally, substituting n, m, and Λ^*, we rewrite condition (2.86) in the form

$$A^{\beta-1} \geq \left(\frac{1}{\beta-1} + \frac{2N}{\sigma}\frac{A^\sigma}{a^2}\right)\left(\frac{1 + 2A^\sigma(N+2/\sigma)/a^2}{\frac{\beta-(\sigma+1)}{\beta-1} + \frac{A^\sigma}{a^2}\frac{4}{\sigma}}\right)^{(\beta+\sigma-1)/\sigma}. \tag{2.87}$$

The system of inequalities (2.85) and (2.87) has a solution (a, A) for all $\sigma > 0$ and $\beta > 1$. Indeed, for $\beta < \sigma+1$, condition (2.85) imposes a restriction on A^σ/a^2. Then, increasing A and a so that this ratio remains constant, we can achieve the fulfillment of condition (2.87). The case where $\beta \geq \sigma+1$ is simpler since condition (2.85) is certainly satisfied. Thus, we have proved the following assertion.

Lemma 2.6. *Let the initial data of the Cauchy problem* (2.78)–(2.79) *satisfy the condition*

$$u_0(x) \geq u_-(0,x) = \frac{1}{T^{1/(\beta-1)}}\theta_-\left(\frac{|x|}{T^{(\beta-\sigma-1)/(2\beta-2)}}\right), \quad x \in \mathbb{R}^N, \tag{2.88}$$

where

$$\theta_-(\xi) = A\left(1 - \frac{\xi^2}{a^2}\right)_+^{1/\sigma}$$

and the positive constants T, a, and A are related by formulas (2.85) *and* (2.87). *Then the solution is unbounded, and its lifetime does not exceed T.*

This lemma immediately implies the following result.

Assertion 2.4. *Let $\beta \in (1, \sigma+1)$ and $u_0(x) \not\equiv 0$. Then the solution of the problem blows up in finite time.*

Proof. Indeed, since $u_0 \not\equiv 0$, there exists a ball

$$\{x \in \mathbb{R}^N;\ |x - x_0| < \rho\}, \quad \rho > 0,$$

in which $u_0(x) \ge \varepsilon > 0$. Then, choosing in the lower solution $u_-(t, x)$ (see (2.80), (2.81)) the value T so large that the inequalities

$$AT^{-1/(\beta-1)} < \varepsilon \quad \text{and} \quad aT^{(\beta-(\sigma+1))/(2(\beta-1))} < \rho$$

hold, we obtain $u_0(x) \ge u_-(0, x - x_0)$. Therefore, by the theorem proved earlier, the solution is unbounded with lifetime T_0:

$$T_0 \le T_*, \quad T_* = \max\{(A/\varepsilon)^{\beta-1},\ (a/\rho)^{2(\beta-1)/(\sigma+1-\beta)}\}.$$

The assertion is proved. □

Now we return to the proof of theorem. It is based on the comparison of the solution $u(x, t)$ with the known self-similar solution of the Cauchy problem for the sourceless equation:

$$v_t = \nabla(v^\sigma \nabla v), \quad v(x, 0) = v_0(x), \quad x \in \mathbb{R}^N. \tag{2.89}$$

By a direct calculation we verify that in the N-dimensional case the self-similar solution (2.89) is of the form

$$v_A(t, x) = \frac{1}{(T_1 + t)^{N/(N\sigma+2)}} f(\eta), \tag{2.90}$$

where

$$f(\eta) = B(\eta_0^2 - \eta^2)_+^{1/\sigma}, \quad B = \left[\frac{\sigma}{2(N\sigma + 2)}\right]^{1/\sigma}, \quad \eta = \frac{|x|}{(T_1 + t)^{1/(N\sigma+2)}},$$

where T_1 and η_0 are arbitrary positive constants. We show that, for any $u_0(x) \not\equiv 0$ and $\beta < \sigma + 1 + 2/N$, after a finite time, the solution of the Cauchy problem (2.78)–(2.79) satisfies condition (2.88) and has a finite lifetime. We use the self-similar solution for estimating the time of "blurring" of the initial perturbation and increasing in time the amplitude of the spatial profile.

Without loss of generality, we assume that

$$u_0(0) > 0, \quad u_0(x) \ge \varepsilon > 0 \quad \text{in the ball } \{x \in \mathbb{R}^N : |x| < \delta\}.$$

We choose $\eta_0 = \eta_0(T_1)$ so that $u_0(x) \ge v_A(0, x)$ in $\mathbb{R}^N$; it suffices to take

$$B\eta_0^{2/\sigma} T_1^{-N/(N\sigma+2)} \le \varepsilon, \quad \eta_0 T_1^{1/(N\sigma+2)} \le \delta.$$

Then by the second comparison theorem we have

$$u(t,x) \geq v_A(t,x), \quad t > 0,\ x \in \mathbb{R}^N.$$

We show that, for $1 < \beta < \sigma + 1 + 2/N$, there exists a time moment t_1 such that, for certain T_1, the function $v_A(t_1,x)$ satisfies condition (2.88), and therefore it is also valid for the solution $u(t_1,x)$.

The condition $v_A(t_1,x) \geq u_-(0,x)$ in $\mathbb{R}^N$ holds if

$$(T_1 + t_1)^{-N/(N\sigma+2)} B\eta_0^{2/\sigma} \geq T^{-1/(\beta-1)} A, \tag{2.91}$$

$$\eta_0 (T_1 + t_1)^{1/(N\sigma+2)} \geq aT^{[\beta-(\sigma+1)]/[2(\beta-1)]}, \tag{2.92}$$

where A and a are arbitrary solutions of inequalities (2.85) and (2.87). We prove that system (2.91)–(2.92) is always solvable with respect to t_1 and T.

Assume that in (2.91) we have the equality

$$T_1 + t_1 = (B\eta_0^{2/\sigma}/A)^{(N\sigma+2)/N} T^{(N\sigma+2)/(N(\beta-1))}, \tag{2.93}$$

where T_1 is fixed, whereas T is sufficiently large. We verify whether inequality (2.92) holds for sufficiently large T. Rewrite it in the form

$$\eta_0 (B\eta_0^{2/\sigma}/A)^{1/N} T^{1/(N(\beta-1))} \geq aT^{(\beta-(\sigma+1))/(2(\beta-1))} \tag{2.94}$$

or, equivalently,

$$T^{\frac{\beta-(\sigma+1+2/N)}{2(\beta-1)}} \leq \frac{1}{a}\left(\frac{B}{A}\right)^{1/N} \eta_0^{2/N\sigma+1}. \tag{2.95}$$

Obviously, in the case where $\beta \in (1, \sigma+1+2/N)$, it holds for sufficiently large T, which completes the proof. □

In fact, in the theorem on blow-up, we proved that, for $1 < \beta < \sigma + 1 + 2/N$, the collapsing time consists of two parts: $T_0 \leq t_1 + T$, where t_1 is the time of "blurring" of the initial perturbation to a certain resonance state without practically noticeable energy release, and T is the time of intensive growth of the resonance solution described in the auxiliary lemma. Conversely, for $\beta > \sigma+1+2/N$, the stage of "blurring" can continue indefinitely; in other words, in this case, nontrivial global solutions are possible. Thus, in this case the Fujita critical exponent is

$$\beta = \sigma + 1 + \frac{2}{N}.$$

The proof of the global solvability based on the construction of upper solutions can be found in [31].

2.7 Bibliographical notes

The most complete description of the method of self-similar solutions of V. A. Galaktionov can be found in the monograph [31] (see also the references therein). Comparison theorems and maximum principles are presented in detail, for example, in [90, 21]. Problems related to self-similar solutions of the heat transfer problem are discussed in [26]. The analysis of collapsing self-similar solutions of the nonlinear heat transfer problem with a source can be found in [93, 52]. Numerical simulations of evolution of *S*-, *HS*-, and *LS*-modes are presented in [13, 113]. Finally, new results concerning the material of this chapter are discussed in [110].

Note that the method of comparison with self-similar unbounded lower solutions can also be used for a wide class of equations for which the maximum principle is valid. For example, it can be applied to higher-order equations of Sobolev type of the form

$$\frac{\partial}{\partial t}(\Delta u - u) + \Delta u + f(x,u) = 0.$$

In this connection, we note the work of Kozhanov [51].

Despite the emergence of such a powerful method as the method of nonlinear capacity, the method based on comparison theorems is extensively used for various nonlinear parabolic equations and systems with nonlocal nonlinearities. The comparison method allows us to get sufficiently subtle results. Many such results have recently been obtained in works of Chinese mathematicians.

3 Method of test functions in combination with method of nonlinear capacity

In this chapter, we discuss the method of test functions developed by V. A. Galaktionov, S. I. Pokhozhaev, and E. L. Mitidieri (see, e. g., [28]) and consisting of a special choice of a test function. Its main advantage is that this method, together with the method of nonlinear capacity, allows obtaining sufficient conditions of the absence of nontrivial solutions for a wide class of problems of mathematical physics.

3.1 Pokhozhaev's identity

Consider the following problem for a nonlinear elliptic partial differential equation in a domain Ω:

$$-\Delta u = |u|^{q-1}u, \tag{3.1}$$

$$u|_{\partial\Omega} = 0. \tag{3.2}$$

Assume the condition

$$\frac{N+2}{N-2} < q. \tag{3.3}$$

We prove that under a certain geometric condition for the domain $\Omega \subset \mathbb{R}^N$, the function $u(x) \equiv 0$ is a unique smooth solution of problem (3.1), whereas restriction (3.3) is natural in a certain sense, so that the exponent

$$q = \frac{N+2}{N-2}$$

can be called *critical*.

Definition 3.1. An open set Ω is said to be *starlike* with respect to the point 0 if for any point $x \in \overline{\Omega}$, the straight line segment $\{\lambda x \mid 0 \le \lambda \le 1\}$ lies in $\overline{\Omega}$.

Obviously, if Ω is convex and $0 \in \Omega$, then Ω is a starlike domain with respect to 0. However, in general, a starlike domain can be nonconvex.

Lemma 3.1. *Let the boundary $\partial\Omega$ belong to the class $\mathbb{C}^1$, and let Ω be a starlike domain with respect to 0. Then*

$$(x, \nu(x)) \ge 0 \quad \forall x \in \partial\Omega,$$

where ν is the unit outer normal.

https://doi.org/10.1515/9783110602074-003

Proof. Since $\partial\Omega$ belongs to the class $\mathbb{C}^1$, for any $x \in \partial\Omega$ and $\varepsilon > 0$, there exists $\delta > 0$ such that, for $|x - y| < \delta$ and $y \in \overline{\Omega}$, we have the inequality

$$\left(\nu(x), \frac{y - x}{|y - x|}\right) \leq \varepsilon,$$

since the value $\frac{y-x}{|y-x|}$ is close to the tangent vector at the point $y \in \partial\Omega$. In particular,

$$\limsup_{\overline{\Omega}\ni y\to x}\left(\nu(x), \frac{y - x}{|y - x|}\right) \leq 0.$$

Let $y = \lambda x$, where $0 < \lambda < 1$. Then $y \in \overline{\Omega}$ since Ω is starlike. Thus we have the equality

$$\left(\nu(x), \frac{x}{|x|}\right) = -\lim_{\lambda\to 1-0}\left(\nu(x), \frac{\lambda x - x}{|\lambda x - x|}\right) \geq 0.$$

The lemma is proved. □

Theorem 3.1. *Let $u \in \mathbb{C}^{(2)}(\overline{\Omega})$ be a solution of problem* (3.1), *and let the exponent q satisfy inequality* (3.3). *Moreover, assume that the set Ω is starlike with respect to the point* 0 *and the boundary $\partial\Omega$ belongs to the class $\mathbb{C}^1$. Then*

$$u \equiv 0 \quad \text{inside } \Omega.$$

Proof. Step 1. Multiplying the equation by $(x, \nabla u)$ and integrating over Ω, we obtain

$$\int_\Omega (-\Delta u)(x, \nabla u)\,dx = \int_\Omega |u|^{q-1}u(x, \nabla u)\,dx. \tag{3.4}$$

We denote the left- and right-hand sides of this equality by A and B, respectively.

Step 2. The left-hand side is of the form

$$\begin{aligned} A &\equiv -\sum_{i,j=1,1}^{N,N}\int_\Omega u_{x_i x_i} x_j u_{x_j}\,dx \\ &= \sum_{i,j=1,1}^{N,N}\int_\Omega u_{x_i}(x_j u_{x_j})_{x_i}\,dx - \sum_{i,j=1,1}^{N,N}\int_{\partial\Omega} u_{x_i}\nu^i x_j u_{x_j}\,dx \equiv A_1 + A_2. \end{aligned} \tag{3.5}$$

Step 3. We have

$$\begin{aligned} A_1 &= \sum_{i,j=1,1}^{N,N}\int_\Omega (u_{x_i}\delta_{ij}u_{x_j} + u_{x_i}x_j u_{x_i x_j})\,dx \\ &= \int_\Omega\left(|\nabla u|^2 + \sum_{j=1}^{N}\left(\frac{|\nabla u|^2}{2}\right)_{x_j} x_j\right)dx \end{aligned}$$

$$= \left(1 - \frac{N}{2}\right) \int_{\Omega} |\nabla u|^2\, dx + \int_{\partial\Omega} \frac{|\nabla u|^2}{2} (\nu, x)\, dS. \tag{3.6}$$

On the other hand, since $u = 0$ on $\partial\Omega$, the gradient ∇u is parallel to the normal ν at each point of the boundary, and we have the following representation:

$$\nabla u \equiv \pm |\nabla u| \nu.$$

Hence we have

$$A_2 = -\int_{\partial\Omega} |\nabla u|^2 (\nu, x)\, dS. \tag{3.7}$$

From (3.5)–(3.7) we conclude that

$$A = \frac{2-N}{2} \int_{\Omega} |\nabla u|^2\, dx - \frac{1}{2} \int_{\partial\Omega} |\nabla u|^2 (\nu, x)\, dS.$$

Step 4. Returning to (3.4), we obtain

$$B \equiv \sum_{j=1}^{N} \int_{\Omega} |u|^{q-1} u x_j u_{x_j}\, dx = \sum_{j=1}^{N} \int_{\Omega} \left(\frac{|u|^{q+1}}{q+1}\right)_{x_j} x_j\, dx = -\frac{N}{q+1} \int_{\Omega} |u|^{q+1}\, dx.$$

Step 5. Using these results and equation (3.4), we have

$$\frac{N-2}{2} \int_{\Omega} |\nabla u|^2\, dx + \frac{1}{2} \int_{\partial\Omega} |\nabla u|^2 (\nu, x)\, dS = \frac{N}{q+1} \int_{\Omega} |u|^{q+1}\, dx. \tag{3.8}$$

Using Lemma 3.1, we arrive at the inequality

$$\frac{N-2}{2} \int_{\Omega} |\nabla u|^2\, dx \le \frac{N}{q+1} \int_{\Omega} |u|^{q+1}\, dx. \tag{3.9}$$

Multiplying equation (3.1) by the function $u(x)$ and integrating by parts, we get

$$\int_{\Omega} |\nabla u|^2\, dx = \int_{\Omega} |u|^{q+1}\, dx.$$

Substituting this into (3.9), we have

$$\left(\frac{N-2}{2} - \frac{N}{q+1}\right) \int_{\Omega} |u|^{q+1}\, dx \le 0.$$

Therefore, if a solution u is not identically zero, then

$$\frac{N-2}{2} - \frac{N}{q+1} \le 0 \quad \Longleftrightarrow \quad q \le \frac{N+2}{N-2}.$$

The theorem is proved. □

3.2 Classical Fujita theorem

In this section, we present a beautiful result obtained in [23]. Consider the following Cauchy problem for a semilinear parabolic equation in $\mathbb{R}^N$:

$$u_t = \Delta u + u^{1+\alpha}, \quad u(x,0) = u_0(x), \quad t > 0,\ x \in \mathbb{R}^N. \tag{3.10}$$

First, we formulate the definition of a classical solution of the Cauchy problem (3.10).

Definition 3.2. A nonnegative function $u(t,x)$ satisfying (3.10) is called a *classical solution* of the Cauchy problem on a segment $[0,T]$ if u, $\nabla_x u$, $\nabla_x \nabla_x u$, and u_t exist and are continuous in the strip $Q_T = [0,T] \times \mathbb{R}^N$.

Definition 3.3. For $T > 0$, let $\mathcal{E}[0,T]$ be the set of all continuous functions $u = u(t,x)$ defined in the strip $[0,T] \times \mathbb{R}^N$ and satisfying the inequality

$$|u(t,x)| \le M \exp(|x|^\beta) \tag{3.11}$$

with a certain constant $M > 0$ and $\beta \in [0,2)$. We also introduce the set $\mathcal{A}$ consisting of functions $v(x)$ such that

$$|v(x)| \le M \exp(|x|^\beta).$$

Theorem 3.2. *Let $0 < \alpha < 2/N$. Assume that $u_0(x) \in \mathcal{A}$ is not identically zero. Then there are no solutions of problem* (3.10) *in $\mathcal{E}[0,T]$ for all $T > 0$.*

Proof. First, we prove an auxiliary lemma.

Lemma 3.2. *Let $u = u(t,x)$ be a classical solution of problem* (3.10) *in $\mathcal{E}[0,T]$ with nontrivial initial condition $u_0(x) \in \mathcal{A}$. Then*

$$J_0^{-\alpha} - u(t,0)^{-\alpha} \ge \alpha t, \quad t \in [0,T], \tag{3.12}$$

where

$$J_0 = J_0(t) = \int_{\mathbb{R}^N} U(t,x) u_0(x)\, dx, \quad U(t,x) = \frac{1}{(4\pi t)^{N/2}} \exp\left(-\frac{|x|^2}{4t}\right). \tag{3.13}$$

Proof. Let $\varepsilon > 0$ be a constant. Fix $t \in [0,T]$ and consider the function

$$J_\varepsilon = J_\varepsilon(s) = \int_{\mathbb{R}^N} v_\varepsilon(s,x) u(s,x)\, dx, \tag{3.14}$$

where

$$v_\varepsilon = v_\varepsilon(s,x) = U(t - s + \varepsilon, x), \quad s \in [0,T],\ x \in \mathbb{R}^N, \tag{3.15}$$

and

$$U(t,x) = \frac{1}{(4\pi t)^{N/2}} \exp\Big(-\frac{|x|^2}{4t}\Big).$$

Clearly, $v_\varepsilon(s,x)$ satisfies the conjugate heat equation

$$\frac{\partial v_\varepsilon}{\partial s} = -\Delta v_\varepsilon.$$

Under the condition $u_0(x) \geq 0$, the function $\mathrm{J}_\varepsilon(s)$ also is nonnegative. Now let $u \in \mathcal{E}[0,T]$; then there exist a positive constant $M > 0$ and a constant $\beta \in (0,2)$ such that

$$0 \leq u(s,x) \leq M \exp(|x|^\beta), \quad s \in [0,T],\ x \in \mathbb{R}^N. \tag{3.16}$$

Applying the change of variables

$$x = 2\eta\sqrt{t-s+\varepsilon},$$

we get the estimate

$$\begin{aligned} 0 \leq \mathrm{J}_\varepsilon(s) &\leq M \int_{\mathbb{R}^N} U(t-s+\varepsilon,x)\exp(|x|^\beta)\,dx \\ &= M\pi^{-N/2} \int_{\mathbb{R}^N} \exp(-|\eta|^2)\exp(|2\sqrt{t-s+\varepsilon}|^\beta|\eta|^\beta)\,d\eta \\ &\leq M\pi^{-N/2} \int_{\mathbb{R}^N} \exp(-|\eta|^2 + \gamma|\eta|^\beta)\,d\eta, \end{aligned} \tag{3.17}$$

where

$$\gamma = 2^\beta(t+\varepsilon)^{\beta/2}.$$

The last inequality shows that the function $\mathrm{J}_\varepsilon(s)$ exists and is continuous with respect to s. Now we prove that $\mathrm{J}_\varepsilon(s)$ is continuously differentiable with respect to s and satisfies the equality

$$\frac{d}{ds}\mathrm{J}_\varepsilon(s) = \int_{\mathbb{R}^N} v_\varepsilon(s,x)u(s,x)^{1+\alpha}\,dx. \tag{3.18}$$

Indeed, let a monotonic function $\rho(x) \in \mathbb{C}_0^\infty(\mathbb{R}^N)$ be such that

$$\rho(x) = \begin{cases} 1 & \text{for } |x| \leq 1, \\ 0 & \text{for } |x| \geq 2. \end{cases}$$

Introduce the notation

$$\rho_n(x) \equiv \rho\left(\frac{x}{n}\right)$$

and set

$$\mathrm{I}^{(n)}(s) = \int_{\mathbb{R}^N} v_\varepsilon(s,x)u(s,x)\rho_n(x)\,dx. \tag{3.19}$$

By the definition of the function $\rho_n(x)$ we have the following equalities:

$$\begin{aligned}\frac{d}{ds}\mathrm{I}^{(N)}(s) &= \int_{\mathbb{R}^N}\left(\frac{\partial v_\varepsilon(s,x)}{\partial s}u + v_\varepsilon\frac{\partial u(s,x)}{\partial s}\right)\rho_n(x)\,dx \\ &= \int_{\mathbb{R}^N}(-u(s,x)\Delta v_\varepsilon(s,x) + v_\varepsilon(s,x)\Delta u(s,x))\rho_n(x)\,dx \\ &\quad + \int_{\mathbb{R}^N} v_\varepsilon(s,x)u^{1+\alpha}(s,x)\rho_n(x)\,dx = \mathrm{I}_1 + \mathrm{I}_2.\end{aligned} \tag{3.20}$$

The condition $u(s,x) \in \mathcal{E}[0,T]$ implies that $u^{1+\alpha}(s,x) \in \mathcal{E}[0,T]$; therefore, we have the following limit equality:

$$\mathrm{I}_2 \to \int_{\mathbb{R}^N} v_\varepsilon(s,x)u(s,x)^{1+\alpha}\,dx \quad \text{as } n \to +\infty.$$

Now we prove that $\mathrm{I}_1 \to 0$ as $n \to +\infty$. Integrating by parts, we have

$$\mathrm{I}_1 = -2\int_{\mathbb{R}^N} v_\varepsilon(s,x)(\nabla_x\rho_n(x), \nabla_x u)\,dx - \int_{\mathbb{R}^N} v_\varepsilon(s,x)u(s,x)\Delta\rho_n(x)\,dx. \tag{3.21}$$

Due to (3.16), the last term in the right-hand side of equation (3.21) can be estimated as follows:

$$p_n = \left|\int_{\mathbb{R}^N} v_\varepsilon(s,x)u(s,x)\Delta\rho_n(x)\,dx\right| \le cn^{-2}\int_{\mathbb{R}^N} v_\varepsilon(s,x)u(s,x)\,dx \le cn^{-2} \to 0$$

as $n \to +\infty$. We can prove that

$$\frac{\partial u}{\partial x_i} \in \mathcal{E}[0,T],$$

and hence

$$\int_{\mathbb{R}^N} v_\varepsilon(s,x)|\nabla_x u(s,x)|\,dx \le c.$$

This implies

$$q_n = \left| \int_{\mathbb{R}^N} v_\varepsilon(s,x)(\nabla_x \rho_n(x), \nabla_x u(s,x))\, dx \right| \le cn^{-1} \to 0$$

as $n \to +\infty$. Similarly, we can verify that

$$\frac{d\mathrm{I}^{(n)}(s)}{ds} \to \frac{d\mathrm{J}_\varepsilon(s)}{ds} \quad \text{as } n \to +\infty,$$

which completes the proof of equation (3.18). Note that $v_\varepsilon(s,x)$ satisfies the equality

$$\int_{\mathbb{R}^N} v_\varepsilon(s,x)\, dx = 1.$$

Therefore by the Jensen inequality we can write

$$\frac{d\mathrm{J}_\varepsilon(s)}{ds} \ge \mathrm{J}_\varepsilon^{1+\alpha}(s), \quad 0 \le s \le t, \tag{3.22}$$

where we have applied the lower convexity of the function $x^{1+\alpha}$ for $x \ge 0$. Solving the differential inequality (3.22), we obtain

$$\mathrm{J}_\varepsilon(0)^{-\alpha} - \mathrm{J}_\varepsilon(t)^{-\alpha} \ge \alpha t. \tag{3.23}$$

Passing to the limit as $\varepsilon \to +0$, we complete the proof of the lemma:

$$\mathrm{J}_\varepsilon(t) \to u(t,0), \quad \mathrm{J}_\varepsilon(0) \to \mathrm{J}_0(t).$$

The lemma is proved. □

Now we return to the proof of Theorem 3.1. By Lemma 3.1 we have the inequality

$$\mathrm{J}_0^{-\alpha}(t) \ge \alpha t. \tag{3.24}$$

Now we estimate $\mathrm{J}_0(t)$ from below. Without loss of generality, we can assume that the initial function $u_0(x)$ is positive in a neighborhood of the origin. Then we can choose positive constants $\gamma > 0$ and $\delta > 0$ such that $u_0(x) \ge \gamma$ for $|x| < 2\delta$.

Assume that $t \ge \delta^2$; then we have the following lower estimate:

$$\mathrm{J}_0(t) = \int_{\mathbb{R}^N} U(x,t)u_0(x)\, dx \ge \int_{|x|\le 2\delta} U(x,t)u_0(x)\, dx \ge \gamma \int_{|x|<2\delta} (4\pi t)^{-N/2} e^{-1}\, dx.$$

Therefore

$$\mathrm{J}_0(t) \ge c_1 t^{-N/2} \quad \text{for } t \ge \delta^2 \tag{3.25}$$

for a certain positive constant $c_1 > 0$. Substituting (3.25) into (3.24), we obtain the inequality

$$t^{N\alpha/2} \ge \alpha c_1 t \quad \text{for } t \ge \delta^2. \tag{3.26}$$

However, (3.26) cannot hold for all $t \ge \delta^2$ since $N\alpha < 2$. The theorem is proved. □

Note that the fact that the Fujita critical exponent is defined by the relation $N\alpha = 2$ was proved later.

3.3 Korteweg–de Vries equation

3.3.1 Introduction

An important role played by the Korteweg–de Vries equation in applications is well known. This equation is one of the fundamental soliton equations; it has the form

$$u_t + uu_x + u_{xxx} = 0.$$

Recently, sufficient conditions for the blow-up in finite time of solutions of initial-boundary-value problem for the Korteweg–de Vries equation were obtained by S. I. Pokhozhaev (see [87–89]); both classical and weak generalized solutions were considered. For this study, the original method of nonlinear capacity developed in [74] was used.

Note that investigation of the blow-up of solutions of the Korteweg–de Vries equation gave impetus to active development of the theory of blow-up for other soliton equations, for example, the Kadomtsev–Petviashvili equation, Rosenau–Burgers equation, Benjamin–Bona–Mahony–Burgers equation, Zakharov–Kuznetsov equation, etc.

3.3.2 Problem on an interval $x \in (0, L)$

Consider the initial-boundary-value problem with zero conditions on the left boundary:

$$u_t - uu_x + u_{xxx} = 0, \quad t > 0,\ x \in (0, L), \tag{3.27}$$

$$u|_{x=0} = u_x|_{x=0} = u_{xx}|_{x=0} = 0, \quad u(x, 0) = u_0(x). \tag{3.28}$$

A solution of problem (3.27)–(3.28) is meant in the classical sense:

$$u(x, t) \in \mathbb{C}^{(1)}([0, T); \mathbb{C}[0, L]) \cap \mathbb{C}([0, T); \mathbb{C}^{(3)}[0, L])$$

for some $T > 0$.

We multiply equation (3.27) by the test function

$$\psi(x) = (L - x)^{\lambda}, \quad \lambda > 5,$$

and integrate by $x \in (0, L)$. Integrating by parts and taking into account the boundary condition (3.28), we obtain the following equality:

$$\frac{dJ}{dt} = \frac{\lambda}{2} \int_0^L (L-x)^{\lambda-1} u^2 \, dx - \lambda(\lambda-1)(\lambda-2) \int_0^L (L-x)^{\lambda-3} u \, dx, \tag{3.29}$$

where

$$J(t) = \int_0^L (L-x)^{\lambda} u \, dx. \tag{3.30}$$

Since

$$1 \geq \frac{L-x}{L} \quad \text{for } x \in [0, L],$$

equation (3.29) implies the inequality

$$\frac{dJ}{dt} \geq \frac{\lambda}{2L} \int_0^L (L-x)^{\lambda} u^2 \, dx - \lambda(\lambda-1)(\lambda-2) \int_0^L (L-x)^{\lambda-3} u \, dx. \tag{3.31}$$

We have the following inequalities:

$$\begin{aligned} \left| \int_0^L (L-x)^{\lambda-3} u \, dx \right| &\leq \left(\int_0^L (L-x)^{\lambda-6} \, dx \right)^{1/2} \left(\int_0^L (L-x)^{\lambda} u^2 \, dx \right)^{1/2} \\ &\leq \left(\frac{L^{\lambda-5}}{\lambda-5} \right)^{1/2} \left(\int_0^L (L-x)^{\lambda} u^2 \, dx \right)^{1/2}. \end{aligned} \tag{3.32}$$

Thus from estimate (3.32) we obtain the inequality

$$\begin{aligned} &\lambda(\lambda-1)(\lambda-2) \left| \int_0^L (L-x)^{\lambda-3} u \, dx \right| \\ &\quad \leq \frac{\lambda}{4L} \int_0^L (L-x)^{\lambda} u^2 \, dx + \frac{\lambda(\lambda-1)^2(\lambda-2)^2}{\lambda-5} L^{4-\lambda}. \end{aligned} \tag{3.33}$$

Using estimates (3.32)–(3.33), we rewrite (3.31) in the form

$$\frac{dJ}{dt} \geq \frac{\lambda}{4L} \int_0^L (L-x)^{\lambda} u^2 \, dx - c_{\lambda} L^{\lambda-4}, \tag{3.34}$$

where

$$c_\lambda = \frac{\lambda(\lambda-1)^2(\lambda-2)^2}{\lambda-5} \quad \text{for } \lambda > 5.$$

By the Jensen inequality we have

$$\frac{\lambda}{4L}\int_0^L (L-x)^\lambda u^2\,dx = \frac{\lambda}{4L}\frac{L^{\lambda+1}}{\lambda+1}\int_0^L \frac{(L-x)^\lambda}{L^{\lambda+1}/(\lambda+1)}u^2\,dx$$
$$\geq \frac{\lambda(\lambda+1)}{4L^{\lambda+1}}\Big(\int_0^L (L-x)^\lambda u\,dx\Big)^2. \tag{3.35}$$

Thus from (3.34) we obtain the ordinary differential inequality

$$\frac{dJ}{dt} \geq b_\lambda L^{-\lambda-2}J^2 - c_\lambda L^{\lambda-4}, \tag{3.36}$$

where

$$b_\lambda = \frac{\lambda(\lambda+1)}{4}, \quad c_\lambda = \frac{\lambda(\lambda-1)^2(\lambda-2)^2}{\lambda-5}.$$

Now we assume that $L \geq 1$; then from inequality (3.36) we can obtain the following rougher estimate:

$$\frac{dJ}{dt} \geq b_\lambda L^{-\lambda-2}J^2 - c_\lambda L^{\lambda+2}. \tag{3.37}$$

We rewrite it in the form

$$\frac{dJ}{dt} \geq k^2J^2 - a^2, \quad k^2 = b_\lambda L^{-\lambda-2}, \quad a^2 = c_\lambda L^{\lambda+2}. \tag{3.38}$$

Thus, integrating the differential inequality (3.38), we obtain the following statement.

Theorem 3.3. *Let*

$$J_0 = J(0) = \int_0^L (L-x)^\lambda u_0\,dx > \frac{a}{k}, \quad L \geq 1,\ \lambda > 5. \tag{3.39}$$

Then there are no global in time classical solutions of problem (3.27)–(3.28). *Moreover, we have the following inequality*:

$$J(t) \geq \frac{a}{k}\frac{1+c_0e^{2akt}}{1-c_0e^{2akt}}, \quad c_0 = \frac{kJ_0-a}{kJ_0+a}, \tag{3.40}$$

and hence

$$\lim_{t\uparrow T_\infty} J(t) = +\infty, \quad T \leq T_\infty \leq -\frac{1}{2ak}\ln\Big(\frac{kJ_0-a}{kJ_0+a}\Big), \tag{3.41}$$

where

$$k = \frac{\sqrt{\lambda(\lambda+1)}}{2} L^{-\lambda/2-1}, \quad a = \frac{\sqrt{\lambda}(\lambda-1)(\lambda-2)}{\sqrt{\lambda-5}} L^{\lambda/2+1}. \tag{3.42}$$

Before proceeding to the following result, we note that the zero boundary conditions at one end are not necessary but sufficient for the applicability of the method. An attentive reader can easily invent nonzero conditions or two-sided conditions that also lead to an inequality of the form (3.38) and, therefore, to the blow-up result. For brevity, we examine the following problem also with zero condition, but now we consider an unbounded domain.

3.3.3 Problem on the half-line $x > 0$

Consider the initial-boundary-value problem

$$u_t - uu_x + u_{xxx} = 0, \quad t > 0,\ x \in (0, +\infty), \tag{3.43}$$

$$u|_{x=0} = u_x|_{x=0} = u_{xx}|_{x=0} = 0, \quad u(x,0) = u_0(x). \tag{3.44}$$

A solution of problem (3.43)–(3.44) is meant in the classical sense:

$$u(x,t) \in \mathbb{C}^{(1)}([0,T); \mathbb{C}[0,+\infty)) \cap \mathbb{C}([0,T); \mathbb{C}^{(3)}[0,+\infty))$$

for some $T > 0$. As before, we multiply equation (3.43) by the power test function

$$\psi(x) = (L-x)^{\lambda} \quad \text{for } \lambda > 5$$

and integrate by $x \in (0, L)$. Now we pass to the limit as $L \to +\infty$ in the formulas of Theorem 3.3. First, we rewrite condition (3.39) in the form

$$\int_0^L (L-x)^{\lambda} u_0(x)\, dx > \left(\frac{c_\lambda}{b_\lambda}\right)^{1/2} L^{\lambda+2}. \tag{3.45}$$

Performing the change of variables $x = Lr$ in the left-hand side of (3.45), we obtain the inequality

$$L \int_0^1 (1-r)^{\lambda} u_0(Lr)\, dr > \left(\frac{c_\lambda}{b_\lambda}\right)^{1/2}. \tag{3.46}$$

Assume the following condition:

$$\lim_{L\to+\infty} L \int_0^1 (1-r)^{\lambda} u_0(Lr)\, dr = b_\infty > \left(\frac{c_\lambda}{b_\lambda}\right)^{1/2}. \tag{3.47}$$

Moreover, we introduce the value

$$J_\infty(t) = \lim_{L\to+\infty} L\int_0^1 (1-r)^\lambda u(Lr,t)\,dr. \tag{3.48}$$

By Theorem 3.1 we obtain the following result.

Theorem 3.4. *Let*

$$J_\infty(0) = b_\infty = \lim_{L\to+\infty} L\int_0^1 (1-r)^\lambda u_0(Lr)\,dr > \left(\frac{c_\lambda}{b_\lambda}\right)^{1/2}, \quad \lambda > 5. \tag{3.49}$$

Then there are no global in time classical solutions of problem (3.43)–(3.44). *Moreover, we have the following inequality*:

$$J_\infty(t) \geq \left(\frac{c_\lambda}{b_\lambda}\right)^{1/2} \frac{1 + c_\infty e^{2\sqrt{c_\lambda b_\lambda}t}}{1 - c_0 e^{2\sqrt{c_\lambda b_\lambda}t}}, \qquad c_\infty = \frac{b_\infty - (c_\lambda/b_\lambda)^{1/2}}{b_\infty + (c_\lambda/b_\lambda)^{1/2}}, \tag{3.50}$$

and therefore

$$\lim_{t\uparrow T_\infty} J_\infty(t) = +\infty, \qquad T \leq T_\infty \leq -\frac{1}{2\sqrt{c_\lambda b_\lambda}} \ln(c_\infty). \tag{3.51}$$

We make the following important remark. In the examples presented, we choose as a test function either an abstract function or a power function. This choice yields sufficient conditions of blow-up but does not provide optimal conditions. To obtain an optimal estimate of the lifetime and the rate of blow-up, we must solve a variation problem for an optimal test function. However, this drawback of the method has not yet been overcome.

3.4 Method of first eigenfunction

Now we analyze the blow-up phenomenon for multidimensional elliptic problems in the case where as a test function we choose a nonnegative eigenfunction of the corresponding eigenvalue problem.

First, we consider a simple example of the initial-boundary-value problem in the ball $\Omega_R \equiv \{x \in \mathbb{R}^N : |x| < R\}$:

$$u_t = \Delta u + |u|^q, \quad x \in \Omega_R,\ q > 1, \tag{3.52}$$

$$u|_{t=0} = u_0(x), \quad u|_{\partial\Omega_R} = \mu(t). \tag{3.53}$$

We find initial and boundary conditions under which the problem has no classical solutions global in time. As a test function $\psi(x)$, we take an eigenfunction of the following problem:

$$\Delta\psi + \lambda\psi = 0 \quad \text{for } x \in \Omega_R, \quad \psi = 0 \quad \text{for } x \in \partial\Omega_R. \tag{3.54}$$

We choose a solution of this eigenvalue problem so that $\psi(x)$ is nonnegative and $\lambda > 0$. To do this, we choose the first eigenfunction

$$\psi_1(x) = c_0 r^{(N-2)/2} J_{(N-2)/2}(\lambda_1^{1/2} r)$$

corresponding to the first eigenvalue

$$\lambda_1 = \left[\frac{z_N^{(1)}}{R}\right]^2,$$

where $z_N^{(1)}$ is the first positive root of the Bessel function $J_{(N-2)/2}(z)$, and the positive constant c_0 is determined by the normalization condition (see, e. g., [94, p. 327]).

Multiplying both sides of equation (3.52) by the test function $\psi_1(x)$ and integrating over the domain Ω_R, we apply the second Green formula:

$$\int_{\Omega_R} \Delta u \psi_1 \, dx = \int_{\Omega_R} u \Delta \psi_1 \, dx + \int_{\partial\Omega_R} \left(\psi_1 \frac{\partial u}{\partial n} - u \frac{\partial \psi_1}{\partial n}\right) dx. \tag{3.55}$$

Since $\partial/\partial n$ is the derivative in the direction of the outer normal to $\partial\Omega_R$ and $\psi_1 = 0$, we see that $\partial\psi_1/\partial n \le 0$ on the boundary due to the positiveness of $\psi_1(x)$ inside the domain. If we assume that $\mu(t) \ge 0$, then from the Green's formula we obtain the inequality

$$\frac{d}{dt} \int_{\Omega_R} u\psi_1 \, dx \ge \int_{\Omega_R} |u|^q \psi_1 \, dx - \lambda_1 \int_{\Omega_R} u\psi_1 \, dx. \tag{3.56}$$

Now we use the Hölder inequality

$$\left|\int_{\Omega_R} u\psi_1 \, dx\right|^q \le \left(\int_{\Omega_R} |u|^q \psi_1 \, dx\right)\left(\int_{\Omega_R} \psi_1 \, dx\right)^{q-1} \tag{3.57}$$

and rewrite (3.56) in the form of an ordinary differential inequality

$$\frac{dJ(t)}{dt} \ge kJ^q - \lambda_1 J, \quad J(t) = \int_{\Omega_R} u\psi_1 \, dx, \quad k = \left(\int_{\Omega_R} \psi_1 \, dx\right)^{1-q}. \tag{3.58}$$

Let the initial conditions satisfy the inequality

$$J^{q-1}(0) > \frac{\lambda_1}{k}.$$

Then, integrating (3.58) by time, we arrive at the inequality

$$\frac{1}{q-1} \ln\left(\frac{kJ^{q-1} - \lambda_1}{kJ_0^{q-1} - \lambda_1}\right) - \ln\left(\frac{J}{J_0}\right) \ge \lambda_1 t, \quad J_0 = J(0), \tag{3.59}$$

which can be rewritten in the form

$$J(t) \geq \frac{\lambda_1 J_0^{q-1}}{kJ_0^{q-1} - (kJ_0^{q-1} - \lambda_1)\exp(\lambda_1(q-1)t)}. \tag{3.60}$$

This result is the content of the following theorem.

Theorem 3.5 (theorem on the global insolvability). *Let the boundary function satisfy the condition $\mu(t) \geq 0$, and let the initial data satisfy the inequality*

$$J_0 \equiv \int_{\Omega_R} u_0(x)\psi_1(x)\,dx \geq \left(\frac{\lambda_1}{k}\right)^{1/(q-1)}.$$

Then there are no global in time solutions of problem (3.52)–(3.53). *Moreover, the lifetime of the solution has the following upper estimate*:

$$T < \frac{1}{\lambda_1(q-1)} \ln\left(\frac{kJ_0^{q-1}}{kJ_0^{q-1} - \lambda_1}\right).$$

Remark 3.1. We note that a similar problem can also be stated for a second-order elliptic operator of a general form and an arbitrary bounded domain with sufficiently smooth boundary. In this case, a unique necessary condition is the existence of a positive eigenfunction of the eigenvalue problem and the uniform ellipticity of the corresponding elliptic operator. Now we present a proof of the existence of a sign-constant eigenfunction of the Laplace operator in an arbitrary bounded domain with an arbitrary smooth boundary (see, e. g., [100]).

Consider the conditional variation extremum problem for the functional

$$H(u) = \frac{1}{2}\int_{\Omega} |\nabla u|^2\,dx$$

on the manifold

$$G(u) = \frac{1}{2}\int_{\Omega} |u|^2\,dx = 1.$$

Based on the completely continuous embedding $\mathbb{H}_0^1(\Omega) \hookrightarrow \mathbb{L}^2(\Omega)$, we can easily prove the following four facts:

(1) the manifold

$$L \equiv \{u \in \mathbb{H}_0^1(\Omega),\ G(u) = 1\}$$

is weakly sequentially closed in $\mathbb{H}_0^1(\Omega)$;

(2) the functional $H(u)$ is weakly sequentially lower semicontinuous on the Hilbert space $\mathbb{H}_0^1(\Omega)$;

(3) the functional $H(u)$ is coercive on the manifold L;
(4) the functionals $H(u)$ and $G(u)$ are Fréchet differentiable.

These properties and results of [100] imply the existence of a unique extremal $u \in \mathbb{H}_0^1(\Omega)$, which provides a minimum of the functional $H(u)$ on the manifold $\{u \in \mathbb{H}_0^1(\Omega), G(u) = 1\}$; moreover, due to the Lagrange theorem (see [100]), the extremal satisfies the equation

$$\langle H_f'(u) - \lambda G_f'(u), \varphi \rangle = 0 \quad \Rightarrow \quad \langle \Delta u + \lambda u, \varphi \rangle = 0 \quad \text{for all } \varphi \in \mathbb{H}_0^1(\Omega),$$

where H_f' and G_f' are the Fréchet derivatives of the functionals H and G, respectively, and $\langle \cdot, \cdot \rangle$ is the duality bracket between $\mathbb{H}_0^1(\Omega)$ and $\mathbb{H}^{-1}(\Omega)$.

Note that, together with u, the function $|u|$ is also a solution of the variation problem. Indeed,

$$\big|\nabla |u|\big| = \big|\operatorname{sign}(u)\nabla u\big| = |\nabla u|, \quad H(|u|) = H(u), \quad G(|u|) = G(u) = 1.$$

Remark 3.2. Here we have used the equality

$$\nabla |u| = \operatorname{sign}(u)\nabla u$$

(in the weak sense) if the operator ∇ is meant in the weak sense.

Note that, for higher-order elliptic equations, a similar conclusion is invalid since the function $|u|$ is not twice differentiable in the weak sense if u is sign-alternating.

Due to the uniqueness, we see that the extremal $u \in \mathbb{H}_0^1(\Omega)$ is sign-constant.

Remark 3.3. In addition, we note that the blow-up result can also be reformulated for generalized solutions. Certainly, in this case, Green's formula has another form, but this does not complicates the proof (see, e. g., [94]).

Remark 3.4. The method presented can be applied for a wide range of problems. In particular, it is applicable to a Sobolev-type equation of the form

$$\frac{\partial}{\partial t}(\Delta u - u) + \Delta u + f(x, u) = 0, \quad f(x, u) \geq c|u|^q, \quad q > 1, \ c > 0.$$

3.5 Bibliographical notes

The material of this chapter is based on [84, 18, 23]. Further results concerning equations of the Korteweg–de Vries type and similar equations can be found in [50, 108, 109].

In addition to test functions of the form $(L - x)^\lambda$, we can also use test functions from the kernel of the corresponding differential operator (see, e. g., [49, 79]).

4 Energy method of H. A. Levine

In this chapter, we consider the well-known classical energy method proposed by Levine [57, 58]. Due to its simplicity and ease of understanding, this method is one of the most popular methods of the study of blow-up of solutions to various hyperbolic and parabolic problems of mathematical physics. Nowadays, this method has many modifications, in particular, for pseudo-hyperbolic, pseudo-parabolic, and integro-differential equations (see the following chapters).

4.1 Blow-up of solutions to nonlinear parabolic equations

4.1.1 Introduction

In the middle of the 20th century the first results appeared showing that the global-in-time solvability is not at all an attribute of initial and initial-boundary-value nonlinear parabolic problems (see [22, 23, 38]). However, specific examples of nonlinear equations whose solutions blow up in finite time were little known. The energy method invented by H. A. Levine was initially intended namely for the search for such nonlinear equations, but due to its conciseness, it became indispensable for the analysis of the blow-up of a wide class of problem of mathematical physics.

We state the most general theorems proved by H. A. Levine. Roughly speaking, the sense of these theorems is that a continuously differentiable solution $u(\cdot)$ exists only in a finite time interval $[0, T)$ and unboundedly increases as $t \to T$, that is,

$$(u, Pu) \quad \text{and} \quad \int_0^t (u, Pu)\, d\eta$$

tend to infinity.

Consider an abstract parabolic problem

$$P\frac{du}{dt} = -Au + F(u(t)), \quad t \in [0, T), \tag{4.1}$$
$$F(0) = 0, \quad u(0) = u_0,$$

where P and A are "positive" linear operators defined in a subdomain D of a real or complex Hilbert space, and the nonlinear term $F(u)$ satisfies the following conditions.

Conditions for $F : D \to H$

(i)$_1$ for any $x \in D$, $F(x)$ has a symmetric Fréchet derivative F_x, and the mapping $x \to F_x$ is strongly continuous;

https://doi.org/10.1515/9783110602074-004

$(\mathrm{ii})_1$ the scalar function $G : D \to R$ defined by the formula

$$G(x) = \int_0^1 (F(\rho x), x)\, d\rho \tag{4.2}$$

for $x \in D$ satisfies the following inequalities for some constant $\alpha > 0$:

$$2(\alpha + 1)G(x) \le (x, F(x)), \quad G(u_0) > \frac{1}{2}(u_0, A(0)u_0). \tag{4.3}$$

Note, however, that for simple equations of the form

$$u_t = u_{xx} + u_x + u^2,$$

the Levine energy method is inapplicable, but the blow-up result can be easily obtained by some less well-known methods (see [22, 38]). The reason for this lies in the fact that the operator $A = -d^2/dx^2 - d/dx$ is not symmetric; since the methods of [22] are based on the maximum principle, problems with d/dx do dot appear. On the contrary, for the equations

$$u_t = -u_{xxxx} + u^2, \quad u_t - u_{xxt} = -u_{xxxx} + u^2,$$

the absence of global-in-time solutions can be easily proved by the energy method, which is impossible by the method of [22] since higher-order parabolic equations do not admit the maximum principle. The last fact can be illustrated by the following problem for the fourth-order parabolic equation:

$$\begin{gathered} u_t = -u_{xxxx}, \quad D = (0, \sqrt{2}\pi) \times [0, T], \\ u(x, 0) = e^{x/\sqrt{2}} \sin \frac{x}{\sqrt{2}}, \quad x \in (0, \sqrt{2}\pi), \\ u(0, t) = u(\sqrt{2}\pi, t) = 0, \end{gathered}$$

whose solution

$$u(x, t) = \exp\left(t + \frac{x}{\sqrt{2}}\right) \sin \frac{x}{\sqrt{2}}$$

has a strict extremum $\exp(3\pi/4 + T)/\sqrt{2}$ at the inner point $(3\sqrt{2}\pi/4, T)$.

Another advantage of the energy method is manifested when comparing with the method used in [38], where the result on the global-in-time insolvability was proved for the equation

$$\frac{\partial u}{\partial t} = Au + G(u, t), \quad A = \sum_{i,j=1}^{n} \frac{\partial}{\partial x_i}\left(a_{ij}(x) \frac{\partial}{\partial x_j}\right),$$

where A is a second-order elliptic operator. The absence of a global solution for an arbitrary initial function was proved for a cylinder $\Omega \times [0, \infty)$, where Ω is a bounded domain, and Dirichlet boundary conditions. The theorem is proved for a wide class of nonlinearities (including $G(u, t) = u^2$). The proof proposed in [38] does not use the maximum principle but is based on the fact that the first eigenfunction of the operator A is simple and sign-constant in Ω. However, similar facts are unknown for the case of higher-order elliptic operators and for multidimensional spaces, where the method from [38] became inapplicable; at the same time, the Levine method is insensitive to the presence or absence of eigenfunctions.

4.1.2 General theorems

We further assume that P and A are symmetric linear operators defined on a dense embedding

$$D \overset{\text{ds}}{\subset} \mathbb{H}$$

(for convenience, we assume that $\mathbb{H}$ is a real Hilbert space). By the conditions for $F : D \to \mathbb{H}$, for any strictly continuous differentiable mapping $v : [0, T) \to D$, we have the following formal equalities:

$$\begin{aligned}
\frac{d}{dt} G(v(t)) &= \frac{d}{dt} \int_0^1 (F(\rho v), v)\, d\rho \\
&= \int_0^1 [\rho(F_{\rho v}(\rho v)v, v_t) + (F(\rho v), v_t)]\, d\rho \\
&= \int_0^1 \left[\rho \frac{d}{d\rho}(F(\rho v), v_t) + (F(\rho v), v_t)\right] d\rho \\
&= \int_0^1 \frac{d}{d\rho} [\rho(F(\rho v), v_t)]\, d\rho = (F(v(t)), v_t(t)).
\end{aligned} \tag{4.4}$$

Theorem 4.1. *Assume that $u : [0, T) \to D$ is a strictly continuous, differentiable with respect to the norm of D, nonextendable solution of problem (4.1). For any t, let the conditions for $F : D \to H$ hold, and let symmetric operators A and P satisfy the inequalities*

(i)$_2$ $(x, Px) > 0$ *for all* $x \in D$, $x \neq 0$;

(ii)$_2$ $(x, A(t)x) \geq 0$ *for all* $x \in D$; *and*

(iii)$_2$ $(x, \dot{A}(t)x) \leq 0$ *for all* $x \in D$.

Then the lifetime T of the solution is bounded:

$$\lim_{t\to T^-}\int_0^t (u(\eta), Pu(\eta))\,d\eta = +\infty, \quad \limsup_{t\to T^-}(u(t), Pu(t)) = +\infty,$$

and we have the following upper estimate:

$$T \le \frac{(2\alpha+1)(u_0, Pu_0)}{\alpha^2(2\alpha+2)}\left[G(u_0) - \frac{1}{2}(u_0, A(0)u_0)\right]^{-1}. \tag{4.5}$$

Proof. We use the convexity property of a power of functional that has a sense of energy. Assume that $T = +\infty$. For all $T_0 > 0$, $\beta > 0$, and $\tau > 0$, we define the functional mentioned as follows:

$$\Phi(t) = \int_0^t (u, Pu)\,d\eta + (T_0 - t)(u_0, Pu_0) + \beta(t+\tau)^2 \tag{4.6}$$

for $t \in [0, T_0]$. Taking into account the equalities

$$\begin{aligned}\Phi'(t) &= (u, Pu) - (u_0, Pu_0) + 2\beta(t+\tau)\\ &= \int_0^t \frac{d}{d\eta}(u, Pu)\,d\eta + 2\beta(t+\tau)\\ &= \int_0^t [(u_\eta, Pu) + (u, Pu_\eta)]d\eta + 2\beta(t+\tau)\\ &= 2\int_0^t (u, Pu_\eta)\,d\eta + 2\beta(t+\tau)\end{aligned} \tag{4.7}$$

for all $t \in [0, T_0]$, we conclude that

$$\Phi'(0) = 2\beta\tau > 0, \quad \Phi(t) > 0.$$

Thus, for any $\alpha > 0$, we can define $\Phi^{-\alpha}(t)$. Assume that $(\Phi^{-\alpha})'' \le 0$; then integrating by time, we obtain

$$\Phi^{-\alpha} \le \Phi^{-\alpha}(0) + [\Phi^{-\alpha}]'(0)t$$

or, extracting the root of a power $-\alpha$,

$$\Phi(t) \ge \Phi^{1+1/\alpha}(0)[\Phi(0) - \alpha t\Phi'(0)]^{-1/\alpha}. \tag{4.8}$$

It is easy to see that, for sufficiently large τ, we can choose T_0 such that

$$T_0 \ge \frac{\Phi(0)}{\alpha\Phi'(0)} \equiv T_{\beta\tau}.$$

Thus, from (4.1) and (4.8) we conclude that the existence interval is embedded in the interval

$$\left[0, \frac{\Phi(0)}{\alpha\Phi'(0)}\right),$$

and the limit equalities of Theorem 4.1 hold under the assumption $(\Phi^{-\alpha})''(t) \leq 0$. Obviously, this assumption is equivalent to the condition

$$\Phi\Phi'' - (\alpha+1)(\Phi')^2 \geq 0. \tag{4.9}$$

Differentiating (4.7) by time, we obtain

$$\begin{aligned}\Phi''(t) &= 2\int_0^t \frac{d}{d\eta}(u_\eta, Pu)\,d\eta + 2(u_t, Pu)(0) + 2\beta \\ &= 4(\alpha+1)\left[\int_0^t (u_\eta, Pu_\eta)\,d\eta + \beta\right] \\ &\quad + 2\int_0^t \left[\frac{d}{d\eta}(u_\eta, Pu) - 2(\alpha+1)(u_\eta, Pu_\eta)\right]d\eta \\ &\quad + 2[(u_t, Pu)(0) - (2\alpha+1)\beta].\end{aligned} \tag{4.10}$$

In the last equality, we have used simple algebraic ideas. Introduce the value S and verify its well-posedness by using the Cauchy–Bunyakovsky inequality:

$$\begin{aligned}S^2 &= \left[\int_0^t (u, Pu)\,d\eta + \beta(t+\tau)^2\right]\left[\int_0^t (u_\eta, Pu_\eta)\,d\eta + \beta\right] \\ &\quad - \left[\int_0^t (u_\eta, Pu)\,d\eta + \beta(t+\tau)\right]^2 \\ &= \left[\int_0^t (u, Pu)\,d\eta \int_0^t (u_\eta, Pu_\eta)\,d\eta - \left(\int_0^t (u_\eta, Pu)\,d\eta\right)^2\right] \\ &\quad + \beta\left[(t+\tau)^2 \int_0^t (u_\eta, Pu_\eta)\,d\eta + \int_0^t (u, Pu)\,d\eta\right. \\ &\quad \left. - 2(t+\tau)\int_0^t (u_\eta, Pu)\,d\eta\right] \geq 0.\end{aligned} \tag{4.11}$$

By the properties of the operator P we have

$$-(\alpha+1)(\Phi')^2 = 4(\alpha+1)S^2 - 4(\alpha+1)\left[\int_0^t (u_\eta, Pu_\eta)\,d\eta + \beta\right](\Phi - (T_0 - t)(u_0, Pu_0))$$

$$\geq 4(\alpha+1)S^2 + 4(\alpha+1)(T_0 - t)(u_0, Pu_0)\left[\int_0^t (u_\eta, Pu_\eta)\,d\eta + \beta\right], \tag{4.12}$$

where the expression in square brackets is nonnegative. Multiplying (4.10) by Φ and substituting expression (4.1), we obtain the equality

$$\begin{aligned}\Phi\Phi'' &\geq 2\Phi\left[-\int_0^t \left(\frac{d}{d\eta}(u, Au) - 2(\alpha+1)(u_\eta, Au)\right)d\eta\right]\\ &\quad + 2\Phi\int_0^t \left[\frac{d}{d\eta}(u, F(u)) - 2(\alpha+1)(u_\eta, F(u))\right]d\eta\\ &\quad + 2\Phi((u_t, Pu)_0 - (2\alpha+1)\beta).\end{aligned} \tag{4.13}$$

Combining (4.12) and (4.13), we obtain the expression that is involved in condition (4.9):

$$\begin{aligned}\Phi\Phi'' - (\alpha+1)(\Phi')^2 &\geq 4(\alpha+1)S^2 + 2\Phi\left[-\int_0^t \left(\frac{d}{d\eta}(u, Au) - 2(\alpha+1)(u_\eta, Au)\right)d\eta\right]\\ &\quad + 2\Phi\int_0^t \left[\frac{d}{d\eta}(u, F(u)) - 2(\alpha+1)(u_\eta, F(u))\right]d\eta\\ &\quad + 4(\alpha+1)(T_0 - t)(u_0, Pu_0)\left[\int_0^t (u_\eta, Pu_\eta)\,d\eta + \beta\right]\\ &\quad + 2\Phi[(u_t, Pu)_0 - (2\alpha+1)\beta].\end{aligned} \tag{4.14}$$

Since the third term in the square brackets is nonnegative, it can be omitted. Using property (iii$_2$), equality (4.4), and the first property in (4.3), respectively, we analyze the expression in the first, second, and third square brackets. Finally, we have

$$\begin{aligned}\Phi\Phi'' - (\alpha+1)(\Phi')^2 &\geq 4\alpha\Phi\int_0^t (u_\eta, Au)\,d\eta + 2\Phi[(u, F(u)) - 2(\alpha+1)G(u)NNN]\\ &\quad + 2\Phi[2(\alpha+1)G(u_0) - (u_0, A(0)u_0) - (2\alpha+1)\beta].\end{aligned} \tag{4.15}$$

Now, substituting the inequality

$$\begin{aligned}4\int_0^t (u_\eta, Au)\,d\eta &= 2\int_0^t \left[\frac{d}{d\eta}(u, Au) - (u, A_\eta u)\right]d\eta\\ &\geq 2(u, Au) - 2(u_0, A(0)u_0) \geq -2(u_0, A(0)u_0)\end{aligned}$$

into (4.15) and applying the conditions for F and (4.15), we obtain, for all $t \in [0, T_0)$, the inequality

$$\Phi\Phi'' - (\alpha+1)(\Phi')^2 \geq 4(\alpha+1)\Phi\left[G(u_0) - \frac{1}{2}(u_0, A(0)u_0) - \frac{(2\alpha+1)\beta}{2(\alpha+1)}\right]. \tag{4.16}$$

Thus, for all $\beta > 0$ satisfying the estimate

$$(2\alpha+1)\beta \leq 2(\alpha+1)\left[G(u_0) - \frac{1}{2}(u_0, A(0)u_0)\right],$$

we obtain that the assumption $(\Phi^{-\alpha})'' \leq 0$ is valid. Therefore, the lifetime of the solution is bounded.

Introduce the notation

$$\beta_0 = \frac{2(\alpha+1)}{2\alpha+1}\left[G(u_0) - \frac{1}{2}(u_0, A(0)u_0)\right],$$
$$T_{\beta\tau} = \frac{\Phi(0)}{\alpha\Phi'(0)} = \frac{T_0(u_0, Pu_0) + \beta\tau^2}{2\alpha\beta\tau}.$$

Note that estimate (4.9) remains valid even for $T_0 = T_{\beta\tau}$; in this case,

$$T_{\beta\tau} = \beta\tau^2[2\alpha\beta\tau - (u_0, Pu_0)]^{-1},$$

that is, we must take τ so large that

$$2\alpha\beta\tau > (u_0, Pu_0).$$

The function

$$T_{\beta\tau} = T_{\beta\tau}(\tau)$$

has the minimal value $T_{\beta\tau} = (u_0, Pu_0)/(\alpha^2\beta)$ at the point

$$\tau = \frac{(u_0, Pu_0)}{(\alpha\beta)}.$$

On the other hand, as a function of β, the value $T_{\beta\tau}$ becomes minimal for $\beta = \beta_0$ since β is bounded by the segment $(0, \beta_0]$. Therefore, the lifetime T of the solution does not exceed $(u_0, Pu_0)\alpha^{-2}\beta_0^{-1}$. □

Theorem 4.1 implies the following important consequence: there exist a number of initial functions u_0 corresponding to solutions of problem (4.1) that blow up in finite time.

Assertion 4.1. *Assume that the nonlinear term $F(u)$ satisfies the equality*

$$F(sx) = s^{1+\delta}F(x)$$

for $x \in D$, where $\delta > 0$ is a certain constant, and $s > 0$. Moreover, assume that there exists an element $x_0 \in D$ such that $(x_0, F(x_0)) > 0$. Then there exists an infinite number of initial functions u_0 for which condition (ii$_1$) *of Theorem 4.1 holds.*

Proof. Indeed, take $u_0 = sx_0$ and choose s so large that

$$s^{\delta+2}G(x_0) = s^{\delta+2}\int_0^1 (F(\rho x_0), x_0)\, d\rho > \frac{s^2}{2}(x_0, A(0)x_0)$$

Then condition (ii$_1$) of Theorem 4.1,

$$G(u_0) > \frac{1}{2}(u_0, A(0)u_0),$$

holds for all

$$s > [2^{-1}(2+\delta)(x_0, A(0)x_0)(x_0, F(x_0))^{-1}]^{1/\delta}.$$

Note also that if F satisfies the conditions of Theorem 4.1 and Assertion 4.1, then $T = T(u_0) = T(sx_0) \to 0$ as $s \to +\infty$. This result is a direct consequence of inequality (4.5):

$$0 < T(sx_0) \le \frac{(2\alpha+1)(x_0, Px_0)}{2\alpha^2(\alpha+1)}\left[G(x_0)s^\delta - \frac{(x_0, A(0)x_0)}{2}\right]^{-1},$$

where the right-hand side tends to zero as $s \to +\infty$. Roughly speaking, this means that the larger the initial data, the smaller the lifetime of the solution. The assertion is proved. □

Remark 4.1. Another feature of the energy method and Theorem 4.1 is the importance of the second condition in (4.3). We clarify this by the following example. Let a function $f \in \mathbb{C}^{(2)}(0,\pi)$, $f \not\equiv 0$, satisfy the boundary-value problem

$$f'' + f^2 = 0, \quad f(0) = f(\pi) = 0.$$

Then $u(x,t) = f(x)$ is a global-in-time solution of the following parabolic initial-boundary-value problem:

$$u_t = u_{xx} + u^2, \quad \Omega = (0,\pi) \times [0,\infty),$$
$$u(x,0) = f(x), \quad x \in (0,\pi),$$
$$u(0,t) = u(\pi,t) = 0.$$

The blow-up of a solution does not occur, although the nonlinearity $F(u) = u^2$ satisfies all conditions of the theorem with $\alpha = 1/2$ and

$$G(f) = \frac{1}{3}\int_0^{\pi} f^3\,dx,$$

except for the second condition in (4.3):

$$G(f) = \frac{1}{3}(f, f^2) = \frac{1}{3}(f', f') < \frac{1}{2}(f', f').$$

A blow-up result similar to Theorem 4.1 can be also obtained for weak solutions. Denote by $D_* \supseteq D$ a dense subdomain of $\mathbb{H}$ on which $P^* \equiv P^{1/2}$, $A^* \equiv A^{1/2}$, and $F(\cdot)$ are defined (here we assume that A is independent of time).

Definition 4.1. We say that a function $u : [0, T) \to D_*$ is a *weak solution* of problem (4.1) if u, P^*u, A^*u, and $F(u)$ are strictly continuous, the function u possesses weak derivatives that are defined on D_* and are locally integrable on $[0, T)$, P^*u possesses the weak derivative $(P^*u)_t$ such that $(P^*u)_t = P^*u_t$, and for all $\varphi : [0, T) \to D_*$, we have the following relation:

$$(P^*\varphi, P^*u) = (P^*\varphi(0), P^*u(0)) + \int_0^t [(P^*\varphi_\eta, P^*u) + (\varphi, F(u)) - (A^*\varphi, A^*u)]\,d\eta. \tag{4.17}$$

Note that the definition of a weak solution implies the following relation:

$$\frac{d}{d\eta}(P^*u, P^*u) = 2(P^*u_\eta, P^*u).$$

In addition to the condition of Theorem 4.1 for the function F, we also assume that the weak solution satisfies the inequality

$$2\int_0^t \|P^*u_\eta\|^2\,d\eta + \|A^*u(t)\|^2 + 2G(u_0) \le \|A^*u_0\|^2 + 2G(u(t)). \tag{4.18}$$

The energy inequality (4.18) can be formally obtained by multiplying (4.1) by u_η, integrating by time from 0 to t, and applying (4.4). For weak solutions, we have the following blow-up result.

Theorem 4.2. *Assume that $u : [0, T) \to D_*$ is a weak nonextendable solution of problem (4.1) satisfying inequality (4.18). If the initial data satisfy the inequality*

$$G(u_0) > \frac{1}{2}\|A^*u_0\|^2, \tag{4.19}$$

then the solution blows up in finite time $T < \infty$, and we have the following limit relations:

$$\lim_{t\to T^-} \int_0^t \|P^* u\|^2 \, d\eta = +\infty, \quad \limsup_{t\to T^-} \|P^* u\| = +\infty.$$

Proof. For arbitrary $T_0, \beta, \tau > 0$, and $t \in [0, T_0)$, we define the function

$$\Phi(t) \equiv \int_0^t \|P^* u\|^2 \, d\eta + (T_0 - t)\|P^* u_0\|^2 + \beta(t+\tau)^2. \tag{4.20}$$

Setting $\varphi = u$ in (4.17), we obtain

$$\begin{aligned}\Phi'(t) &= \|P^* u\|^2 - \|P^* u_0\|^2 + 2\beta(t+\tau) \\ &= 2\int_0^t [(u, F(u)) - \|A^* u\|^2] \, d\eta + 2\beta(t+\tau) \\ &= 2\int_0^t (P^* u, P^* u_\eta) \, d\eta + 2\beta(t+\tau).\end{aligned} \tag{4.21}$$

Differentiating the second line by time, we obtain

$$\begin{aligned}\Phi''(t) = 4(\alpha+1)&\left[\int_0^t \|P^* u_\eta\|^2 \, d\eta + \beta\right] \\ &+ 2\left[(u, F(u)) - \|A^* u\|^2 - 2(\alpha+1)\int_0^1 \|P^* u_\eta\|^2 \, d\eta - (2\alpha+1)\beta\right].\end{aligned} \tag{4.22}$$

Now from (4.18) and the condition for F we conclude that

$$\begin{aligned}\Phi''(t) \geq 4(\alpha+1)&\left[\int_0^t \|P^* u_\eta\|^2 \, d\eta + \beta\right] \\ &+ 2[2(\alpha+1)G(u_0) - (\alpha+1)\|A^* u_0\|^2 \\ &+ \alpha\|A^* u(t)\|^2 - (2\alpha+1)\beta].\end{aligned} \tag{4.23}$$

Combining the conditions of the theorem, (4.20), (4.22), and (4.23), we arrive at the ordinary differential inequality

$$\Phi\Phi'' - (\alpha+1)(\Phi')^2 \geq 0$$

under the condition

$$0 < \beta \leq \frac{2(\alpha+1)}{2\alpha+1}\left[G(u_0) - \frac{\|A^* u_0\|^2}{2}\right].$$

Since $\Phi'(0) = 2\beta\tau > 0$, the lifetime of the solution is finite, and the limit relations of Theorem 4.2 hold. Finally, following the scheme of the proof of Theorem 4.1, we can obtain the following estimate of the blow-up time:

$$T \leq \frac{2\alpha + 1}{2\alpha^2(\alpha + 1)} \frac{\|P^* u_0\|^2}{G(u_0) - \|A^* u_0\|^2/2}.$$

The theorem is proved. □

4.1.3 Examples

We present examples of several equations to which the classical energy method can be applied. This list proposed by H. A. Levine demonstrates wide possibilities of the method although it is not exhaustive and complete. In particular, we consider only examples in real functional spaces. Moreover, to avoid excessively long calculations, we omit proofs of certain assumptions for P, A, and F; the main attention will be paid to the positiveness property and conditions for F. For a more detailed acquaintance with applicability of the method, we refer the reader to [57, 58]. Solutions are assumed to be classical.

Example 4.1. Let $\Omega \subset \mathbb{R}^N$ be a bounded domain with sufficiently smooth boundary. Consider the Hilbert space $\mathbb{H} = \mathbb{L}^2(\Omega)$ with the scalar product

$$(f, g) = \int_\Omega f(x)g(x)\,dx.$$

Consider the operator

$$A = \sum_{|\beta|\leq 2m} a_\beta(x)D^\beta = \sum_{k=0}^{2m} \sum_{\beta_1+\dots+\beta_n=k} a_{\beta_1\dots\beta_n}(x)D_1^{\beta_1}\dots D_n^{\beta_n}. \tag{4.24}$$

By definition we set $P = I$ and assume that the nonlinear function $F(u)(x,t) = F(u(x,t))$ possesses the properties

$$F : \mathbb{R}^1 \to \mathbb{R}^1, \quad F(0) = 0.$$

Consider the parabolic initial-boundary-value problem

$$\begin{aligned} u_t &= -Au(x,t) + F\big(u(x,t)\big), \quad \Omega \times [0,T), \\ u(x,0) &= f(x), \quad f \in D_A,\ x \in \Omega, \\ (B_i u)(x,t) &= 0, \quad (x,t) \in \partial\Omega \times [0,T), \end{aligned} \tag{4.25}$$

where B_i are homogeneous boundary conditions for $i = 0, \dots, m-1$, and the set of initial functions belongs to the space

$$D = D_A = \{f \in \mathbb{C}^{2m}(\bar{\Omega}) \mid B_i f \equiv 0,\ i = 0, 1, \dots, m-1\}.$$

We assume that, for all $f, g \in D_A$,
(i)$_3$ $(f, Ag) = (Af, g)$;
(ii)$_3$ $(f, Af) \geq 0$.

In this case, definition (4.2) of the functional $G(f)$ for all $f \in D_A$ implies the relation

$$G(f) \equiv \int_0^1 \int_\Omega F(\rho f(x)) f(x)\, d\rho\, dx = \int_\Omega \left[\int_0^1 F(\rho f) f\, d\rho \right] dx$$
$$= \int_\Omega \left[\int_0^1 \frac{d}{d\rho} \left[\int_0^{\rho f} F(z)\, dz \right] d\rho \right] dx = \int_\Omega \left[\int_0^{f(x)} F(z)\, dz \right] dx. \tag{4.26}$$

Thus the necessary conditions for F, stated in Theorem 4.1, are fulfilled for all $f \in D_A$ and certain $\alpha > 0$ if the following inequality holds:

$$\int_\Omega \left[f(x) F(f(x)) - 2(\alpha + 1) \int_0^{f(x)} F(z) dz \right] dx \geq 0. \tag{4.27}$$

Note that

$$yF(y) = \int_0^y [zF'(z) + F(z)]\, dz; \tag{4.28}$$

then we can rewrite (4.27) in the form

$$\int_\Omega \int_0^{f(x)} [zF'(z) - (2\alpha + 1)F(z)]\, dz\, dx \geq 0. \tag{4.29}$$

In turn, the last inequality holds if the expression in the square brackets is nonnegative for $z > 0$ and nonpositive for $z < 0$. Both these conditions are satisfied if the nonlinearity F is of the form

$$F(z) = |z|^{2\alpha+1} \varphi(z), \tag{4.30}$$

where $\varphi : \mathbb{R}^1 \to \mathbb{R}^1$ is a nondecreasing function. Finally, we assume that the initial function $f(x)$ satisfies the inequality

$$\int_\Omega \left(\int_0^{f(x)} F(z) dz \right) dx > \frac{1}{2}(f, Af). \tag{4.31}$$

Applying Theorem 4.1, we conclude that there exists $T < \infty$ such that

$$\lim_{t \to T} \int_0^t \int_\Omega |u(x, \eta)|^2\, dx\, d\eta = +\infty \tag{4.32}$$

and the estimate for the lifetime

$$T \leq \frac{2\alpha+1}{2\alpha^2(\alpha+1)} \int_\Omega |f|^2\, dx \times \left[\int_\Omega \left(\int_0^{f(x)} F(z)dz - \frac{1}{2} f(x) \sum_{|\beta|\leq 2m} a_\beta(x) D^\beta f(x) \right) dx \right]^{-1} \tag{4.33}$$

is valid. Thus we have obtained the following sufficient condition of blow-up of solutions of problem (4.25):

$$\limsup_{t\to T} \max_{x\in\bar{\Omega}} |u(x,t)| = +\infty. \tag{4.34}$$

Note that if we take a second-order operator A of a particular form,

$$A = -\sum_{i,j=1}^{n} \frac{\partial}{\partial x_i}\left(a_{ij}(x) \frac{\partial}{\partial x_j} \right), \quad B_0 u = u, \tag{4.35}$$

then we can compare our sufficient conditions of blow-up with the results of [22, 38] and make sure that conditions imposed on the nonlinearity F are neither equivalent nor one implies the other. However, roughly speaking, both conditions require that $F(z) > 0$ for sufficiently large z and, for some z_0,

$$\int_{z_0}^{\infty} \frac{dz}{F(z)} < \infty, \quad F(z) > 0 \text{ for } z \in [z_0, \infty).$$

As noted before, the results of [22, 38] can be easily extended by using the energy method, for example, to the problem

$$\begin{gathered} \frac{\partial u}{\partial t} = -\left(\frac{\partial^2}{\partial x^2} + \frac{\partial^2}{\partial y^2} \right)^2 u + u^2, \quad (x,y,t) \in \Omega \times [0,T), \\ u(x,y,0) = f(x,y), \quad (x,y) \in \Omega, \\ u(x,y,t) = \frac{\partial u}{\partial n}(x,y,t) = 0, \quad (x,y) \in \partial\Omega,\ t \in [0,\infty), \end{gathered} \tag{4.36}$$

where the bounded domain $\Omega \subset \mathbb{R}^2$ has a smooth boundary. It is easy to see that the blow-up of a solution is a direct consequence of Theorem 4.1 if the initial function $f \in C_0^2(\Omega)$ satisfies the inequality

$$\int_\Omega f^3(x,y)\, dx\, dy > \frac{3}{2} \int_\Omega \left(\frac{\partial^2 f}{\partial x^2} + \frac{\partial^2 f}{\partial y^2} \right)^2 dx\, dy.$$

Example 4.2. Obviously, in the differential equation of Example 4.1, we can change the operator P and replace u_t by $\rho(x)u_t$, where $\rho(x) > 0$ in Ω. Then, instead of (4.32) and (4.34), we have the following limit relations:

$$\lim_{t\to T}\int_0^t\int_\Omega \rho(x)u^2(x,\eta)\,dx\,d\eta = +\infty, \tag{4.37}$$

$$\limsup_{t\to T}\int_\Omega \rho(x)u^2(x,t)\,dx = +\infty. \tag{4.38}$$

However, another form of the operator P is more important. In [11], a problem for the equation

$$a\Delta u_t + bu_t = c\Delta u$$

was considered, where Δ is the n-dimensional Laplacian, and a, b, and c are constants. For a constant $\lambda > 0$, consider a similar nonlinear problem:

$$\begin{aligned} \lambda u_t - u_{xxt} &= u_{xx} + u^2, \quad (x,t)\in(0,\pi)\times[0,T),\\ u(x,0) &= f(x), \quad x\in(0,\pi),\\ u(0,t) = u(\pi,t) &= 0, \quad t\in[0,\infty). \end{aligned} \tag{4.39}$$

Applying Theorem 4.1, we easily see that, for the operator

$$P = \lambda I - \frac{d^2}{dx^2}$$

and an initial function satisfying the condition

$$\int_0^\pi f^3(x)\,dx > \frac{3}{2}\int_0^\pi f'(x)^2\,dx, \tag{4.40}$$

we have the following limit relation for some $T < \infty$:

$$\lim_{t\to T}\left[\int_0^t\int_0^\pi \lambda u^2\,dx\,d\eta + \int_0^t\int_0^\pi u_x^2\,dx\,d\eta\right] = +\infty.$$

By the Poincaré inequality this is equivalent to

$$\lim_{t\to T}\int_0^t\int_0^\pi u_x^2\,dx\,d\eta = +\infty.$$

Thus, a global-in-time solution $u(x,t)$ of problem (4.39) does not exist since its gradient on a finite time interval becomes infinite.

Example 4.3 (integral nonlinear operator F). Let $K(x,y)$ be a symmetric real-valued function defined in the square $[0,\pi] \times [0,\pi]$. Consider the initial-boundary-value problem

$$\begin{aligned} u_t = u_{xx} + u(x,t) \int_0^\pi K(x,y) u^2(y,t)\,dy, \quad (x,t) \in (0,\pi) \times [0,T), \\ u(x,0) = f(x), \quad x \in [0,\pi], \\ u(0,t) = u(\pi,t) = 0, \quad t \in [0,\infty). \end{aligned} \tag{4.41}$$

We can consider more complicated problems, for example, for the equations $u_t = -u_{xxxx} + uKu^2$ or $u_t - u_{xxt} = uKu^2$, but we would like to make the example clearer.

For the case

$$F(f)(x) = f(x) \int_0^\pi K(x,y) f^2(y)\,dy,$$

the Fréchet derivative F_f is a linear operator on $\mathbb{H} = \mathbb{L}^2(0,\pi)$ acting on a function $h \in \mathbb{H}$ as follows:

$$(F_f h)(x) = h(x) \int_0^\pi K(x,y) f^2(y)\,dy + 2f(x) \int_0^\pi K(x,y) f(y) h(y)\,dy.$$

Obviously, the derivative F_f is symmetric, so that the potential G (see (4.2)) exists and is defined by the formula

$$\begin{aligned} G(f) &= \int_0^1 \int_\Omega \rho f^2(x) \int_\Omega K(x,y) \rho^2 f^2(y)\,dy\,dx\,d\rho \\ &= \frac{1}{4} \int_\Omega \int_\Omega K(x,y) f^2(x) f^2(y)\,dx\,dy. \end{aligned}$$

This implies that the limit equality (4.32) is valid ($\Omega = (0,\pi)$) if the initial function satisfies the inequality

$$\int_0^\pi \int_0^\pi K(x,y) f^2(x) f^2(y)\,dx\,dy > 2 \int_0^\pi (f'(x))^2\,dx; \tag{4.42}$$

moreover, the lifetime T can be estimated as follows:

$$T \le 3 \int_0^\pi f^2(x)\,dx \left[\int_0^\pi \int_0^\pi K(x,y) f^2(x) f^2(y)\,dx\,dy - 2 \int_0^\pi (f'(x))^2\,dx \right]^{-1}.$$

We emphasize that, for a positive kernel $K(x,y)$, the existence in an open subset of the set $[0,\pi] \times [0,\pi]$ of an appropriate function $f \in \mathbb{C}_0^2(0,\pi)$ satisfying the condition (4.42) is guaranteed.

Example 4.4. Consider the system of n scalar equations

$$\frac{\partial u_i}{\partial t} = \sum_{j=1}^{n} \frac{\partial}{\partial x}\left(a_{ij}\frac{\partial u_j}{\partial x}\right) + F_i(u_1,\dots,u_n), \quad i = 1,\dots,n, \tag{4.43}$$

where $a_{ij} = a_{ji}$, and the initial-boundary-value problem for system (4.43) in the strip $(x,t) \in [-a,a] \times [0,T)$. We assume that

$$\inf_{x\in(-a,a)} \lambda_1(x) \geq 0,$$

where λ_1 is the smallest eigenvalue of the matrix $[a_{ij}]$. We work in the space

$$\mathbb{H} = \mathbb{L}^2(-a,a) \otimes \cdots \otimes \mathbb{L}^2(-a,a)$$

with the scalar product

$$(f,g) = \sum_{i=1}^{n} \int_{-a}^{a} f_i(x)g_i(x)\,dx.$$

We take

$$D = \{f \in \mathbb{H} \,|\, f_i \in \mathbb{C}^{(2)}([-a,a]),\ f_i(-a) = f_i(a) = 0\}.$$

Let a solution $u = (u_1(x,t),\dots,u_n(x,t))$ of equation (4.43) satisfy the following initial and boundary conditions:

$$u(a,t) = u(-a,t) = 0, \quad t \in [0,\infty), \tag{4.44}$$

$$u(x,0) = f = (f_1,f_2,\dots)(x), \quad x \in (-a,a),\ f \in D. \tag{4.45}$$

Let the nonlinearity $F(u)$ possess a symmetric Fréchet derivative such that

$$\frac{\partial F_i}{\partial u_j} = \frac{\partial F_j}{\partial u_i};$$

in other words, we assume that it can be represented in the form $F = \nabla\Psi(u)$, where Ψ is a scalar function of real arguments $u_1,\dots,u_n$. In this case, the potential G (see (4.2)) is defined by the formula

$$\begin{aligned} G(f) &= \int_0^1 (F(\rho f),f)\,d\rho \\ &= \int_0^1 \int_{-a}^{a} \sum_{i=1}^{n} f_i(x)F_i(\rho f(x))\,dx\,d\rho \end{aligned}$$

$$= \int_{-a}^{a}\int_{0}^{1}\sum_{i=1}^{n} f_i(x)\frac{\partial\Psi}{\partial(\rho f_i)}(\rho f(x))\,d\rho\,dx$$

$$= \int_{-a}^{a}\int_{0}^{1}\frac{\partial\Psi}{\partial\rho}(\rho f(x))\,d\rho\,dx = \int_{-a}^{a}\Psi(f(x))\,dx. \tag{4.46}$$

Thus, the condition of Theorem 4.1 for the function F is fulfilled if there exists a constant $\alpha > 0$ such that

$$2(\alpha+1)\Psi(z) \le \sum_{i=1}^{n} z_i\frac{\partial\Psi}{\partial z_i}. \tag{4.47}$$

Note that if Ψ is a homogeneous function of degree $2\alpha+2$, that is,

$$\Psi(sx) = s^{2\alpha+2}\Psi(x),$$

then condition (4.47) holds. In the general case where $n \ge 2$, inequality (4.47) is difficult to verify.

We apply Theorem 4.1. The condition for the initial function of system (4.43) can be rewritten in the form

$$\int_{-a}^{a}\Psi(f)\,dx > \frac{1}{2}\int_{-a}^{a}\sum_{i,j=1}^{n} a_{ij}(x)f_i'(x)f_j'(x)\,dx. \tag{4.48}$$

If this inequality is fulfilled, then a global-in-time solution cannot exist, and we have the following relation:

$$\lim_{t\to T}\sum_{i=1}^{n}\int_{-a}^{a}|u_i(x,t)|^2\,dx = +\infty. \tag{4.49}$$

We have proved the fact of blow-up although it is unknown which element of the sum has led to the collapse.

Example 4.5. Note that the energy method allows observing the influence of the size of the domain on the stability of solutions. Indeed, consider the problem

$$\begin{aligned} u_t = u_{xx} + u^2, &\quad (x,t) \in (0,a)\times[0,\infty),\\ u(x,0) = f(x), &\quad x \in (0,a),\\ u(0,t) = u(a,t) = 0, &\quad t \in [0,\infty). \end{aligned} \tag{4.50}$$

As the initial function, we take

$$f_0(x) = \sqrt{\frac{2}{a}}\sin\frac{\pi x}{a}$$

and set $f(x) = rf_0(x)$, $r > 0$. For $F(u) = u^2$ and $\alpha = 1/2$, condition (ii_1) holds only if

$$\int_0^a f^3(x)\,dx > \frac{3}{2}\int_0^a |f'(x)|^2\,dx \quad \Rightarrow \quad ra^{3/2} > K,$$

where K is a constant independent of a and r. Therefore, if we fix $\|f\| = r$ and take a sufficiently large domain a, then we can obtain a blow-up of a solution of the corresponding problem.

Example 4.6. In his classical paper [23, Chap. 3, Sec. 2], Fujita examined the blow-up of solutions of the initial-boundary-value problem

$$\begin{gathered} u_t = \Delta u + u^{1+\alpha}, \quad \mathbb{R}^m \times [0, T), \\ u(x,t) \geq 0, \quad u(x,0) = a(x), \quad x \in \mathbb{R}^m, \end{gathered} \tag{4.51}$$

where Δa, ∇a, and a are subject to the continuity and boundedness conditions, and, moreover, a is nonnegative. The main result of [23] is as follows.

Theorem 4.3. *If $0 < m\alpha < 2$, then problem (4.51) with nonzero initial conditions does not possess global-in-time solutions of the class*

$$E(0,\infty) \equiv \{u : \mathbb{R}^m \times [0,\infty) \to \mathbb{R}^1 \mid u(x,t) \textit{ is bounded in } E(0,T) \textit{ for all } T\},$$

where

$$E(0,T) \equiv \{u : \mathbb{R}^m \times [0,T) \to \mathbb{R}^1 \mid \exists M > 0,\ \beta \in [0,2) : |u(x,t)| \leq M \exp(|x|^\beta)\}.$$

If $m\alpha > 2$, then a global solution exists for sufficiently small initial data.

Certainly, if we apply Levine's energy method to problem (4.51), then we can only prove a weaker result that, under the condition

$$\int_{\mathbb{R}^m} (a(x))^{2+\alpha}\,dx > \frac{2+\alpha}{2}\int_{\mathbb{R}^m} |\nabla a|^2\,dx, \quad a,\ \nabla a,\ \Delta a \in \mathbb{L}^2(\mathbb{R}^1), \tag{4.52}$$

there are no global-in-time nonnegative solutions of the class $u(\cdot,t) \in \mathbb{L}^2(\mathbb{R}^m)$. However, to obtain a blow-up result, Fujita used very strong assumptions regarding the properties of the fundamental solution $u_t = \Delta u$ (positiveness, integrability, etc.), which makes his method inapplicable to the study of the Cauchy problem of the following form:

$$\begin{gathered} u_t - \mu\Delta u_t = -\Delta^2 u + u^{1+\alpha}, \quad \mathbb{R}^m \times [0,T), \quad \mu > 0,\ \alpha > 0, \\ u(x,0) = a(x) \geq 0, \quad x \in \mathbb{R}^m, \end{gathered} \tag{4.53}$$

due to the higher complexity of the question on the explicit form and properties of the Green function of this equation. On the contrary, using Levine's energy method, we can easily obtain that, under the condition

$$\int_{\mathbb{R}^m} (a(x))^{2+\alpha}\,dx \geq \frac{2+\alpha}{2}\int_{\mathbb{R}^m} |\Delta a|^2\,dx, \tag{4.54}$$

the lifetime $T > 0$ of solutions of problem (4.53) is finite:

$$\limsup_{t\to T}\int_{\mathbb{R}^m} [u^2(x,t) + \mu|\nabla u(x,t)|^2]\,dx = \infty. \tag{4.55}$$

Thus, the examples presented show that the energy method is applicable to a wide class of parabolic problems, easy for verification of sufficient conditions, and allows us to obtain estimates of the blow-up time.

4.2 Blow-up of solutions of nonlinear wave equation

4.2.1 Introduction

Even in the earliest classical studies on the solvability of nonlinear hyperbolic problems, special attention was paid to the fact that solutions of these problems are not stable with respect to arbitrary changes in the nonlinear terms and initial data, and in certain cases, they become infinite in finite time (see [41, 96]).

Most of the original methods for the investigation of such phenomena were developed for specific tasks and were based on comparison principles and the Huygens principle, until the appearance in 1974 of Levine's energy method (see [58]) for the study of a wide class of abstract initial and initial-boundary-value problems of the form

$$\begin{aligned} P\frac{d^2u}{dt^2} &= -A(t)u + F(u), \quad t \in [0,T),\\ u(0) &= u_0, \quad u_t(0) = v_0, \end{aligned} \tag{4.56}$$

where u is a function of time in the Hilbert space $\mathbb{H}$, $A(t)$ is a symmetric linear operator, which is defined and nonnegative for all $t \geq 0$, P is strictly positive symmetric operator, and F is a given nonlinearity.

Roughly speaking, the basic result of the energy method consists of the following. Let a function $u(x,t)$ be a strictly continuous, twice differentiable, and nonextendable solution of problem (4.56) on the corresponding maximal interval time $[0,T)$. Assume that F has a symmetric Fréchet derivative F_x and defines the "potential" energy of the nonlinearity G:

$$G(x) \equiv \int_0^1 (F(\rho x), x)\,d\rho. \tag{4.57}$$

Moreover, assume that there exists a constant $\alpha > 0$ such that, for all x, we have the inequality

$$(x, F(x)) \geq 2(2\alpha + 1)G(x).$$

If the initial data satisfy the condition

$$G(u_0) > \frac{1}{2}(u_0, A(0)u_0) + \frac{1}{2}(v_0, Pv_0) \equiv E(0),$$

then the lifetime of the nonextendable solutions is bounded, $T < \infty$, and we have the following limit relation:

$$\lim_{t \to T}(u(t), Pu(t)) = +\infty. \tag{4.58}$$

In other words, if the initial potential energy of the nonlinearity is greater than the total initial energy of the linear problem, then (4.56) has no global-in-time solutions. If $G(u_0) < E(0)$, then the question on the blow-up within the framework on Levine's method cannot be solved unambiguously. However, note that a modification of Levine's method developed by M. O. Korpusov and A. G. Sveshnikov easily overcame this difficulty (see Chapter 8).

In addition to the fact that Levine's energy method yields a unified solution scheme (this fact distinguishes this method from the methods used in [41, 96]), it is also interesting for the following reasons (this will be shown further on specific examples):

(1) To solve the blow-up question, special properties of classical wave solution (for example, the finite propagation speed or the Huygens principle) are not used. In fact, these special conditions for problem (4.56) may not be fulfilled at all.
(2) The results can be generalized for higher-orders equations, systems of equations, and equations in which the "density" can take zero values on some curves or surfaces for various initial and initial-boundary-value problems with unbounded operator P.
(3) Nonlinearity can be very "soft." In addition to the power law, the nonlinearity can take the form of a nonlinear integral operator.
(4) As in [96], the energy method makes it possible to study the influence of the size of the domain on the stability of the problem.
(5) The method is applicable for problems with damping. In particular, we further consider a problem for equation

$$\frac{d^2u}{dt^2} + a\frac{du}{dt} = -Au + F(u), \quad a > 0. \tag{4.59}$$

For a more detailed introduction to the application of the energy method to hyperbolic equations, we refer the reader to the original paper [58].

4.2.2 General theorems

Let $\mathbb{H}$ be a real Hilbert space. Let D be another Hilbert space, and let $D \subset \mathbb{H}$ be a dense linear embedding considered as a mapping of Hilbert spaces. We denote by $(\cdot,\cdot)$ the scalar product in $\mathbb{H}$, by $\|\cdot\|$ the corresponding norm, and by $(\cdot,\cdot)_D$ the scalar product in D. We assume that, for all $x \in D$, there exists a constant $c > 0$ such that $\|x\| \le c\|x\|_D$.

Consider the problem for equation (4.56) and state conditions for the operators A and P and the nonlinearity F. Namely, we assume that, for all $t \ge 0$:

(i_A) $A(t) : D \to \mathbb{H}$ is a symmetric linear operator;

(ii_A) $(x, A(t)x) \ge 0$ for all $x \in D$;

(iii_A) if $v : [0,\infty) \to H$ is a strictly continuous differentiable mapping and $v(t)$, $dv/dt \in D$ for all $t \ge 0$, then the function $(v(t), A(t)v(t))$ is continuously differentiable, and we have the following inequality:

$$Q(v,v) \equiv \frac{d}{dt}(v, Av) - 2\left(\frac{dv}{dt}, Av\right) \le 0;$$

(i_P) $P : D_P \to \mathbb{H}$, $D \subseteq D_P \subseteq \mathbb{H}$, is a symmetric linear operator;

(ii_P) $(x, Px) > 0$ for all $x \in D_P$, $x \neq 0$.

(i_F) $F : D \to \mathbb{H}$ is continuously differentiable as a function from D into $\mathbb{H}$ possessing a symmetric Fréchet derivative F_x, which is a bounded linear operator on $\mathbb{H}$, and the mapping $x \to F_x$ is strictly continuous;

(ii_F) denote by $G(x) : D \to \mathbb{R}^1$ the potential related to the nonlinearity F by the formula

$$G(x) \equiv \int_0^1 (F(\rho x), x)\, d\rho$$

and such that the action of its Fréchet derivative for all x and y is defined by the formula $G_x y = (F(x), y)$. Moreover, there exists a constant $\alpha > 0$ such that

$$(x, F(x)) \ge 2(2\alpha + 1)G(x), \quad x \in D. \tag{4.60}$$

In the sequel, we need the following formula, which is valid for functions $v : [0,T) \to D$ with strictly continuous derivatives v_t (here D is a Banach space):

$$G(v(t)) - G(v(0)) = \int_0^t (F(v(\eta)), v_\eta(\eta))\, d\eta. \tag{4.61}$$

This formula is obtained by integration by time of the expression of the similar equality (4.4) for nonlinear parabolic problems. The solution of (4.56) is meant in the following strong generalized sense.

Definition 4.2. A function $u : [0,T) \to \mathbb{H}$ is called a *strong generalized solution* of equation (4.56) if, for all t, the functions $u(t)$ and $u_t(t)$ belong to the space D, u_{tt} exists, takes its values in D_T, and is strictly continuous with respect to the norm of $\mathbb{H}$, and the differential equality is fulfilled in the classical sense.

Theorem 4.4. *Let $u : [0,T) \to \mathbb{H}$ be a solution of the initial-boundary-value problem of the form*

$$P\frac{d^2u}{dt^2} = -A(t)u(t) + F(u(t)), \quad t \in [0,T), \tag{4.62}$$

$$u(0) = u_0, \quad u_t(0) = v_0, \tag{4.63}$$

and let P, F, and A satisfy the conditions stated before. Then the following two assertions hold:

(A) *If the initial data satisfy the inequality*

$$\beta_0 = 2G(u_0) - [(u_0, A(0)u_0) + (v_0, Pv_0)] > 0, \tag{4.64}$$

then the solution blows up in a finite time T, which can be estimated from above as follows:

$$T \le T_{\beta_0} = (u_0, Pu_0)\alpha^{-1}[(\beta_0(u_0, Pu_0) + (u_0, Pv_0)^2)^{1/2} + (u_0, Pv_0)]^{-1};$$

(B) *If the initial data satisfy the equality*

$$G(u_0) = \frac{1}{2}[(u_0, A(0)u_0) + (v_0, Pv_0)], \quad \frac{(u_0, Pv_0)}{u_0, Pu_0} = \lambda > 0, \tag{4.65}$$

then the solution blows up in a finite time T, which satisfies the following upper estimate:

$$T \le (2\alpha\lambda)^{-1}.$$

The blow-up of a solution is understood as the following limit relation:

$$\lim_{t \to T}(u(t), Pu(t)) = +\infty. \tag{4.66}$$

Proof. Introduce the notation

$$\Phi(t) = (u(t), Pu(t)) + q^2 + \beta(t+\tau)^2, \tag{4.67}$$

where q, β, and τ are arbitrary nonnegative constants to be specified later. Differentiating (4.67) by time and taking into account the symmetry of the operator P, we obtain

$$\Phi'(t) = 2(u_t, Pu) + 2\beta(t+\tau), \tag{4.68}$$

$$\Phi''(t) = 2(u_t, Pu_t) + 2(u, Pu_{tt}) + 2\beta. \tag{4.69}$$

Introduce the following function $S(t)$ and verify that it is well-defined (cf. (4.11)):

$$S^2 = [(u, Pu) + \beta(t+\tau)^2][(u_t, Pu_t) + \beta] - [(u, Pu_t) + \beta(t+\tau)]^2 \geq 0.$$

We have

$$-(\alpha+1)(\Phi')^2 = 4(\alpha+1)S^2 - 4(\alpha+1)(\Phi - q^2)[(u_t, Pu_t) + \beta]. \tag{4.70}$$

Therefore, combining (4.70) and (4.69), we obtain

$$\begin{aligned}\Phi\Phi'' - (\alpha+1)(\Phi')^2 &= 4(\alpha+1)S^2 + 4(\alpha+1)q^2[(u_t, Pu_t) + \beta] \\ &\quad + 2\Phi[(u, Pu_{tt}) - (2\alpha+1)((u_t, Pu_t) + \beta)]. \end{aligned} \tag{4.71}$$

Assume that we were able to prove the nonnegativeness of (4.71); then the function $\Phi^{-\alpha}$ is concave since

$$(\Phi^{-\alpha})'' = -\alpha\Phi^{-\alpha-2}[\Phi\Phi'' - (\alpha+1)(\Phi')^2] \leq 0.$$

Integrating by time the last differential inequality under the conditions

$$\Phi(0) > 0, \quad \Phi'(0) > 0,$$

we obtain the following inequalities:

$$\begin{aligned} \Phi^{-\alpha} &\leq \Phi^{-\alpha}(0) - \alpha t\Phi'(0)\Phi^{-\alpha-1}(0), \\ \Phi^{\alpha}(t) &\geq \frac{\Phi^{\alpha+1}(0)}{\Phi(0) - \alpha t\Phi'(0)}. \end{aligned} \tag{4.72}$$

Therefore, we obtain the blow-up $\Phi(t) \to +\infty$ as $t \to T \leq \Phi(0)/(\alpha\Phi'(0))$. The main difficulty is to prove the concavity of the function $\Phi^{-\alpha}$.

We return to expression (4.71). Note that

$$4(\alpha+1)S^2 + 4(\alpha+1)q^2[(u_t, Pu_t) + \beta] > 0.$$

Introduce the functional

$$\begin{aligned} H(t) &\equiv (u, Pu_{tt}) - (2\alpha+1)((u_t, Pu_t) + \beta) \\ &= -(u, Au) - (2\alpha+1)((u_t, Pu_t) + \beta) + (u, F(u)). \end{aligned} \tag{4.73}$$

Using property (iii$_A$) and the definition of $Q(u,u)$ and differentiating (4.73) by time, we obtain

$$H'(t) = -[Q(u,u) + 2(u_t, Au) + 2(2\alpha+1)(u_t, Pu_{tt})] + \frac{d(u, F(u))}{dt}$$

$$= -Q(u,u) + 4\alpha(u_t, Au) + \frac{d(u, F(u))}{dt} - 2(2\alpha+1)(u_t, F(u))$$
$$= -(2\alpha+1)Q(u,u) + 2\alpha\frac{d(u,Au)}{dt}$$
$$+ \frac{d(u,F(u))}{dt} - 2(2\alpha+1)(u_t, F(u)). \tag{4.74}$$

Integrating (4.74) by time from 0 to t and taking into account the positive definiteness of $A(t)$, properties (4.60) and (4.61), and the definition

$$H(0) - (u_0, F(u_0)) = -(u_0, A(0)u_0) - (2\alpha+1)((v_0, Pv_0) + \beta)$$

of the functional H, we obtain the following relations:

$$H(t) = H(0) + 2\alpha(u, Au) - 2\alpha(u_0, A(0)u_0) - (2\alpha+1)\int_0^t Q(u,u)\,d\eta$$
$$+ (u, F(u)) - (u_0, F(u_0)) - 2(2\alpha+1)(G(u) - G(u_0))$$
$$\geq 2(2\alpha+1)\Big(G(u_0) - \frac{1}{2}((u_0, A(0)u_0) + (v_0, Pv_0) + \beta)\Big). \tag{4.75}$$

Assume that the initial data satisfy condition (4.64). Then, for $q^2 = 0$ and $\beta = \beta_0$, from (4.75) we obtain that $H(t) \geq 0$ and, therefore,

$$(\Phi^{-\alpha})'' \leq 0.$$

Moreover, for sufficiently large τ, we have

$$\Phi'(0) = 2(u_0, Pv_0) + 2\beta_0\tau > 0.$$

Thus, we see that the lifetime of the solution cannot exceed

$$T = T(\beta) = \frac{\Phi(0)}{\alpha\Phi'(0)} = f(\tau),$$

where

$$f(\tau) = \frac{(u_0, Pu_0) + \beta_0\tau^2}{2\alpha((u_0, Pv_0) + \beta_0\tau)^2}.$$

It is easy to verify that the function $f(\tau)$ attains the minimal value at the point

$$\tau = \tau_0 \equiv \beta_0^{-1}(-(u_0, Pv_0) + \sqrt{(u_0, Pv_0)^2 + \beta_0(u_0, Pu_0)}).$$

Thus the estimate of the lifetime has the form

$$T \leq T_\beta = (u_0, Pu_0)\alpha^{-1}[(\beta_0(u_0, Pu_0) + (u_0, Pv_0)^2)^{1/2} + (u_0, Pv_0)]^{-1}.$$

Now we prove the second assertion of the theorem. Assume that the condition (4.65) holds; then we set $q^2 = 0$ and $\beta = 0$, so that we have the required inequality:

$$(\Phi^{-\alpha})'' \le 0, \quad \Phi(t) = (u(t), Pu(t)).$$

Therefore, the solution blows in finite time:

$$\lim_{t\to T}(u(t), Pu(t)) = +\infty, \quad T \le \frac{\Phi(0)}{\alpha\Phi'(0)} = \frac{(u_0, Pu_0)}{2\alpha(u_0, Pv_0)} = \frac{1}{2\alpha\lambda}.$$

The theorem is proved. □

An important question on the validity of the theorem is the question on the existence of initial data satisfying conditions (4.64) and (4.65) simultaneously. First, we note that if the initial data satisfy the inequality

$$G(u_0) > \frac{1}{2}(u_0, A(0)u_0), \tag{4.76}$$

then we can find a function v_0 such that the corresponding solution of the problem blows up in finite time. Second, we verify the validity of the following assertion.

Lemma 4.1. *Assume that, for all $x \in D$, the nonlinearity has the form $F(sx) = s^{1+\delta}F(x)$, where $\delta > 0$, and let $(x_0, F(x_0)) > 0$ for some $x_0 \in D$. Then there exist infinitely many initial data u_0 such that inequality (4.76) holds.*

Indeed, we choose $s = s_0$ so large that

$$s^{\delta+2}G(x_0) = s^{\delta+2}\int_0^1 (F(\rho x_0), x_0)\, d\rho > \frac{s^2}{2}(x_0, A(0)x_0). \tag{4.77}$$

Then, for $s \ge s_0$, the inequality

$$G(u_0) = s^{\delta+2}G(x_0) > \frac{s^2}{2}(x_0, A(0)x_0) \tag{4.78}$$

is fulfilled for all $u_0 = sx_0$. Thus, we obtain a wide class of nonlinearities for which there exist a set of initial data u_0 satisfying (4.76) and the corresponding set of functions v_0 satisfying (4.64) and (4.65).

4.2.3 Examples

Now we consider elementary applications of Levine's energy method to certain nonlinear wave equations. Certainly, this list of examples in not exhaustive; we only present the simplest cases that illustrate various advantages of the method.

To shorten the calculations, we leave a strict check of certain properties of the properties of the operators P, F, and A to the reader, referring him/her to [58]. In particular, among the conditions imposed on the nonlinearity F, we mainly focus on (4.60) and (4.61). The discussion of the regular properties (i_F) is omitted since they follow from the corresponding Sobolev inequalities and in most cases can be proved for the Hilbert space D with the norm

$$\|x\|_D = \sqrt{\|x\|^2 + \lambda_1(Px, Px) + \lambda_2(Ax, Ax)}, \quad \lambda_1, \lambda_2 > 0.$$

Example 4.7. Let $\Omega \subset \mathbb{R}^N$ be a bounded domain with sufficiently smooth boundary $\partial\Omega$ (in this case, all transformations below are valid). For all $(x,t) \in \Omega \times (0,\infty)$, let the symmetric $(n \times n)$-matrix $A(x,t) = [a_{ij}(x,t)]$ consist of continuously differentiable real elements such that

$$\sum_{i,j=1}^{n} a_{ij}(x,t)\xi_i\xi_j \geq 0, \tag{4.79}$$

$$\sum_{i,j=1}^{n} \left(\frac{\partial}{\partial t} a_{ij}(x,t)\right)\xi_i\xi_j \leq 0 \tag{4.80}$$

for all $\xi = (\xi_1, \dots, \xi_n) \in \mathbb{R}^n$. Let the boundary of the domain Ω consist of two parts, $\partial\Omega = \Gamma_1 \cup \Gamma_2$, where Γ_1 and Γ_2 are nonintersecting "smooth" submanifolds of Ω. Assume that $A(x,t)$ is independent of time on Γ_2 and consider the problem

$$\frac{\partial^2 u}{\partial t^2} = \sum_{i,j=1}^{n} \frac{\partial}{\partial x_i}\left(a_{ij}(x,t)\frac{\partial u}{\partial x_j}\right) + F(u), \quad F(0) = 0, \tag{4.81}$$

in the domain $\Omega \times [0,\infty)$ with boundary conditions

$$u(x,t) = 0 \quad \text{on } \Gamma_1 \times [0,\infty)$$

and

$$\sum_{i,j=1}^{n} a_{ij}(x)\nu_j \frac{\partial u}{\partial x_i} = 0 \quad \text{on } \Gamma_2 \times [0,\infty),$$

where $\nu = (\nu_1, \dots, \nu_n)$ is the outer normal of the boundary Γ_2. Moreover, we complement problem (4.81) with the initial data

$$u(x,0) = u_0(x), \quad \frac{\partial u}{\partial t}(x,0) = v_0(x).$$

Assume that the nonlinearity depends only on the solution of the problem, that is, $F(u)(x,t) = F(u(x,t))$. We denote the scalar product in the Hilbert space $\mathbb{H} = \mathbb{L}^2(\Omega)$ by

$$(f,g) = \int_\Omega fg\,dx$$

and introduce the space

$$D = \left\{ f \in \mathbb{H} \,|\, f \in \mathbb{C}^{(2)}(\bar{B}),\ f = 0 \text{ on } \Gamma_1,\ \sum_{i,j=1} a_{ij}(x) v_j f_{x_i} = 0 \text{ on } \Gamma_2 \right\}.$$

Moreover, for all $f \in D$, let

$$A(t)f(x) = -\sum_{i,j=1}^{n} \frac{\partial}{\partial x_i}\left(a_{ij}(x,t) \frac{\partial}{\partial x_j} \right) f(x).$$

It is easy to prove that $A(\cdot)$ satisfies all conditions for the operator A of Theorem 4.1, including the condition for Q. Indeed, for all smooth $v(x,t)$ such that $v(\cdot,t) \in D$ for all t, we have the following equality:

$$Q(v,v) = \int_{\Omega} \sum_{i,j=1}^{n} \frac{\partial a_{ij}}{\partial t} \frac{\partial v}{\partial x_i} \frac{\partial v}{\partial x_j}\, dx.$$

Since $P = I$, we take $D_P = \mathbb{H}$. The most frequent cases encountered in applications refer to $\Gamma_2 = \emptyset$ or to the operator $A(t)$ that is independent of time. As for parabolic equations discussed in the previous section, from the definition of the functional G we obtain the following equality for $f \in D$:

$$G(f) = \int_{\Omega} \int_{0}^{1} F(\rho f(x)) f(x)\, d\rho\, dx = \int_{\Omega} \left(\int_{0}^{f(x)} F(z)\, dz \right) dx.$$

Thus condition (4.60) holds for arbitrary $f \in D$ if and only if

$$\int_{\Omega} \left(\int_{0}^{f(x)} \left(\frac{d}{dz}(zF(z)) - (4\alpha + 2)F(z) \right) dz \right) dx \geq 0.$$

This implies that (4.60) holds if

$$zF'(z) - (4\alpha + 1)F(z) \geq 0 \quad \text{for all } z > 0,$$
$$zF'(z) - (4\alpha + 1)F(z) \leq 0 \quad \text{for all } z < 0.$$

It is easy to verify that the nonlinearity satisfying these conditions is of the form

$$F(z) = |z|^{4\alpha+1} \varphi(z), \quad \varphi : \mathbb{R}^1 \to \mathbb{R}^1, \tag{4.82}$$

where φ is an increasing (or nondecreasing) function. Moreover, condition (4.76) holds only if

$$G(u_0) = \int_{S_+} \left[\int_{0}^{u_0} F(z) dz \right] dx$$

$$- \int_{S_-} \left[\int_{u_0}^{0} F(z)\, dz \right] dx$$

$$> \frac{1}{2}(u_0, A(0)u_0), \tag{4.83}$$

where

$$S_+ = \{x \in \Omega \mid u_0 > 0\}, \quad S_- = \{x \in \Omega \mid u_0 < 0\}.$$

Obviously, condition (4.83) does not hold for all $u_0 \in D$ if $\varphi \le 0$ for $z \ge 0$ and $\varphi \ge 0$ for $z \le 0$. Since φ is nondecreasing, condition (4.83) does not hold only if $\varphi = 0$. Therefore, if for some time $T < \infty$, inequality (4.83) holds, then the solution blows up:

$$\lim_{t \to T} \int_B u^2(x, t)\, dx = +\infty.$$

Example 4.8. In the previous example, we have considered the nonlinearity

$$F(g)(x) = f(g(x))g(x), \quad f(s) = O(s^\delta) \quad \text{as } |s| \to +\infty,\ \delta > 0.$$

Here for problem (4.81), the nonlinearity is of the form

$$F(g)(x) = g(x) \int_\Omega K(x, y) g^2(y)\, dy, \tag{4.84}$$

where the kernel $K(x, y)$ is a square-integrable real-valued function such that $K(x, y) = K(y, x)$. It is easy to calculate the Fréchet derivative with respect to $g \in \mathbb{H}$:

$$F_g(h(x)) = h(x) \int_\Omega K(x, y) g^2(y)\, dy + 2g(x) \int_\Omega K(x, y) g(y) h(y)\, dy. \tag{4.85}$$

This implies that conditions (i_F) and (ii_F) are fulfilled since F_g is symmetric and bounded. Moreover,

$$G(g) = \frac{1}{4} \int_\Omega \int_\Omega K(x, y) g^2(y) g^2(x)\, dy\, dx, \tag{4.86}$$

and therefore, for $\alpha = 1/2$, condition (4.60) holds:

$$2(2\alpha + 1)G(g) = (F(g), g).$$

Now we can write the condition for the initial function (4.76) in the form

$$\int_\Omega \int_\Omega u_0^2(x) u_0^2(y) K(x, y)\, dy\, dx > 2 \int_\Omega \sum_{i,j}^{n} a_{ij}(x, 0) \frac{\partial u_0}{\partial x_i} \frac{\partial u_0}{\partial x_j}\, dx. \tag{4.87}$$

Therefore, if K is positive on a subset of Ω of size of the Lebesgue n^2-measure, then we can choose an initial function $u_0 \in \mathbb{C}_0^{(2)}(\Omega)$ such that (4.87) is fulfilled. This example of application of the energy method to the proof of the blow-up of solutions to nonlinear hyperbolic integro-differential problems can be extended to general integral operators.

Example 4.9. Consider a particular case of Example 4.7, in which we are interested in the influence of the size of the domain on the blow-up. We consider the problem on the segment

$$\begin{gathered}\frac{\partial^2 u}{\partial t^2} = \frac{\partial^2 u}{\partial x^2} + \varepsilon F(u), \quad (x,t) \in (0,a) \times [0,\infty), \\ u(x,0) = u_0(x), \quad u_t(x,0) = v_0(x), \quad x \in [0,a], \\ u(0,t) = u(a,t) = 0, \quad t \in [0,\infty).\end{gathered} \tag{4.88}$$

We study the connection of the blow-up with parameters a and ε. If $F(u) = u$, then, applying the Fourier transformation, we can easily prove that, for a solution of problem (4.88) for $\varepsilon \le \pi^2/a^2$, there exists a constant $K > 0$ such that the following estimate holds:

$$\|u_\varepsilon(t)\|_a \le K(\|u_0\|_a + \|v_0\|_a), \quad \|f\|_a = \left|\int_0^a |f|^2\,dx\right|^{1/2}. \tag{4.89}$$

Otherwise, if $\varepsilon > \pi^2/a^2$, then solutions can grow in time, but this growth is not faster than exponential. Note that although the domain of stable solutions diminishes as $a \to +\infty$, for $\varepsilon < 0$ and all $a > 0$, a solution exists and is stable. On the contrary, we can show that even in the case of the quadratic nonlinearity $F(u) = u^2$, for all $\varepsilon \neq 0$, there exist initial data such that the corresponding solutions blow up in finite time.

In the condition of Theorem 4.1, we take $\alpha = 1/4$; then the "potential" on the nonlinearity takes the form

$$G(f) = \frac{1}{3}\int_0^a f^3(x)\,dx.$$

We take a function

$$f \in \mathbb{C}^{(1)}[0,a], \quad f(x) > 0 \text{ for } x \in (0,a), \quad f(0) = f(a) = 0,$$

and set $u_0(x) = rf(x)$. Then condition (4.76) is fulfilled if

$$\varepsilon r\int_0^a f^3(x)\,dx > \frac{3}{2}\int_0^a f'(x)^2\,dx. \tag{4.90}$$

Thus, for any ε, choosing sufficiently large $|r|$ so that $r\varepsilon > 0$, we can achieve that condition (4.90) holds, and therefore there exists $T > 0$ such that under the initial conditions $u_0(x) = rf(x)$, the solution is pointwise unbounded on $[0, a] \times [0, T)$.

Consider the particular case where

$$f(x) = \sqrt{\frac{2}{a}} \sin \frac{\pi x}{a}.$$

Fix $\varepsilon > 0$; then we can choose $r > 0$ such that

$$\varepsilon r \int_0^a f^3(x)\, dx > \frac{3\pi^2}{2a^2}. \tag{4.91}$$

Thus, for fixed r and $\|u_0\|_a = r$, the solution blows up if the size of domain increases, that is, $a \to \infty$. We have obtained the well-known result of Sattinger: “large domains are less stable than small” (see [96]).

Example 4.10. The following example demonstrates the application of the energy method to nonlinear higher-order equations. Consider the following problem in a bounded domain $\Omega \subset \mathbb{R}^2$:

$$\begin{gathered} \frac{2\rho h}{d} \frac{\partial^2 w}{\partial t^2} = -\left(\frac{\partial^2}{\partial x^2} + \frac{\partial^2}{\partial y^2}\right)^2 w + F(w), \quad (x, y, t) \in \Omega \times [0, \infty), \\ w(x, y, 0), \ \frac{\partial w}{\partial t} \quad \text{are defined for } (x, y) \in \Omega, \\ w(x, y, t) = \frac{\partial w}{\partial \nu}(x, y, t) = 0, \quad (x, y, t) \in \partial\Omega \times [0, \infty). \end{gathered} \tag{4.92}$$

This equation describes oscillations of a clamped beam (see [67]); here F is the vertical load, $\rho(x, y)$ is the density of the beam as the point (x, y), d is the rigidity, and h is the half-thickness of the beam. We assume that $\rho h/d > 0$ at each point.

To apply Theorem 4.1, we introduce the Hilbert space $\mathbb{H} = \mathbb{L}^2(\Omega)$ with the standard scalar product and the operator $A = \Delta^2$ with domain

$$D_A = D = \left\{ f \in \mathbb{C}^{(4)}(\bar{\Omega}) \;\middle|\; f = \frac{\partial f}{\partial \nu} = 0 \text{ on } \partial\Omega \right\}.$$

Note that $(f, Af) \geq 0$ for all $f \in D$. As a nonlinearity, we choose the function $F(w) = \varepsilon w^2$. Then for any function $f(x, y) = w(x, y, 0)$ satisfying the conditions of Theorem 4.1,

$$\varepsilon \int_\Omega f^3(x, y)\, dx\, dy \geq \frac{3}{2} \int_\Omega (f_{xx} + f_{yy})^2\, dx\, dy, \tag{4.93}$$

the corresponding solution blows up for a finite time:

$$\lim_{t \to T} \int_\Omega \rho h d^{-1} w^2(x, y, t)\, dx\, dy = +\infty.$$

However, a similar linear problem is solvable globally in time. We note that characteristics of the equations $w_{tt} = -\Delta^2 w$ are planes parallel to the plane xy; therefore analogs of the Huygens principle are invalid, and hence methods based on this principle do not operate (see [41]).

Example 4.11. In this example, we consider the application of the energy method to the study of influence of nonlinear forces on the behavior of a linear elastic system

$$\rho(x)\frac{\partial^2 u_i}{\partial t^2} = \frac{\partial}{\partial x_k}\left(c_{ijkl}(x)\frac{\partial u_j}{\partial x_i}\right) + F_i(u_1, u_2, u_3), \quad i = 1, 2, 3. \tag{4.94}$$

We rewrite (4.94) in the domain $\Omega \times [0, \infty)$ in the vector form:

$$\rho(x)\frac{\partial^2 u}{\partial t^2} = \frac{\partial}{\partial x_k}\left(C_{kl}(x)\frac{\partial u}{\partial x_l}\right) + F(u). \tag{4.95}$$

Assume that the matrix $C_{kl} = (c_{ikjl})$ is symmetric ($C_{lk} = C_{kl}^t$, $c_{ikjl} = c_{jlik}$) and satisfies the inequality $c_{ikjl}\eta_{jl}\eta_{ik} \geq 0$ for all real $[\eta_{ij}]_{3\times 3}$ and all $x \in \bar{\Omega}$. Moreover, we assume that the solution satisfies the standard homogeneous boundary conditions, which, in the notation of Example 4.7, can be written in the form

$$u(x, 0) = 0 \quad \text{on } \Gamma_1 \times [0, \infty), \quad \nu_k c_{ikjl} u_{jl} = 0 \quad \text{on } \Gamma_2 \times [0, \infty).$$

These assumptions, together with the requirement of sufficient smoothness of u, satisfy the conditions of Theorem 4.1 for the operator A in the Hilbert space

$$\mathbb{H} = \mathbb{L}^2(\Omega) \otimes \mathbb{L}^2(\Omega) \otimes \mathbb{L}^2(\Omega)$$

equipped with the scalar product

$$(u, v) = \int_\Omega \sum_{i=1}^{3} u_i v_i \, dx.$$

As a nonlinearity, we take the vector-valued function

$$F(u)(x, t) = F(u(x, t)) = (F_1, F_2, F_3). \tag{4.96}$$

Assume that there exists a scalar potential of the nonlinearity H related to F by the expression

$$F = \nabla H(u_1, u_2, u_3), \quad H(0, 0, 0) = 0.$$

It is necessary to require that

$$\frac{\partial F_i}{\partial u_j} = \frac{\partial F_j}{\partial u_i}.$$

Then by definition the functional G is of the form

$$G(f) = G(f_1, f_2, f_3) = \int_0^1 \left(\int_\Omega F_i(\rho f(x)) f_i(x)\, dx \right) d\rho$$
$$= \int_\Omega \left(\int_0^1 \frac{\partial H}{\partial(\rho f_i)} (\rho f) f_i \, d\rho \right) dx = \int_\Omega H(f(x))\, dx. \tag{4.97}$$

Therefore condition (4.60) holds for any solution of problem (4.94) if

$$(4\alpha + 2) H(\xi_1, \xi_2, \xi_3) \le \sum_{i=1}^{3} \xi_i \frac{\partial H}{\partial \xi_i}(\xi_1, \xi_2, \xi_3), \tag{4.98}$$

and condition (4.76) is fulfilled if the initial data satisfy the inequality

$$\int_\Omega H(u(x,0))\, dx > \frac{1}{2} \int_\Omega c_{ikjl} \frac{\partial u_i(x,0)}{\partial x_k} \frac{\partial u_j(x,0)}{\partial x_l}\, dx. \tag{4.99}$$

Unfortunately, it is impossible to give a simple interpretation of all scalar functions H satisfying (4.98), in contrast to Example 4.7. However, for example, if H is a homogeneous function of degree $4\alpha + 2$, then in (4.98) an equality holds. As a consequence of Theorem 4.1, for the initial value u_0, we can find an appropriate value of the velocity $u'(x,0)$ such that the solution blows up in finite time in the following limit relation:

$$\lim_{t \to T} \int_\Omega \rho h d^{-1} u_i(x,t)\, dx = +\infty; \tag{4.100}$$

however, we cannot say with certainty which of the three integrals (4.100) is infinite.

Example 4.12. In the last example of application of Levine's energy method, we consider the problem on spatial oscillations of a rod with nonconstant cross section, one of whose ends is fixed and the other is free. This problem is described by the wave equation $\rho w_{tt} = E w_{xx}$, where ρ is the density, E is the elastic modulus, and w is the displacement. With account of the inertia of the lateral motion, the equation takes the form (see [67])

$$\rho \frac{\partial^2}{\partial t^2} \left(w - \sigma^2 K^2 \frac{\partial^2 w}{\partial x^2} \right) = E \frac{\partial^2 w}{\partial x^2}, \tag{4.101}$$

where σ and K are positive constants. The boundary conditions are $w = 0$ at the free end and $\partial w / \partial x = 0$ at the fixed end. Thus, we arrive at the following problem:

$$\frac{\partial^2}{\partial t^2} \left(\rho - \frac{\partial}{\partial x_i} \left(a_{ij}(x) \frac{\partial}{\partial x_j} \right) \right) u = \frac{\partial}{\partial x_i} \left(b_{ij}(x) \frac{\partial}{\partial x_j} \right) u + F(u), \tag{4.102}$$

where $[a_{ij}(x)]$ and $[b_{ij}(x)]$ are positive definite matrices for all $x \in \mathbb{R}^n$. If we take the nonlinearity as in Examples 4.7 and 4.8 and an initial condition satisfying (4.76), then we can apply Theorem 4.1 to the corresponding initial-boundary-value problem for equation (4.102) and prove the blow-up of a solution in finite time. Equations of such type are discussed in detail in Chap. 8.

4.2.4 Wave equation with damping

Consider a nonlinear hyperbolic problem with linear damping. Let $P = I$, and let the operator $A(t) = A \geq 0$ be independent of time. Applying Levine's method, we obtain sufficient conditions of blow-up for the initial-boundary-value problem

$$\begin{gathered} u_{tt} + au_t = -Au + F(u), \quad a > 0, \\ u(0) = u_0, \quad u_t(0) = v_0. \end{gathered} \tag{4.103}$$

Theorem 4.5. *Assume that F and A satisfy all conditions of Theorem 4.1. We treat nonextendable in time solutions $u : [0, T) \to \mathbb{H}$ of problem (4.103) in the same sense as solutions of equation (4.62). We assume that the function u_0 satisfies the inequality*

$$G(u_0) > \frac{1}{2}((u_0, Au_0) + (a/2\alpha)^2(u_0, u_0)). \tag{4.104}$$

Then, for all initial functions $v_0 \in D$ satisfying the inequalities

$$\|v_0\| \leq r(u_0), \quad \|v_0\| \cos\theta > \frac{a\|u_0\|}{2\alpha},$$

where

$$\cos\theta = \frac{(u_0, v_0)}{\|u_0\|\,\|v_0\|}, \quad r(u_0) = \sqrt{2G(u_0) - (u_0, Au_0)},$$

the corresponding solution blows up in finite time:

$$\lim_{t\to T} \|u(t)\| = +\infty, \quad T \leq \frac{1}{a}\ln\left(\frac{2\alpha\|v_0\|\cos\theta}{2\alpha\|v_0\|\cos\theta - a\|u_0\|}\right). \tag{4.105}$$

Proof. Introduce the notation $\Phi(t) = (u, u)$. Differentiating this function twice by time, we have

$$\Phi'(t) = 2(u, u_t), \quad \Phi''(t) = 2(u, u_{tt}) + 2(u_t, u_t).$$

Using the Cauchy–Bunyakovsky inequality, we verify the inequality

$$S^2 = (u, u)(u_t, u_t) - (u, u_t)^2 \geq 0,$$

which can be rewritten in the form

$$4(\alpha+1)S^2 = 2\Phi(2\alpha+2)(u_t,u_t) - (\alpha+1)(\Phi')^2 \geq 0.$$

It is easy to prove the estimate

$$\Phi\Phi'' - (\alpha+1)(\Phi')^2 \geq 4(\alpha+1)S^2 + 2\Phi((u,u_{tt}) - (2\alpha+1)(u_t,u_t)). \tag{4.106}$$

Introduce the notation

$$H(t) \equiv (u,u_{tt}) - (2\alpha+1)(u_t,u_t) \tag{4.107}$$

and consider the last term of (4.106) separately. Differentiate $H(t)$ by time:

$$\begin{aligned} H'(t) &= (u,u_{ttt}) - (4\alpha+1)(u_t,u_{tt}) \\ &= -a(u,u_{tt}) + 4\alpha(u_t,Au) + a(4\alpha+1)(u_t,u_t) + \frac{d}{dt}(u,F(u)) - (4\alpha+2)(u_t,F(u)) \\ &= a(4\alpha+2)(u_t,u_t) - (4\alpha+2)(u_t,F(u)) \\ &\quad + \frac{d}{dt}\big[2\alpha(u,Au) - a(u,u_t) + (u,F(u))\big]. \end{aligned} \tag{4.108}$$

Now we integrate (4.108) from 0 to t and apply the properties of A and G and the definition of H:

$$\begin{aligned} H(t) &\geq H(0) - a(u_0,u_{t,0}) + (u_0,Au_0) - (u_0,F(u_0)) - a(u,u_t) \\ &\quad - (u,F(u)) - (4\alpha+2)\int_0^t (u_\eta,F(u))d\eta - (2\alpha+1)(u_0,Au_0) \\ &\geq -a(u,u_t) + 2(2\alpha+1)G(u_0) - (2\alpha+1)((u_0,Au_0) + (v_0,v_0)) \\ &\geq -a\Phi'(t)/2. \end{aligned} \tag{4.109}$$

In the last inequality, we have applied condition (4.104). Substituting (4.109) into (4.106), we obtain the ordinary differential inequality

$$\Phi\Phi'' - (\alpha+1)(\Phi')^2 \geq -a\Phi\Phi' \quad \Rightarrow \quad (\Phi^{-\alpha})'' + a(\Phi^{-\alpha})' \leq 0.$$

Integrating this inequality twice by time, we arrive at the lower estimate

$$\Phi^\alpha \geq \frac{\Phi^\alpha(0)}{1 - \alpha(1-e^{-at})\Phi'(0)/\Phi(0)}. \tag{4.110}$$

Under the condition

$$\frac{\alpha\Phi'(0)}{\Phi(0)} > 1,$$

the denominator of (4.110) at some time moment $t = t_0$ vanishes. Thus, we have obtained sufficient conditions of the blow-up of a solution for a finite time,

$$\lim_{t \to T} \|u(t)\| = +\infty,$$

and the upper estimate of the lifetime,

$$T \le t_0 = \frac{1}{a} \ln\left(\frac{2\alpha(u_0, v_0)}{2\alpha(u_0, v_0) - a(u_0, u_0)}\right).$$

The theorem is proved. □

As for the nonlinear problem without damping, we can prove that if the nonlinearity F satisfies the conditions of Lemma 4.1, then there exist initial conditions u_0 and v_0 under which solution (4.103) blows up in finite time.

4.3 Bibliographical notes

The classical energy method of H. A. Levine for the study of the blow-up of solutions of parabolic and hyperbolic equations is presented in detail in [57, 61] and [58, 60], respectively. In [60], we can find a geometric interpretation of blow-up theorems presented here. For further development of the method for problems with nonlinear damping and for pseudo-hyperbolic and pseudo-parabolic equations, see Chapters 5–8.

Unfortunately, Levine's energy method and its modifications have not yet been applied to the calculation of the critical exponent for the Cauchy problem

$$u_t = \Delta u + |u|^{q-1}u, \quad u(x, 0) = u_0(x), \quad q > 1.$$

Modifications of the energy method (for example, the modification proposed by Pokhozhaev) allow us to prove the blow-up only for "large" initial functions $u_0(x)$.

By the way, there are not yet results about the applicability of the energy method for the problem on the critical exponent for the simpler problem

$$u_t = \Delta u + |u|^q, \quad u(x, 0) = u_0(x), \quad q > 1,$$

which can be naturally examined by the method of nonlinear capacity of S. I. Pokhozhaev. Permanent intensive work is being done in this direction.

5 Energy method of G. Todorova

Energy method proposed by G. Todorova is applicable to equations that contain, in addition to a nonlinear source, the term $|u_t|^m u_t$, which describes nonlinear damping. Levine's classical method is inapplicable for equations with such terms. However, a modification invented by G. Todorova allow us to take into account such terms without changing the essence of the energy method (see [29]).

In this chapter, we consider two problems, namely Todorova's result on the blow-up of solutions for the initial-boundary-value problem for a hyperbolic differential equation (see [29]) and Messaoudi's result for an integro-differential equation with the same nonlinear damping term (see [73]).

5.1 Statement of the problem and local solvability

Consider the following initial-boundary-value problem for the hyperbolic nonlinear equation:

$$u_{tt} - \Delta u + au_t|u_t|^{m-1} = bu|u|^{p-1}, \quad (t,x) \in [0,T] \times \Omega, \tag{5.1}$$

$$u(t,x) = 0, \quad (t,x) \in [0,T] \times \partial\Omega, \tag{5.2}$$

$$u(0,x) = \varphi(x), \quad u_t(0,x) = \psi(x), \tag{5.3}$$

where $a, b > 0$, $p, m > 1$, and the bounded domain $\Omega \subset \mathbb{R}^N$, $N \geq 1$, has a smooth boundary $\partial\Omega$. The case $a = 0$ is well studied (see, e. g., [31, 103]); in particular, it is known that, for sufficiently large initial data, the nonlinear term $bu|u|^{p-1}$ guarantees the blow-up of a solution in finite time. In the case $b = 0$, the term $u_t|u_t|^{m-1}$ provides the global solvability for arbitrary initial data due to a priori estimates obtained in [33].

Setting $a = b = 1$, we examine the mutual influence of the nonlinear damping term $au_t|u_t|^{m-1}$ and exterior sources $bu|u|^{p-1}$. We define the values of the parameters p and m for which the solution blows up and for which it exists globally in time. Before the proof of basic results, we justify that problem (5.1)–(5.3) is well posed and there exists a local-in-time solution.

The local solvability of the problem with nonlinear terms $u|u|^{p-1}$ and $u_t|u_t|^{m-1}$ was studied by Lions [66], but only for equations containing one of these terms. In our case, we consider both these terms with $a = b = 1$. Applying an appropriate approximation and the solvability result obtained by Lions in the space $\mathbb{L}^\infty((0,T);\mathbb{X})$, we prove the local solvability in a "less regular" space.

The result on the local solvability can be formulated as follows.

Theorem 5.1. *Let $m > 1$, and let the following inequalities hold:*

$$1 < p \leq \frac{N}{N-2} \quad \textit{for } N \geq 3, \tag{5.4}$$

https://doi.org/10.1515/9783110602074-005

$$p > 1 \quad for\, N \leq 2. \tag{5.5}$$

Then, for sufficiently small T and any initial data $\varphi \in \mathbb{H}_0^1(\Omega)$, $\psi \in \mathbb{L}^2(\Omega)$, there exists a unique solution $u(t,x)$ of problem (5.1)–(5.3) *of the class*

$$U = \begin{pmatrix} u \\ u_t \end{pmatrix} \in \mathbb{Y}_T = \left\{ U = \begin{pmatrix} u_1 \\ u_2 \end{pmatrix},\ u_1 \in \mathbb{C}([0,T];\mathbb{H}),\ u_2 \in \mathbb{L}^{m+1}((0,T)\times\Omega) \right\}.$$

Proof. We rewrite problem (5.1)–(5.3) for the following abstract evolution equation:

$$\partial_t U - AU + g(U) = f(U), \quad U(x,0) = \begin{pmatrix} \varphi \\ \psi \end{pmatrix}, \tag{5.6}$$

where

$$U(x,t) = \begin{pmatrix} u \\ u_t \end{pmatrix}, \quad A = \begin{pmatrix} 0 & 1 \\ \Delta & 0 \end{pmatrix},$$

$$g\begin{pmatrix} u_1 \\ u_2 \end{pmatrix} = \begin{pmatrix} 0 \\ u_2|u_2|^{m-1} \end{pmatrix}, \quad f\begin{pmatrix} u_1 \\ u_2 \end{pmatrix} = \begin{pmatrix} 0 \\ u_1|u_1|^{p-1} \end{pmatrix}.$$

Applying a result of [66], we prove the following assertion.

Assertion 5.1. *Assume that the conditions of Theorem* 5.1 *hold. Then, for any $U \in \mathbb{C}([0,T];\mathbb{H})$, there exists a unique solution $V \in \mathbb{Y}_T$ of the problem*

$$\partial_t V - AV + g(V) = f(U), \quad V(x,0) = \begin{pmatrix} \varphi \\ \psi \end{pmatrix}, \tag{5.7}$$

satisfying the following energy inequality for all $t \geq s \geq 0$:

$$\frac{1}{2}\|V(t)\|_{\mathbb{H}}^2 - \frac{1}{2}\|V(s)\|_{\mathbb{H}}^2 + \int_s^t (g(V),V)_{\mathbb{H}}\, d\tau = \int_s^t (f(U),V)_{\mathbb{H}}\, d\tau. \tag{5.8}$$

Proof. We will work in the natural energy Hilbert space $\mathbb{H} = \mathbb{H}_0^1(\Omega) \otimes \mathbb{L}^2(\Omega)$, in which the scalar product of any two elements $U = (u_1,u_2)^T$ and $V = (v_1,v_2)^T$ is defined as follows:

$$(U,V)_H = (\nabla u_1, \nabla v_1)_{L_2(\Omega)} + (u_2,v_2)_{L_2(\Omega)}. \tag{5.9}$$

Due to the dense embedding

$$\mathbb{C}_0^\infty(\Omega) \otimes \mathbb{C}_0^\infty(\Omega) \overset{ds}{\subset} \mathbb{H},$$

the initial data $(\varphi,\psi)^T$ of the problem can be approximated by elements

$$\begin{pmatrix} \varphi^\nu \\ \psi^\nu \end{pmatrix}, \quad \varphi^\nu, \psi^\nu \in \mathbb{C}_0^\infty(\Omega).$$

Moreover, any element $U \in \mathbb{C}([0,T];\mathbb{H})$ can be approximated by elements $U^\nu \in \mathbb{C}([0,T];\mathbb{C}_0^\infty(\Omega) \otimes \mathbb{C}_0^\infty(\Omega))$. Then by [66, Chap. 1, Thm. 3.1] we conclude the existence of a solution $V^\nu = (v^\nu, v_t^\nu)^T$ of the abstract approximated problem

$$\partial_t V^\nu - AV^\nu + g(V^\nu) = f(U^\nu), \quad V^\nu(0,x) = \begin{pmatrix} \varphi^\nu \\ \psi^\nu \end{pmatrix}. \tag{5.10}$$

Moreover, this solution possesses the following properties:

$$v^\nu \in \mathbb{L}^\infty((0,T);\ \mathbb{H}^2(\Omega) \cap \mathbb{H}_0^1(\Omega)), \quad v_{tt}^\nu \in \mathbb{L}^\infty((0,T);\mathbb{L}^2(\Omega)),$$
$$v_t^\nu \in \mathbb{L}^\infty((0,T);\ \mathbb{H}_0^1(\Omega)), \quad v_t^\nu \in \mathbb{L}^{m+1}((0,T)\times\Omega).$$

We show that the sequence $\{V^\nu\}$ is a Cauchy sequence in the space $\mathbb{Y}_T$ with the norm

$$\|U\|_{\mathbb{Y}_T} = \left\| \begin{pmatrix} u_1 \\ u_2 \end{pmatrix} \right\|_{\mathbb{Y}_T} = \sup_{t\in[0,T]} \|U(t)\|_{\mathbb{H}} + \|u_2\|_{\mathbb{L}^{m+1}((0,T)\times\Omega)}. \tag{5.11}$$

Thus, it becomes clear that all reasonings can be divided into two stages: first, we prove that the sequence $\{V^\nu\}$ is a Cauchy sequence in the space $\mathbb{C}([0,T];\mathbb{H})$, and then we prove that the sequence $\{v_t^\nu\}$ is a Cauchy sequence in the space $\mathbb{L}^{m+1}((0,T)\times\Omega)$.

For given μ and ν, we introduce the notation $W = V^\nu - V^\mu$. Then W is a solution of the following Cauchy problem:

$$\partial_t W - AW + g(V^\nu) - g(V^\mu) = f(U^\nu) - f(U^\mu), \tag{5.12}$$

where

$$W(0,x) = \begin{pmatrix} \varphi^\nu - \varphi^\mu \\ \psi^\nu - \psi^\mu \end{pmatrix}.$$

Using the definitions of A and the scalar product in $\mathbb{H}$, we can easily prove that

$$(AW, W)_{\mathbb{H}} = 0,$$

and therefore W satisfies the energy equality

$$\frac{1}{2}\|W(t)\|_{\mathbb{H}}^2 - \frac{1}{2}\|W(0)\|_{\mathbb{H}}^2 + \int_0^t (g(V^\nu) - g(V^\mu), W)_{\mathbb{H}}\, d\tau$$
$$= \int_0^t (f(U^\nu) - f(U^\mu), W)_{\mathbb{H}}\, d\tau. \tag{5.13}$$

We analyze the terms of this equality separately. The norm $\|W(0)\|_{\mathbb{H}^2}$ is small for sufficiently large ν and μ since the sequence

$$\left\{ \begin{pmatrix} \varphi^\nu \\ \psi^\nu \end{pmatrix} \right\}$$

is a Cauchy sequence in $\mathbb{H}$. Due to the monotonicity of the power function $|x|^m x$, we can verify that the integrand in the left-hand side is nonnegative:

$$(g(V^\nu) - g(V^\mu), W)_{\mathbb{H}} = (v_t^\nu |v_t^\nu|^m - v_t^\mu |v_t^\mu|^m, v_t^\nu - v_t^\mu)_{\mathbb{L}^2(\Omega)}. \tag{5.14}$$

We estimate the integrand in the right-hand side. For the four functions

$$U = \begin{pmatrix} u_1 \\ u_2 \end{pmatrix}, \quad \overline{U} = \begin{pmatrix} \overline{u}_1 \\ \overline{u}_2 \end{pmatrix}, \quad V = \begin{pmatrix} v_1 \\ v_2 \end{pmatrix}, \quad \overline{V} = \begin{pmatrix} \overline{v}_1 \\ \overline{v}_2 \end{pmatrix}$$

of class $\mathbb{H}$, by the Hölder inequality we have the following estimate:

$$\begin{aligned} |(f(U) - f(\overline{U}), V - \overline{V})_{\mathbb{H}}| &= |(u_1|u_1|^{p-1} - \overline{u}_1|\overline{u}_1|^{p-1}, v_2 - \overline{v}_2)_{\mathbb{L}^2(\Omega)}| \\ &\le c\|u_1 - \overline{u}_1\|_{2N/(N-2)} \|v_2 - \overline{v}_2\|_2 (\|u_1\|_{N(p-1)}^{p-1} + \|\overline{u}_1\|_{N(p-1)}^{p-1}). \end{aligned} \tag{5.15}$$

Sobolev embeddings imply the following inequalities:

$$\|u_1 - \overline{u}_1\|_{2N/(N-2)} \le c\|\nabla(u_1 - \overline{u}_1)\|_2 \le c_1 \|U - \overline{U}\|_{\mathbb{H}}. \tag{5.16}$$

To estimate the norm $\|u_1\|_{N(p-1)}$ through $\|U\|_{\mathbb{H}}$, we must require that

$$N(p-1) \le \frac{2N}{N-2} \quad \Rightarrow \quad p \le \frac{N}{N-2}.$$

Then the Sobolev embedding $\mathbb{L}^{2N/(N-2)}(\Omega) \subset \mathbb{H}_0^1(\Omega)$ leads to the estimate

$$\|u\|_{N/(p-1)}^{p-1} \le c\|U\|_{\mathbb{H}}^{p-1}. \tag{5.17}$$

Thus, we have proved the inequality

$$\begin{aligned} &|(f(U) - f(\overline{U}), V - \overline{V})_{\mathbb{H}}| \\ &\quad \le c(\|U\|_{\mathbb{H}} + \|\overline{U}\|_{\mathbb{H}})^{p-1} \|U - \overline{U}\|_{\mathbb{H}} \|V - \overline{V}\|_{\mathbb{H}}, \end{aligned} \tag{5.18}$$

where the constant $c = c(\Omega, T)$ depends on the domain Ω and time T but is independent of the functions U, $\overline{U}$, V, and $\overline{V} \in \mathbb{H}$.

Combining formulas (5.13), (5.18), and (5.15), we obtain the energy estimate

$$\|W(t)\|_{\mathbb{H}}^2 \le \|W(0)\|_{\mathbb{H}}^2 + c \int_0^t \|U^\nu - U^\mu\|_{\mathbb{H}} \|W(\tau)\|_{\mathbb{H}} \, d\tau, \tag{5.19}$$

which by the Gronwall–Bellman–Bihari lemma implies the inequality

$$\begin{aligned} \|W(t)\|_{\mathbb{H}} &= \|V^\nu(t) - V^\mu(t)\|_{\mathbb{H}} \\ &\le c\|V^\nu(0) - V^\mu(0)\|_{\mathbb{H}} + cT\|U^\nu - U^\mu\|_{\mathbb{C}([0,T];\mathbb{H})}. \end{aligned} \tag{5.20}$$

Since the sequences $\{U^\nu\}$ and $\{V^\nu(0)\}$ are Cauchy sequences in the spaces $\mathbb{C}([0,T];\mathbb{H})$ and $\mathbb{H}$, respectively, we conclude that the sequence $\{V^\nu\}$ is a Cauchy sequence in $\mathbb{C}([0,T];\mathbb{H})$.

Now we return to (5.11). To prove that $\{V^\nu\}$ is a Cauchy sequence in $\mathbb{Y}_T$, it remains to verify that $\{v_t^\nu\}$ is a Cauchy sequence in the space $\mathbb{L}^{m+1}((0,T)\times\Omega)$. From the estimate

$$(\alpha|\alpha|^{m-1} - \beta|\beta|^{m-1})(\alpha-\beta) \geq c|\alpha-\beta|^{m+1}$$

for all real α and β we have

$$(g(V^\nu) - g(V^\mu),\ W)_{\mathbb{H}} \geq c\|v_t^\nu - v_t^\mu\|^{m+1}_{\mathbb{L}^{m+1}}. \tag{5.21}$$

Integrating (5.21) by time from 0 to T and using (5.13), where we set $t = T$, we can continue the inequality (5.21):

$$\begin{aligned} c\|v_t^\nu - v_t^\mu\|^{m+1}_{\mathbb{L}^{m+1}((0,T)\times\Omega)} &\leq \frac{1}{2}\|W(0)\|^2_{\mathbb{H}} - \frac{1}{2}\|W(T)\|^2_{\mathbb{H}} \\ &\quad + \int_0^T (f(U^\nu) - f(U^\mu), W)_{\mathbb{H}}\, d\tau \leq c\|V^\nu(0) - V^\mu(0)\|^2_{\mathbb{H}} \\ &\quad + c\int_0^T \|U^\nu(\tau) - U^\mu(\tau)\|_{\mathbb{H}}\|V^\nu(\tau) - V^\mu(\tau)\|_{\mathbb{H}}\, d\tau. \end{aligned} \tag{5.22}$$

Inequality (5.22) implies that $\{v_t^\nu\}$ is s Cauchy sequence; therefore, $\{V^\nu\}$ is also a Cauchy sequence in $\mathbb{Y}_T$.

Let a function V be the limit of $\{V^\nu\}$ in $\mathbb{Y}_T$. We prove that V is a *weak solution* of problem (5.7) in the sense of [66], that is, for all $\kappa \in \mathbb{H}_0^1(\Omega) \cap \mathbb{L}^{m+1}(\Omega)$ and almost all $t \in [0,T]$, we have the following equality:

$$\frac{d}{dt}(v_t, \kappa) + (\nabla v, \nabla\kappa) + (v_t|v_t|^{m-1}, \kappa) = (u|u|^{p-1}, \kappa). \tag{5.23}$$

Indeed, multiplying the equation

$$v_{tt}^\nu - \Delta v^\nu + v_t^\nu|v_t^\nu|^{m-1} = u^\nu|u^\nu|^{p-1}$$

by $\kappa \in \mathbb{H}_0^1(\Omega) \cap \mathbb{L}^{m+1}(\Omega)$, we obtain

$$\frac{d}{dt}(v_t^\nu, \kappa) + (\nabla v^\nu, \nabla\kappa) + (v_t^\nu|v_t^\nu|^{m-1}, \kappa) = (u^\nu|u^\nu|^{p-1}, \kappa). \tag{5.24}$$

Pass to the limit in the Banach space $\mathbb{C}([0,T])$ as $\nu \to \infty$ with respect to the standard norm:

$$(\nabla v^\nu, \nabla \kappa) \to (\nabla v, \nabla \kappa),$$
$$(v_t^\nu |v_t^\nu|^{m-1}, \kappa) \to (v_t |v_t|^{m-1}, \kappa),$$
$$(u^\nu |u^\nu|^{p-1}, \kappa) \to (u|u|^{p-1}, \kappa).$$

Then from (5.24) we conclude that

$$\lim_{\nu\to\infty} (v_t^\nu, \kappa) = (v_t, \kappa)$$

is a uniformly continuous function and relation (5.23) holds. Similarly, from the energy equality for v^ν we obtain equation (5.8) for v.

Finally, to prove the uniqueness of solution (5.7), we assume that V and $\overline{V}$ are two distinct solutions of problem (5.7). For the function $W = V - \overline{V}$, we have

$$\frac{1}{2}\|W(t)\|_{\mathbb{H}^2} + \int_0^t (g(V) - g(\overline{V}),\ W)_{\mathbb{H}}\, d\tau$$
$$= \int_0^t (f(U) - f(\overline{U}),\ W)_{\mathbb{H}}\, d\tau, \quad U, \overline{U} \in \mathbb{C}([0,T];\mathbb{H}). \tag{5.25}$$

Therefore, if $U = \overline{U}$, then $W = 0$, and the solution is unique. The assertion is proved. □

We have proved the existence and uniqueness of a solution of problem (5.7). Now we return to the proof of the theorem. For an arbitrary function $U \in \mathbb{C}([0,T];\mathbb{H})$, we define a function $V = \Phi(U)$ such that V is a solution of equation (5.7). The assertion proved before implies that Φ maps the space

$$\mathbb{X}_T = \mathbb{X}(T, \varphi, \psi) = \left\{U \in \mathbb{C}([0,T];\mathbb{H}) : U(x,0) = \begin{pmatrix}\varphi\\ \psi\end{pmatrix}\right\} \quad \text{into } \mathbb{Y}_T.$$

For sufficiently large $R > 0$ depending on the initial data and sufficiently small $T > 0$, the following properties hold:
(i) Φ maps a ball B_R of radius R from the space $\mathbb{X}_T$ into itself;
(ii) Φ is a contraction mapping on the ball B_R.

Using these properties and the Banach fixed-point theorem, we complete the proof of the theorem on the local existence of a solution.

To prove property (i), we choose

$$U \in B_R = \left\{U \in \mathbb{X}_T : \sup_{0\le t\le T} \|U(t)\|_{\mathbb{H}} \le R\right\}. \tag{5.26}$$

Then $V = \Phi(U)$ is a solution of the equation

$$\partial_t V - AV + g(V) = f(U),$$

and the assertion proved before guarantees the fulfillment of equation (5.8):

$$\frac{1}{2}\|V(t)\|_{\mathbb{H}^2} - \frac{1}{2}\|V(0)\|_{\mathbb{H}^2} + \int_0^t (g(V), V)_{\mathbb{H}}\, d\tau = \int_0^t (f(U), V)_{\mathbb{H}}\, d\tau. \tag{5.27}$$

Since the integrand is nonnegative, we have

$$(g(V), V)_{\mathbb{H}} = (v_2|v_2|^{m-1}, v_2)_{\mathbb{L}^2} \geq 0, \tag{5.28}$$

and then we obtain the inequality

$$\frac{1}{2}\|V(t)\|^2_{\mathbb{H}} - \frac{1}{2}\|V(0)\|^2_{\mathbb{H}} \leq \int_0^t |(f(U), V)_{\mathbb{H}}|\, d\tau. \tag{5.29}$$

As in the case of estimate (5.18), we can easily prove the inequality

$$|(f(U), V)_{\mathbb{H}}| \leq c\|U\|^p_{\mathbb{H}}\|V\|_{\mathbb{H}}, \tag{5.30}$$

where $c > 0$ is some constant. Substituting (5.30) into (5.29), we obtain

$$\|V(t)\|^2_{\mathbb{H}} - \|V(0)\|^2_{\mathbb{H}} \leq CR^p \int_0^t \|V(\tau)\|\, d\tau. \tag{5.31}$$

For another constant c independent of t, T, and R, we have the following inequality:

$$\|V(t)\|_{\mathbb{H}} \leq c\|V(0)\|_{\mathbb{H}} + cR^pT. \tag{5.32}$$

Now, choosing sufficiently large R and sufficiently small T, we arrive at the estimate

$$\sup_{0\leq t\leq T} \|V(t)\|_{\mathbb{H}} \leq R.$$

Therefore, $V \in B_R$, and property (i) is proved.

To prove property (ii), we take $U, \overline{U} \in \mathbb{X}_T$. Let

$$V = \begin{pmatrix} v_1 \\ v_2 \end{pmatrix} = \Phi(U), \quad \overline{V} = \begin{pmatrix} \overline{v}_1 \\ \overline{v}_2 \end{pmatrix} = \Phi(\overline{U}).$$

Then $W = V - \overline{V}$ satisfies the equation

$$\partial_t W - AW + g(V) - g(\overline{V}) = f(U) - f(\overline{U}). \tag{5.33}$$

Applying the energy equality (5.25), from the monotonicity of g and estimate (5.18) we obtain

$$\|W(t)\|^2_{\mathbb{H}} \leq c \int_0^t (\|U\|_{\mathbb{H}} + \|\overline{U}\|_{\mathbb{H}})^{p-1}\|W(\tau)\|_{\mathbb{H}}\|U(\tau) - \overline{U}(\tau)\|_{\mathbb{H}}\, d\tau. \tag{5.34}$$

Thus, we get the following inequality:

$$\begin{aligned}\|W\|_{\mathbb{C}([0,T];\mathbb{H})} \leq{}& cT(\|U\|_{\mathbb{C}([0,T];\mathbb{H})} \\ &+ \|\overline{U}\|_{\mathbb{C}([0,T];\mathbb{H})})^{p-1}\|U - \overline{U}\|_{\mathbb{C}([0,T];\mathbb{H})}.\end{aligned} \tag{5.35}$$

If $U, \overline{U} \in B_R$, then we choose $T > 0$ so small that $cR^{p-1}T \leq 1/2$. Then from (5.35) we see that Φ is a contraction mapping on the ball B_R. Applying the Banach fixed-point theorem, we conclude that there exists a unique solution U of the problem $U = \Phi(U)$ in the space $\mathbb{C}([0,T];\mathbb{H})$. The fact that Φ maps $\mathbb{C}([0,T];\mathbb{H})$ to $\mathbb{Y}_T$, guarantees that $U \in \mathbb{Y}_T$. The theorem is proved. □

5.2 Global solvability and blow-up

In the previous section, we have proved the local solvability, and now we can use the conventional continuation method (see, e. g., [97, 49, 79]). Using this method, we can easily conclude that either the upper boundary of the lifetime T of the solution is infinite, $T = \infty$, or T is finite, and the following limit equality holds:

$$\lim_{t \to T-0} \|U(t)\|_{\mathbb{H}} = \infty,$$

that is, the blow-up of the solution occurs. We use Todorova's energy method and prove the results on the global solvability and insolvability.

Theorem 5.2 (on the global solvability). *Assume that $a = b = 1$, the initial data belong to the classes $\varphi \in \mathbb{H}_0^1(\Omega)$ and $\psi \in \mathbb{L}^2(\Omega)$, and the following inequalities are fulfilled:*

$$1 < p \leq \frac{N}{N-2} \quad \text{for } N \geq 3, \quad p > 1 \quad \text{for } N \leq 2.$$

Then, for $p \leq m$, problem (5.1)–(5.3) has a unique solution $u(t,x)$ such that, for any $T > 0$,

$$u(t,x) \in \mathbb{C}^{(1)}([0,T];\mathbb{H}_0^1(\Omega)), \quad u_t(t,x) \in \mathbb{C}^{(1)}([0,T];\mathbb{L}^2(\Omega)) \cap \mathbb{L}^{m+1}((0,T)\times\Omega).$$

Theorem 5.3 (on the blow-up). *Assume that $1 < m < p$ and conditions (5.4) and (5.5) hold. Then a solution of problem (5.1)–(5.3) for an appropriate initial data blows up in finite time in the norm of $\mathbb{L}^\infty(\Omega)$.*

Here under "appropriate initial data," we mean data that provide a sufficiently large (by absolute value) negative value of the energy functional

$$E = \frac{1}{2}\|\psi\|_2^2 + \frac{1}{2}\|\nabla\varphi\|_2^2 - \frac{1}{p+1}\|\varphi\|_{p+1}^{p+1}.$$

Proof. Levine's classical method with the conventional energy functional

$$E_{kl} = \frac{1}{2}\|u_t\|_2^2 + \frac{1}{2}\|\nabla u\|_2^2$$

does not allow us to analyze the mutual influence of the source and the nonlinear damping. G. Todorova proposed to use the following modified energy functional to prove the global solvability:

$$E(t) = E_{kl}(t) + \frac{1}{p+1}\|u\|_{p+1}^{p+1}. \tag{5.36}$$

We show that the functional $E(t)$ is exponentially bounded, and therefore the classical energy cannot grow unboundedly. We differentiate the energy equality obtained in (5.8) by time and use the definition of the scalar product in $\mathbb{H}$; then we obtain

$$E'_{kl} = -\|u_t\|_{m+1}^{m+1} + \int_\Omega uu_t|u|^{p-1}dx. \tag{5.37}$$

On the other hand, since

$$\frac{1}{p+1}\frac{d}{dt}\|u\|_{p+1}^{p+1} = \int_\Omega uu_t|u|^{p-1}dx, \tag{5.38}$$

adding (5.38) to both sides of (5.37), we have

$$E' = -\|u_t\|_{m+1}^{m+1} + 2\int_\Omega uu_t|u|^{p-1}dx. \tag{5.39}$$

By the Hölder inequality

$$\left|\int_\Omega uu_t|u|^{p-1}dx\right| \le \|u\|_{p+1}^{p}\|u_t\|_{p+1}, \tag{5.40}$$

and the three-parameter Young inequality

$$ab \le \varepsilon a^\alpha + c(\varepsilon)b^\beta, \quad \frac{1}{\alpha} + \frac{1}{\beta} = 1, \tag{5.41}$$

for positive a, b, α, β, ε, and $c(\varepsilon)$, from (5.40) we obtain the estimate

$$\left|\int_\Omega uu_t|u|^{p-1}dx\right| \le c(\varepsilon)\|u\|_{p+1}^{p+1} + \varepsilon\|u_t\|_{p+1}^{p+1}. \tag{5.42}$$

Since $m+1 \ge p+1$, in the inequality

$$E'(t) \le -\|u_t\|_{m+1}^{m+1} + 2c(\varepsilon)\|u\|_{p+1}^{p+1} + 2\varepsilon\,\|u_t\|_{p+1}^{p+1}, \tag{5.43}$$

we can choose ε sufficiently small; this leads to the estimate

$$E'(t) \le cE(t). \tag{5.44}$$

Applying the Gronwall lemma, we obtain

$$E(t) \le E(0)\exp(ct). \tag{5.45}$$

This upper estimate for $E(t)$ and the continuation principle imply the absence of blow-up of solutions, that is, the global solvability occurs.

Now we prove the blow-up theorem. Introduce the notation $F(t) = \|u\|_2^2$. The proof of blow-up of the functional $F(t)$ in finite time is sufficiently difficult even in the absence of damping (see [2]). Applying the energy approach, we can prove the blow-up only in a certain subdomain $m < p$, for example, for $m < 2p/(p+1)$. To obtain the desired result for all $m < p$, we consider the other functional

$$H^{1-\alpha}(t) + \varepsilon F'(t), \tag{5.46}$$

where

$$H(t) = -\frac{1}{2}\|u_t\|_2^2 - \frac{1}{2}\|\nabla u\|_2^2 + \frac{1}{p+1}\|u\|_{p+1}^{p+1},$$

where $\varepsilon > 0$ and $\alpha > 0$ are small parameters (we choose their values further). Note that the form of $H(t)$ allows us to separate terms related to nonlinear damping in the energy equality (5.8) and to rewrite (5.8) in the form

$$H'(t) = \|u_t\|_{m+1}^{m+1}. \tag{5.47}$$

Thus $H(t)$ is an increasing function:

$$0 < H(0) \le H(t) \le \frac{1}{p+1}\|u\|_{p+1}^{p+1}. \tag{5.48}$$

We start with the equality

$$\frac{d}{dt}(H^{1-\alpha} + \varepsilon F') = (1-\alpha)H^{-\alpha}H' + \varepsilon F''. \tag{5.49}$$

Using the definition of H, we have

$$2(u,\Delta u)_2 = 4H + 2\int_\Omega u_t^2\,dx + \frac{4}{p+1}\|u\|_{p+1}^{p+1}.$$

Calculate the second time derivative $F''(t)$:

$$\begin{aligned} F''(t) &= 2\int_\Omega u_t^2\,dx + 2(u,u_{tt})_2 \\ &= 2\int_\Omega u_t^2\,dx + 2(u,\Delta u - |u_t|^{m-1}u_t + |u|^{p-1}u) \end{aligned}$$

$$= 4\int_\Omega u_t^2\,dx + 4H + 2\frac{p-1}{p+1}\int_\Omega |u|^{p+1}dx - 2\int_\Omega uu_t|u_t|^{m-1}dx. \tag{5.50}$$

Substituting (5.50) into (5.49), we obtain

$$\begin{aligned}\frac{d}{dt}(H^{1-\alpha} + \varepsilon F') &= (1-\alpha)H^{-\alpha}H' + 4\varepsilon\int_\Omega u_t^2\,dx \\ &\quad + 4\varepsilon H + 2\varepsilon\frac{p-1}{p+1}\int_\Omega |u|^{p+1}dx - 2\varepsilon\int_\Omega uu_t|u_t|^{m-1}dx.\end{aligned} \tag{5.51}$$

The last term in the right-hand side of (5.51) can be estimated by using (5.48) and the Hölder inequality for $m < p$:

$$\begin{aligned}\left|2\varepsilon\int_\Omega uu_t|u_t|^{m-1}dx\right| &\le c\varepsilon\|u\|_{m+1}\|u_t\|_{m+1}^m \\ &\le c\varepsilon\|u\|_{p+1}^{1-(p+1)/(m+1)}\|u\|_{p+1}^{(p+1)/(m+1)}\|u_t\|_{m+1}^m \\ &\le c\varepsilon H(t)^{1/(p+1)-1/(m+1)}\|u\|_{p+1}^{(p+1)/(m+1)}\|u_t\|_{m+1}^m.\end{aligned} \tag{5.52}$$

The Young inequality for $m < p$ implies the following estimate for the last two factors in the right-hand side of (5.52):

$$\|u\|_{p+1}^{(p+1)/(m+1)}\|u_t\|_{m+1}^m \le c(\|u\|_{p+1}^{p+1} + \|u_t\|_{m+1}^{m+1}) \le c_1(\|u\|_{p+1}^{p+1} + H'). \tag{5.53}$$

We choose α so that

$$0 < \alpha < \frac{1}{2}, \quad \alpha \le \frac{1}{m+1} - \frac{1}{p+1} \tag{5.54}$$

and assume that $H(0) > 1$; then (5.48) implies the upper estimate

$$H(t)^{-1/(m+1)+1/(p+1)} \le H(t)^{-\alpha} \le H(0)^{-\alpha}.$$

Substituting it into (5.52), we obtain

$$\left|2\varepsilon\int_\Omega uu_t|u_t|^{m-1}dx\right| \le \varepsilon cH(0)^{-\alpha}(\|u\|_{p+1}^{p+1} + H'). \tag{5.55}$$

Using (5.55), we return to (5.51) and rewrite it in the form

$$\begin{aligned}\frac{d}{dt}(H^{1-\alpha} + \varepsilon F') &\ge (1-\alpha-\varepsilon c)H(0)^{-\alpha}H' + 4\varepsilon\int_\Omega u_t^2\,dx \\ &\quad + 4\varepsilon H + \varepsilon\left(2\frac{p-1}{p+1} - cH(0)^{-\alpha}\right)\int_\Omega |u|^{p+1}dx.\end{aligned} \tag{5.56}$$

Now for fixed α (to be specified later), we choose sufficiently small ε and sufficiently large $H(0)$ so that the following inequalities hold:

$$\varepsilon d \leq 1-\alpha, \quad 2\frac{p-1}{p+1} - cH(0)^{-\alpha} > 0. \tag{5.57}$$

Due to (5.47) and estimates (5.57), from inequality (5.56) we obtain the following lower estimate:

$$\frac{d}{dt}(H^{1-\alpha} + \varepsilon F') \geq c_1(\|u_t\|_2^2 + H + \|u\|_{p+1}^{p+1}), \tag{5.58}$$

which implies that the function $H^{1-\alpha} + \varepsilon F'$ increases. Therefore, setting $F'(0) > 0$, we see that $H^{1-\alpha} + \varepsilon F' > 0$.

Finally, we show that estimate (5.58) implies the inequality

$$\frac{d}{dt}(H^{1-\alpha} + \varepsilon F') \geq c(H^{1-\alpha} + \varepsilon F')^{\beta}, \quad \beta > 1, \tag{5.59}$$

where $c > 0$. By the theory of ordinary differential inequalities estimate (5.59) completes the proof of the blow-up in finite time of the functional $H^{1-\alpha} + \varepsilon F'$ and hence of the function u. We separately consider the following two cases.
Case 1: $F'(t) \leq 0$. Obviously,

$$(H^{1-\alpha} + \varepsilon F')^{-(1-\alpha)} \leq H;$$

this inequality, together with (5.48) and (5.58), implies (5.59).
Case 2: $F'(t) > 0$. By the Hölder and Young inequalities, we have the following estimates:

$$\begin{aligned}(\varepsilon F')^{1/(1-\alpha)} &\leq (c\varepsilon\|u\|_{p+1}\|u_t\|_2)^{1/(1-\alpha)} \\ &\leq (c_1\varepsilon)^{1/(1-\alpha)}(\|u\|_{p+1}^{\lambda/(1-\alpha)} + \|u_t\|_2^{\mu/(1-\alpha)}), \quad \frac{1}{\lambda} + \frac{1}{\mu} = 1.\end{aligned} \tag{5.60}$$

We take $\mu = 2(1-\alpha)$; then, for sufficiently small α, the constants μ and λ are close to 2. Since $\lambda/(1-\alpha)$ is also close to 2, we can choose α so small that $\lambda/(1-\alpha) < p+1$. The final choice of the minimal value of α is determined by the simultaneous fulfillment of conditions (5.54). Having chosen α, we further select appropriate ε and $H(0)$ from conditions (5.57). Thus (5.60) implies that

$$\varepsilon F' \leq \varepsilon c(\|u\|_{p+1}^{p+1} + \|u_t\|_2^2)^{1-\alpha}.$$

Combining this estimate with the obvious inequality

$$H^{1-\alpha} \leq c(\|u\|_{p+1}^{p+1} + \|u_t\|_2^2)^{1-\alpha},$$

which follows from the definition of H, we arrive at the inequality

$$(H^{1-\alpha} + \varepsilon F')^{1/(1-\alpha)} \le c(\|u\|_{p+1}^{p+1} + \|u_t\|_2^2). \tag{5.61}$$

Substituting estimate (5.61) into (5.58), we obtain (5.59), which completes the proof of the theorem. □

Note that, as is usual in various modifications of energy methods, a particular polynomial form of nonlinearities is not necessary in the proof of the blow-up result, which can also be obtained for a wider class of nonlinearities under appropriate assumptions (see [29]).

5.3 Integro-differential problem with nonlinear damping

In this section, we present the result of S. A. Messaoudi on sufficient conditions of the blow-up of solutions of an initial-boundary-value problem, which differs from Todorova's problem by an additional viscoelastic integral term:

$$u_{tt} - \Delta u + \int_0^t g(t-s)\Delta u(x,s)\,ds + u_t|u_t|^{m-2} = |u|^{p-2}u, \quad x \in \Omega, \tag{5.62}$$

$$u(x,t)|_{\partial\Omega} = 0, \quad u(x,0) = u_0(x), \quad u_t(x,0) = u_1(x), \tag{5.63}$$

where $p > 2$, $m \ge 1$, g is a bounded, positive, and continuously differentiable function $g \in \mathbb{C}^{(1)}(\mathbb{R}_+^1)$, and $\Omega \subset \mathbb{R}^N$ is a bounded domain with smooth boundary $\partial\Omega \in \mathbb{C}^\infty$, $N \ge 1$.

From the physical standpoint, the integral term characterizes the "memory" and elasticity of the medium; it often occurs in similar problems (see, e. g., Chap. 8). For equations (5.62), the influence of the integral term in the case of a linear localized damping was examined in [10]. We present the result of S. A. Messaoudi (see [73]) obtained by Todorova's method; in addition, this result is also valid in the case of positive initial energies.

First, we formulate the theorem on the local solvability.

Theorem 5.4. *Assume that the initial data belong to the class $(u_0, u_1) \in \mathbb{H}_0^1(\Omega) \times \mathbb{L}^2(\Omega)$, and, moreover, that the numbers $m > 1$ and $p > 2$ and the function $g \in \mathbb{C}^{(1)}(\mathbb{R}_+^1)$ satisfy the following conditions:*

$$\max\{m,p\} \le \frac{2(N-1)}{N-2} \quad \text{for } N \ge 3, \quad 1 - \int_0^\infty g(s)\,ds = k > 0. \tag{5.64}$$

Then, for certain $T_m > 0$, problem (5.62)–(5.63) has a unique local solution of the class

$$u \in \mathbb{C}([0,T_m); \mathbb{H}_0^1(\Omega)), \tag{5.65}$$

$$u_t \in \mathbb{C}([0,T_m); \mathbb{L}^2(\Omega)) \cap \mathbb{L}^m(\Omega \times (0,T_m)). \tag{5.66}$$

We omit the proof of this theorem but note that it is similar to that of Theorem 5.1 but is based on the integral character of the additional term in equation (5.62).

For convenience, we introduce the following notation:

$$E(t) = \frac{1}{2}\|u_t\|_2^2 + \frac{1}{2}\Big(1 - \int_0^t g(s)\,ds\Big)\|\nabla u\|_2^2 + \frac{1}{2}(g \circ \nabla u)(t) - \frac{1}{p}\|u\|_p^p, \tag{5.67}$$

$$(g \circ v) = \int_0^t g(t-\tau)\|v(t) - v(\tau)\|_2^2\,d\tau,$$

$$\alpha = \Big(\frac{c_1}{\sqrt{k}}\Big)^{-p/(p-2)}, \quad E_1 = \Big(\frac{1}{2} - \frac{1}{p}\Big)\alpha^2,$$

where c_1 is the best constant of embedding of $\mathbb{H}_0^1(\Omega)$ in $\mathbb{L}^p(\Omega)$. We apply Todorova's method and prove the following blow-up result.

Theorem 5.5 (on the blow-up). *Assume that $m > 1$, $p > \max\{2, m\}$, and $g(s) \in \mathbb{C}^{(1)}(\mathbb{R}_+^1)$ satisfy conditions (5.64) and the inequalities*

$$g(s) \ge 0, \quad g'(s) \le 0, \tag{5.68}$$

$$\int_0^\infty g(s)\,ds < \frac{p/2 - 1}{p/2 - 1 + 1/(2p)}. \tag{5.69}$$

Then the solution of problem (5.62)–(5.63) corresponding to the initial data such that

$$E(0) < E_1, \quad \|\nabla u_0\|_2 > \alpha \tag{5.70}$$

blows up in finite time.

Proof. First, we prove several auxiliary assertions.

Lemma 5.1. *Let $u(x,t)$ be a solution of problem (5.62)–(5.63) satisfying all the conditions of the theorem on the blow-up. Then the energy $E(t)$ does not increase that is, $E'(t) \le 0$.*

Proof. Multiplying equation (5.62) by u_t and integrating over the domain Ω, we conclude that any regular solution satisfies the equality

$$\frac{d}{dt}\Big(\frac{1}{2}\int_\Omega |\nabla u|^2\,dx + \frac{1}{2}\int_\Omega |u_t|^2\,dx - \frac{1}{p}\int_\Omega |u|^p\,dx\Big) - \int_0^t g(t-\tau)\int_\Omega (\nabla u_t(t), \nabla u(\tau))\,dx\,d\tau = -\int_\Omega |u_t|^m\,dx. \tag{5.71}$$

This result remains valid for weak solutions. We carefully examine the last term in the left-hand side of (5.71):

$$\begin{aligned}
&\int_0^t g(t-\tau)\int_\Omega (\nabla u_t(t), \nabla u(\tau))\,dx\,d\tau \\
&\quad = \int_0^t g(t-\tau)\int_\Omega (\nabla u_t(t), \nabla u(\tau)-\nabla u(t))\,dx\,d\tau \\
&\qquad + \int_0^t g(t-\tau)\int_\Omega (\nabla u_t(t), \nabla u(t))\,dx\,d\tau \\
&\quad = -\frac{1}{2}\int_0^t g(t-\tau)\frac{d}{dt}\int_\Omega |\nabla u(\tau)-\nabla u(t)|^2\,dx\,d\tau \\
&\qquad + \frac{1}{2}\int_0^t g(\tau)\frac{d}{dt}\int_\Omega |\nabla u(t)|^2\,dx\,d\tau \\
&\quad = -\frac{1}{2}\frac{d}{dt}\int_0^t g(t-\tau)\int_\Omega |\nabla u(\tau)-\nabla u(t)|^2\,dx\,d\tau \\
&\qquad + \frac{1}{2}\frac{d}{dt}\int_0^t g(\tau)\int_\Omega |\nabla u(t)|^2\,dx\,d\tau \\
&\qquad + \frac{1}{2}\int_0^t g'(t-\tau)\int_\Omega |\nabla u(\tau)-\nabla u(t)|^2\,dx\,d\tau - \frac{1}{2}g(t)\int_\Omega |\nabla u(t)|^2\,dx. \qquad (5.72)
\end{aligned}$$

Substituting (5.72) into (5.71) and collecting the time derivative in the left-hand side, we arrive at the inequality

$$\begin{aligned}
&\frac{d}{dt}\left(\frac{1}{2}\int_\Omega |\nabla u|^2\,dx + \frac{1}{2}\int_\Omega |u_t|^2\,dx - \frac{1}{p}\int_\Omega |u|^p\,dx\right) \\
&\qquad + \frac{1}{2}\frac{d}{dt}\int_0^t g(t-\tau)\int_\Omega |\nabla u(\tau)-\nabla u(t)|^2\,dx\,d\tau - \frac{1}{2}\frac{d}{dt}\int_0^t g(\tau)\|\nabla u(t)\|^2\,d\tau \\
&\quad = -\int_\Omega |u_t|^m\,dx + \frac{1}{2}\int_0^t g'(t-\tau)\int_\Omega |\nabla u(\tau)-\nabla u(t)|^2\,dx\,d\tau \\
&\qquad - \frac{1}{2}g(t)\|u\|_2^2 \le 0. \qquad (5.73)
\end{aligned}$$

Due to the conditions for $g(s)$, the right-hand side is not positive, whereas the left-hand side contains the derivative $E'(t)$. The lemma is proved. □

The following result is based on the conditions for the initial data stated in Theorem 5.5.

Lemma 5.2. *Let $u(x,t)$ be a solution of problem (5.62)–(5.63) satisfying all the conditions of Theorem 5.5. Then there exists a constant $\beta > \alpha$ such that, for all $t \in [0,T)$, we have the estimates*

$$\left|\left(1-\int_0^t g(s)\,ds\right)\|\nabla u\|_2^2 + (g\circ\nabla u)(t)\right|^{1/2} \geq \beta, \quad \|u\|_p \geq \frac{c_1\beta}{\sqrt{k}}. \tag{5.74}$$

Proof. From definition (5.67) and the Sobolev embedding theorem we obtain the following inequalities:

$$\begin{aligned}
E(t) &\geq \frac{1}{2}\left(1-\int_0^t g(s)\,ds\right)\|\nabla u\|_2^2 + \frac{1}{2}(g\circ\nabla u)(t) - \frac{1}{p}\|u\|_p^p \\
&\geq \frac{1}{2}\left(1-\int_0^t g(s)\,ds\right)\|\nabla u\|_2^2 + \frac{1}{2}(g\circ\nabla u)(t) \\
&\quad - \frac{1}{p}c_1^p\|\nabla u\|_2^p \geq \frac{1}{2}\left(1-\int_0^t g(s)\,ds\right)\|\nabla u\|_2^2 \\
&\quad + \frac{1}{2}(g\circ\nabla u)(t) - \frac{c_1^p}{pk^{p/2}}(k\|\nabla u\|_2^2 + (g\circ\nabla u)(t))^{p/2} \\
&= \frac{1}{2}\xi^2 - \frac{C_1^p}{pk^{p/2}}\left(\left(1-\int_0^t g(s)\,ds\right)\|\nabla u\|_2^2 + (g\circ\nabla u)(t)\right)^{p/2} \\
&= \frac{1}{2}\xi^2 - \frac{C_1^p}{pk^{p/2}}\xi^p = h(\xi),
\end{aligned} \tag{5.75}$$

where we have applied the substitution

$$\xi = \left[\left(1-\int_0^t g(s)\,ds\right)\|\nabla u\|_2^2 + (g\circ\nabla u)(t)\right]^{1/2}.$$

It is easy to verify that the function $h(\xi)$ increases on the interval $0 < \xi < \alpha$ and attains the maximal value at $\xi = \alpha$:

$$h(\alpha) = \left(\frac{1}{2} - \frac{1}{p}\right)\alpha^2 = E_1. \tag{5.76}$$

Then it infinitely decreases for $\xi > \alpha$, that is, $h(\xi) \to -\infty$ as $\xi \to +\infty$. Since, by the assumption $E(0) < E_1$ there exists a constant $\beta > \alpha$ such that $h(\beta) = E(0)$, at $t = 0$, we have

$$\xi = \|\nabla u_0\|_2 \equiv \alpha_0.$$

Since by the condition $\|\nabla u_0\|_2 > \alpha$ we obtain from (5.75) the inequality

$$h(\alpha_0) \le E(0) = h(\beta)$$

at $t = 0$. This inequality implies that $\alpha_0 > \beta$. In other words, at the initial time moment, the point $\xi(0)$ lies to the right of the point $\xi = \beta$. We show that there is no time moment at which $\xi(t)$ becomes equal to β, that is, we prove the first inequality in (5.74). Assume the contrary, that is, let for some $t_0 > 0$ the following inequality hold:

$$\left|\left(1 - \int_0^{t_0} g(s)\,ds\right)\|\nabla u\|_2^2 + (g \circ \nabla u)(t_0)\right|^{1/2} < \beta. \tag{5.77}$$

Using the continuity of the radicand and the fact that $\beta > \alpha$, we conclude that there exists a close time moment (we denote it by the same symbol t_0) for which we have the inequality

$$\left|\left(1 - \int_0^{t_0} g(s)\,ds\right)\|\nabla u\|_2^2 + (g \circ \nabla u)(t_0)\right|^{1/2} > \alpha. \tag{5.78}$$

Taking into account the behavior of the function $h(\xi)$ (see (5.75)), we obtain the inequality

$$E(t_0) > h(\beta) = E(0). \tag{5.79}$$

We have arrived at a contradiction since Lemma 5.1 implies that $E(t) \le E(0)$ for all $t \in [0, T)$. The first inequality is proved.

Now we prove the second inequality (5.74) of the lemma. Using the definition of $E(t)$ and the property $E(t) \le E(0)$, we write

$$\frac{1}{2}\left(\left(1 - \int_0^{t} g(s)\,ds\right)\|\nabla u\|_2^2 + (g \circ \nabla u)(t)\right) \le E(0) + \frac{1}{p}\|u\|_p^p. \tag{5.80}$$

Applying the first inequality from (5.74), we rewrite (5.80) in the form

$$\begin{aligned}\frac{1}{p}\|u\|_p^p &\ge \frac{1}{2}\left(1 - \int_0^{t} g(s)\,ds\right)\|\nabla u\|_2^2 + \frac{1}{2}(g \circ \nabla u)(t) - E(0)\\ &\ge \frac{1}{2}\beta^2 - E(0) \ge \frac{1}{2}\beta^2 - h(\beta) = \frac{C_1^p}{pk^{p/2}}\beta^p.\end{aligned} \tag{5.81}$$

The lemma is proved. □

Lemma 5.3. *Let condition* (5.64) *hold. Then there exists a positive constant* $c > 1$ *such that, for all* $u \in \mathbb{H}_0^1(\Omega)$ *and* $l\, 2 \le s \le p < 2^*$*, we have the estimate*

$$\|u\|_p^s \le c\,(\|\nabla u\|_2^2 + \|u\|_p^p). \tag{5.82}$$

Proof. Indeed, if $\|u\|_p \le 1$, then, due to the continuous embedding $\mathbb{H}_0^1(\Omega) \subset \mathbb{L}^p(\Omega)$, we have

$$\|u\|_p^s \le \|u\|_p^2 \le c\|\nabla u\|_2^2$$

since $p < 2^*$. If $\|u\|_p > 1$, then $\|u\|_p^s \le \|u\|_p^p$. The lemma is proved. □

Lemma 5.4. *Let condition* (5.64) *hold. Then, for any* $2 \le s \le p$, *the solution* $u(x,t)$ *of problem* (5.62)–(5.63) *satisfies the inequality*

$$\|u\|_p^s \le c(-H(t) - \|u_t\|_2^2 - (g \circ \nabla u)(t) + \|u\|_p^p), \tag{5.83}$$

where $H(t) = E_1 - E(t)$.

Proof. Indeed, applying the definitions of k and $E(t)$ (see (5.64) and (5.67)), we have

$$\begin{aligned} \frac{1}{2}k\|\nabla u\|_2^2 &\le \frac{1}{2}\Big(1 - \int_0^t g(s)\,ds\Big)\|\nabla u\|_2^2 \\ &= E(t) - \frac{1}{2}\|u_t\|_2^2 - \frac{1}{2}(g \circ \nabla u)(t) + \frac{1}{p}\|u\|_p^p \\ &= E_1 - H(t) - \frac{1}{2}\|u_t\|_2^2 - \frac{1}{2}(g \circ \nabla u)(t) + \frac{1}{p}\|u\|_p^p. \end{aligned} \tag{5.84}$$

From the definition of E_1 and the second inequality in (5.74) we obtain the estimate

$$\begin{aligned} \|u\|_p^p &\ge B_1^p \beta^p > B_1^p B_1^{-p^2/(p-2)} \\ &= B_1^{-2p/(p-2)} = \frac{2p}{p-2}E_1, \quad B_1 = \frac{c_1}{\sqrt{k}}. \end{aligned} \tag{5.85}$$

Substituting this estimate into (5.84) and applying the estimate for $\|\nabla u\|_2^2$ obtained before and Lemma 5.3, we arrive at (5.83). The lemma is proved. □

Now we return to the proof of Theorem 5.5. From the definitions of $E(t)$ and $H(t)$ and the inequality $E'(t) \le 0$ we obtain the estimate

$$\begin{aligned} 0 &< H(0) \le H(t) \\ &\le E_1 - \frac{1}{2}\Big[\|u_t\|_2^2 + \Big(1 - \int_0^t g(s)\,ds\Big)\|\nabla u\|_2^2 + (g \circ \nabla u)(t)\Big] + \frac{1}{p}\|u\|_p^p. \end{aligned} \tag{5.86}$$

Due to estimate (5.74), we conclude that, for all $t \ge 0$, the following inequalities hold:

$$E_1 - \frac{1}{2}\Big[\|u_t\|_2^2 + \Big(1 - \int_0^t g(s)\,ds\Big)\|\nabla u\|_2^2 + (g \circ \nabla u)(t)\Big]$$

$$< E_1 - \frac{\beta^2}{2} = -\frac{\beta^2}{p} < 0. \tag{5.87}$$

Thus, for all $t \geq 0$, we have the following two-sided estimate of the functional $H(t)$:

$$0 < H(0) \leq H(t) \leq \frac{1}{p}\|u\|_p^p. \tag{5.88}$$

As in the previous section, for some small constants σ and ε, we introduce the functional

$$L(t) = H^{1-\sigma}(t) + \varepsilon \int_\Omega u u_t \, dx, \quad 0 < \sigma < \min\left\{\frac{p-2}{2p}, \frac{p-m}{p(m-1)}\right\}. \tag{5.89}$$

Differentiating (5.89) by time and substituting the expression for u_{tt} from (5.62), we obtain the inequality

$$\begin{aligned}
L'(t) &= (1-\sigma)H^{-\sigma}(t)\left\{\|u_t\|_m^m - \frac{1}{2}(g' \circ \nabla u)(t) + \frac{1}{2}g(t)\|\nabla u\|_2^2\right\} \\
&\quad + \varepsilon \int_\Omega [u_t^2 - |\nabla u|^2] \, dx \\
&\quad + \varepsilon \int_0^t g(t-\tau) \int_\Omega (\nabla u(t), \nabla u(\tau)) \, dx \, d\tau \\
&\quad + \varepsilon \int_\Omega |u|^p \, dx - \varepsilon \int_\Omega |u_t|^{m-2} u_t u \, dx \\
&\geq (1-\sigma)H^{-\sigma}(t)\|u_t\|_m^m + \varepsilon \int_\Omega [u_t^2 - |\nabla u|^2] \, dx \\
&\quad + \varepsilon \int_\Omega |u|^p \, dx - \varepsilon \int_\Omega |u_t|^{m-2} u_t u \, dx \\
&\quad + \varepsilon \int_0^t g(t-\tau)\|\nabla u(t)\|_2^2 \, d\tau \\
&\quad + \varepsilon \int_0^t g(t-\tau) \int_\Omega (\nabla u(t), \nabla u(\tau) - \nabla u(t)) \, dx \, d\tau.
\end{aligned} \tag{5.90}$$

Two terms in the curly brackets can be omitted due to their nonnegativeness. Applying the Cauchy–Bunyakovsky inequality, we rewrite (5.90) in the form

$$\begin{aligned}
L'(t) &\geq (1-\sigma)H^{-\sigma}(t)\|u_t\|_m^m \\
&\quad + \varepsilon \int_\Omega [u_t^2 - |\nabla u|^2] \, dx + \varepsilon \int_\Omega |u|^p \, dx - \varepsilon \int_\Omega |u_t|^{m-2} u_t u \, dx
\end{aligned}$$

$$+\varepsilon\int_0^t g(t-\tau)\|\nabla u(t)\|_2^2\,d\tau$$
$$-\varepsilon\int_0^t g(t-\tau)\|\nabla u(t)\|_2\|\nabla u(\tau)-\nabla u(t)\|_2\,dx\,d\tau. \tag{5.91}$$

We estimate the last term in the right-hand side by using the Young inequality and the third term by using the definition of $H(t)$ through $E(t)$. Finally, we obtain

$$\begin{aligned}
L'(t) &\geq (1-\sigma)H^{-\sigma}(t)\|u_t\|_m^m + \varepsilon\int_\Omega u_t^2\,dx - \varepsilon\Big(1-\int_0^t g(s)\,ds\Big)\|\nabla u(t)\|_2^2 \\
&\quad + \varepsilon\Big(pH(t) + \frac{p}{2}(g\circ\nabla u)(t) + \frac{p}{2}\|u_t\|_2^2 + \frac{p}{2}\Big(1-\int_0^t g(s)\,ds\Big)\|\nabla u(t)\|_2^2\Big) \\
&\quad - \varepsilon\int_\Omega |u_t|^{m-2}u_t u\,dx - \varepsilon\tau(g\circ\nabla u)(t) - \frac{\varepsilon}{4\tau}\int_0^t g(s)\,ds\|\nabla u(t)\|_2^2 \\
&\geq (1-\sigma)H^{-\sigma}(t)\|u_t\|_m^m + \varepsilon\Big(1+\frac{p}{2}\Big)\int_\Omega u_t^2\,dx + \varepsilon pH(t) \\
&\quad + \varepsilon\Big(\frac{p}{2}-\tau\Big)(g\circ\nabla u)(t) - \varepsilon\int_\Omega |u_t|^{m-2}u_t u\,dx \\
&\quad + \varepsilon\Big(\Big(\frac{p}{2}-1\Big) - \Big(\frac{p}{2}-1+\frac{1}{4\tau}\Big)\int_0^\infty g(s)\,ds\Big)\|\nabla u(t)\|_2^2
\end{aligned} \tag{5.92}$$

for some $\tau \in (0, p/2)$. Using condition (5.68), we can rewrite (5.92) as follows:

$$\begin{aligned}
L'(t) &\geq (1-\sigma)H^{-\sigma}(t)\|u_t\|_m^m + \varepsilon\Big(1+\frac{p}{2}\Big)\int_\Omega u_t^2\,dx \\
&\quad + \varepsilon pH(t) + \varepsilon a_1(g\circ\nabla u)(t) + \varepsilon a_2\|\nabla u(t)\|_2^2 - \varepsilon\int_\Omega |u_t|^{m-2}u_t u\,dx,
\end{aligned} \tag{5.93}$$

where

$$a_1 = \frac{p}{2}-\tau > 0, \quad a_2 = \Big(\frac{p}{2}-1\Big) - \Big(\frac{p}{2}-1+\frac{1}{4\tau}\Big)\int_0^\infty g(s)\,ds > 0.$$

To estimate the last term in (5.93), we apply the three-parameter Young inequality:

$$ab \leq \frac{\delta^r}{r}a^r + \frac{\delta^{-q}}{q}b^q, \quad a,b \geq 0, \quad \frac{1}{r}+\frac{1}{q} = 1, \quad \forall\delta > 0. \tag{5.94}$$

We take $r = m$ and $q = m/(m-1)$ and apply the estimate

$$\int_\Omega |u_t|^{m-2}u_t u\,dx \le \frac{\delta^m}{m}\|u\|_m^m + \frac{m-1}{m}\delta^{-m/(m-1)}\|u_t\|_m^m \tag{5.95}$$

to inequality (5.93). We arrive at the inequalities

$$\begin{aligned} L'(t) &\ge \left((1-\sigma)H^{-\sigma}(t) - \frac{m-1}{m}\varepsilon\delta^{-m/(m-1)}\right)\|u_t\|_m^m \\ &\quad + \varepsilon\left(1+\frac{p}{2}\right)\int_\Omega u_t^2\,dx + \varepsilon a_1(g\circ\nabla u)(t) \\ &\quad + \varepsilon p H(t) + \varepsilon a_2\|\nabla u(t)\|_2^2 - \varepsilon\frac{\delta^m}{m}\|u\|_m^m. \end{aligned} \tag{5.96}$$

Since integration in estimate (5.95) is performed over the space, the parameter δ can be a function of time; we choose it as follows:

$$\delta^{-m/(m-1)} = k_1 H^{-\sigma}(t),$$

where $k_1 > 0$ is a sufficiently large constant to be specified further. Then expression (5.96) can be rewritten in the form

$$\begin{aligned} L'(t) &\ge \left((1-\sigma) - \frac{m-1}{m}\varepsilon k_1\right)H^{-\sigma}(t)\|u_t\|_m^m + \varepsilon\left(1+\frac{p}{2}\right)\int_\Omega u_t^2\,dx \\ &\quad + \varepsilon a_1(g\circ\nabla u)(t) + \varepsilon a_2\|\nabla u(t)\|_2^2 \\ &\quad + \varepsilon\left[pH(t) - \frac{k_1^{1-m}}{m}H^{\sigma(m-1)}(t)\|u\|_m^m\right]. \end{aligned} \tag{5.97}$$

From (5.88) and the embedding inequality

$$\|u\|_m^m \le c\|u\|_p^m$$

we obtain the following estimate for the second term in the square brackets:

$$H^{\sigma(m-1)}(t)\|u\|_m^m \le c\left(\frac{1}{p}\right)^{\sigma(m-1)}\|u\|_p^{m+\sigma p(m-1)}. \tag{5.98}$$

Substituting this estimate into (5.97), we obtain

$$\begin{aligned} L'(t) &\ge \left((1-\sigma) - \frac{m-1}{m}\varepsilon k_1\right)H^{-\sigma}(t)\|u_t\|_m^m + \varepsilon\left(1+\frac{p}{2}\right)\int_\Omega u_t^2\,dx \\ &\quad + \varepsilon a_1(g\circ\nabla u)(t) + \varepsilon a_2\|\nabla u(t)\|_2^2 \\ &\quad + \varepsilon\left(pH(t) - c_3\frac{k_1^{1-m}}{m}\left(\frac{1}{p}\right)^{\sigma(m-1)}\|u\|_p^{m+\sigma p(m-1)}\right). \end{aligned} \tag{5.99}$$

Due to estimate (5.83), we conclude that inequality (5.99) for the value $s = m+\sigma p(m-1)$ can be rewritten in the following form:

$$\begin{aligned}
L'(t) &\geq \left((1-\sigma) - \frac{m-1}{m}\varepsilon k_1\right) H^{-\sigma}(t)\|u_t\|_m^m \\
&\quad + \varepsilon\left(1+\frac{p}{2}\right)\int_\Omega u_t^2\,dx + \varepsilon a_1(g\circ\nabla u)(t) + \varepsilon a_2\|\nabla u(t)\|_2^2 \\
&\quad + \varepsilon(pH(t) - c_5 k_1^{1-m}[-H(t) - \|u_t\|_2^2 - (g\circ\nabla u)(t) + \|u\|_p^p]) \\
&\geq \left((1-\sigma) - \frac{m-1}{m}\varepsilon k\right) H^{-\sigma}(t)\|u_t\|_m^m + \varepsilon\left(1+\frac{p}{2} + c_4 k_1^{1-m}\right)\|u_t\|_2^2 \\
&\quad + \varepsilon(a_1 + c_5 k_1^{1-m})(g\circ\nabla u)(t) + \varepsilon a_2\|\nabla u(t)\|_2^2 \\
&\quad + \varepsilon(p + c_5 k_1^{1-m})H(t) - \varepsilon c_5 k^{1-m}\|u\|_p^p,
\end{aligned} \tag{5.100}$$

where we have introduced the notation $c_5 = (1/p)^{\sigma(m-1)}c/m$. Now substituting the estimate for $H(t)$ into (5.100), we obtain the inequality

$$H(t) \geq \frac{1}{p}\|u\|_p^p - \frac{1}{2}\|u_t\|_2^2 - \frac{1}{2}\|\nabla u\|_2^2 - \frac{1}{2}(g\circ\nabla u)(t),$$

where

$$p = 2a_3 + (p - 2a_3) \quad \text{for } a_3 < \min\{a_1, a_2, p/2\}.$$

Omitting nonnegative terms, we obtain

$$\begin{aligned}
L'(t) &\geq \left((1-\sigma) - \frac{m-1}{m}\varepsilon k_1\right) H^{-\sigma}(t)\|u_t\|_m^m \\
&\quad + \varepsilon\left(1+\frac{p}{2} + c_5 k^{1-m} - a_3\right)\|u_t\|_2^2 \\
&\quad + \varepsilon(a_1 + c_5 k^{1-m} - a_3)(g\circ\nabla u)(t) + \varepsilon(a_2 - a_3)\|\nabla u(t)\|_2^2 \\
&\quad + \varepsilon(p - 2a_3 + c_5 k_1^{1-m})H(t) + \varepsilon\left(\frac{2a_3}{p} - c_5 k_1^{1-m}\right)\|u\|_p^p.
\end{aligned} \tag{5.101}$$

Introduce the constant

$$\gamma = \varepsilon\min\left\{(p - 2a_3 + c_5 k_1^{1-m}), \left(1 + \frac{p}{2} + c_5 k_1^{1-m} - a_3\right), \left(\frac{2a_3}{p} - c_5 k_1^{1-m}\right)\right\}.$$

Taking sufficiently large $k_1 > 0$ for the positive constant γ, we simplify (5.101):

$$\begin{aligned}
L'(t) &\geq \left((1-\sigma) - \frac{m-1}{m}\varepsilon k_1\right) H^{-\sigma}(t)\|u_t\|_m^m \\
&\quad + \varepsilon\gamma(H + \|u_t\|_2^2 + \|u\|_p^p + (g\circ\nabla u)).
\end{aligned} \tag{5.102}$$

For fixed $k_1 > 0$ and $\gamma > 0$, we choose $\varepsilon > 0$ so small that the following inequality holds:

$$(1-\sigma) - \varepsilon k_1 \frac{m-1}{m} \geq 0. \tag{5.103}$$

Moreover, we assume that the initial data satisfy the estimate

$$L(0) = H^{1-\sigma}(0) + \varepsilon \int_\Omega u_0 u_1(x)\,dx > 0.$$

Then from (5.102) we obtain the inequality

$$L'(t) \geq \varepsilon\gamma(H(t) + \|u_t\|_2^2 + \|u\|_p^p + (g \circ \nabla u)(t)), \tag{5.104}$$

which yields the following lower estimate for $L(t)$:

$$L(t) \geq L(0) > 0 \quad \text{for all } t \geq 0.$$

Finally, we use the estimate

$$\left|\int_\Omega u u_t\,dx\right| \leq \|u\|_2 \|u_t\|_2 \leq c\|u\|_p \|u_t\|_2. \tag{5.105}$$

Raising it to the power $1/(1-\sigma)$ and applying the Young inequality, we obtain

$$\left|\int_\Omega u u_t\,dx\right|^{1/(1-\sigma)} \leq c(\|u\|_p^{\mu/(1-\sigma)} + \|u_t\|_2^{\nu/(1-\sigma)}), \quad \frac{1}{\mu} + \frac{1}{\nu} = 1. \tag{5.106}$$

In (5.106), we set $\nu = 2(1-\sigma)$. Then

$$\frac{\mu}{1-\sigma} = s = \frac{2}{1-2\sigma} \leq p,$$

and we have

$$\left|\int_\Omega u u_t\,dx\right|^{\frac{1}{1-\sigma}} \leq c(\|u\|_p^s + \|u_t\|_2^2). \tag{5.107}$$

Now, applying Lemma 5.4 and (5.83), we obtain

$$\left|\int_\Omega u u_t\,dx\right|^{\frac{1}{1-\sigma}} \leq c(H(t) + \|u\|_p^p + \|u_t\|_2^2 + (g \circ \nabla u)(t)). \tag{5.108}$$

It remains to verify that the following inequalities hold:

$$L^{1/(1-\sigma)}(t) = \left[H^{(1-\sigma)} + \varepsilon \int_\Omega u u_t\,dx\right]^{1/(1-\sigma)}$$

$$\le 2^{1/(1-\sigma)}\left[H(t) + \left|\int_\Omega uu_t\,dx\right|^{1/(1-\sigma)}\right]$$

$$\le C[H(t) + \|u\|_p^p + \|u_t\|_2^2 + (g \circ \nabla u)(t)]. \tag{5.109}$$

Combining this with (5.104), we obtain the following ordinary differential inequality:

$$L'(\tau) \ge gL^{1/(1-\sigma)}(\tau) \quad \forall t \ge 0. \tag{5.110}$$

Obviously, the constant $g > 0$ can depend only on c, ε, and γ. Then, integrating (5.110) by τ on the interval $(0, t)$, we arrive at the following lower estimate:

$$L^{\sigma/(1-\sigma)}(t) \ge \frac{1}{L^{-\sigma/(1-\sigma)}(0) - tg\sigma/(1-\sigma)}, \tag{5.111}$$

which implies the blow-up of the solution in finite time T such that

$$T < T_0 = \frac{1-\sigma}{g\sigma(L(0))^{\sigma/(1-\sigma)}}. \tag{5.112}$$

Thus, inequality (5.111) completes the proof of the theorem. □

Note that the result obtained is valid in the case of damping depending only on the viscoelastic term $m = 1$. Moreover, the third inequality of condition (5.68) shows a strong relationship between the nonlinear source and the viscosity integral, that is, the greater p, the closer to 1 the coefficient

$$\int_0^\infty g(s)\,ds.$$

We also note the quite expected result that the greater $L(0)$, the faster the blow-up occurs (see (5.112)).

5.4 Bibliographical notes

In this chapter, we follow the papers [29] and [73]. An extension of results obtained by G. Todorova to the case of Cauchy problems with nonlinear sources, damping, and negative initial energy can be found in [101]. The proof of blow-up of solutions of the initial-boundary-value problem with positive energy discussed in Section 5.3, and for problems without integral terms, it can be found in [60].

The papers [29] of Georgiev and Todorova and [73] of Messaoudi aroused great interest in the study of equations with nonlinear damping. Moreover, there exist generalizations of Todorova's method to equations with nonlinear damping of the general form $f(u_t)$ (see, e. g., http://www.ams.org/mathscinet/).

6 Energy method of S. I. Pokhozhaev

The approach proposed by S. I. Pokhozhaev is based on the energy method, which was modified for problems in unbounded domains that cannot admit application of the "classical" energy method due to the absence in $\Omega = \mathbb{R}^N$ of a "coercive" estimate of the form

$$\int_{\mathbb{R}^N} u^2\,dx \le c_p \left| \int_{\mathbb{R}^N} |u|^p\,dx \right|^{2/p}, \quad p > 2,\ c_p < \infty. \tag{6.1}$$

Pokhozhaev developed an idea of H. Levine on the using of the concavity argument for the corresponding functionals with respect to time; this allowed him to consider a wide class of problems for quasilinear parabolic and hyperbolic equations in unbounded domains. Moreover, to analyze the fundamental nonlinear differential equation, Pokhozhaev used the method of nonlinear capacity described in Chapter 1 and based on an appropriate choice of optimal test functions. This allows us to write joint (not isolated) conditions for the nonlinearity and the initial data.

6.1 Blow-up for parabolic equations

6.1.1 Abstract results

Consider an abstract nonlinear problem for an evolution equation in a Banach space $\mathbb{X}$:

$$\frac{du}{dt} = g(t,u), \quad t \ge 0, \tag{6.2}$$

$$u|_{t=0} = u_0 \in \mathbb{H} \cap \mathbb{X}, \tag{6.3}$$

where the operator $g : \mathbb{R}_+ \times \mathbb{X} \to \mathbb{X}^*$ in the right-hand side is potential with respect to $u(x,t)$, that is, there exists a functional $G \in \mathbb{C}^1 : \mathbb{R}_+ \times \mathbb{X} \to \mathbb{R}$ such that

$$G'_u(t,u) = \frac{\partial G}{\partial u} = g(t,u), \quad G(t,0) \equiv 0,$$

where the derivative is meant in the Fréchet space, and $\mathbb{X}^*$ is the dual space to $\mathbb{X}$. Let $\mathbb{H}$ be a Hilbert space with scalar product $(\cdot,\cdot)$ and the corresponding norm $\|\cdot\|$.

A solution of problem (6.2)–(6.3) is a function

$$u : \mathbb{R}_+ \to \mathbb{X}$$

satisfying equation (6.2) in the classical sense and the following conditions for all $t \in \mathbb{R}_+$:

(i)$_1$ there exists $du/dt \in \mathbb{X}^*$;

(ii)$_1$ $u(t), u_t(t) \in \mathbb{H}$, where $\mathbb{H}$ is a Hilbert space such that $\|u(t)\|, \|u'(t)\| \in \mathbb{L}^2_{\mathrm{loc}}(\mathbb{R}_+)$.

https://doi.org/10.1515/9783110602074-006

Moreover, we assume that the dual relations are compatible on solutions:

$$(u_t, u)_{\mathbb{H}} = (G'(t, u), u)_{\mathbb{X}}, \quad (u_t, u_t)_{\mathbb{H}} = (G'(t, u), u_t)_{\mathbb{X}}. \tag{6.4}$$

We divide our study into several steps.

Step I. Levine-type inequality

Introduce the following notation:

$$J(t) = \int_0^t \|u(\eta)\|^2 \, d\eta + c_0^2, \quad c_0^2 > 0. \tag{6.5}$$

Differentiate this function by time:

$$J'(t) = \|u(t)\|^2 = 2 \int_0^t (u, u_\eta) \, d\eta + \|u_0\|^2.$$

By the Hölder inequality for positive $\varepsilon > 0$,

$$ab \leq \frac{\varepsilon}{2} a^2 + \frac{1}{2\varepsilon} b^2,$$

we can estimate the square of $J'(t)$ as follows:

$$\begin{aligned} J'(t)^2 &= 4\left|\int_0^t (u, u_\eta) \, d\eta\right|^2 + 4\|u_0\|^2 \int_0^t (u, u_\eta) \, d\eta \\ &+ \|u_0\|^4 \leq 4 \int_0^t \|u(\eta)\|^2 \, d\eta \int_0^t \|u'(\eta)\|^2 \, d\eta \\ &+ 2\varepsilon\|u_0\|^2 \int_0^t \|u(\eta)\|^2 \, d\eta + 2\frac{\|u_0\|^2}{\varepsilon} \int_0^t \|u'(\eta)\|^2 \, d\eta + \|u_0\|^4. \end{aligned} \tag{6.6}$$

Substituting the derivative from equation (6.2), we write $J''(t)$ in the form

$$J''(t) = 2(u_t, u) = 2(G'(t, u), u). \tag{6.7}$$

Combining (6.6) and (6.7), for all positive $\alpha > 0$, we obtain the inequality

$$\begin{aligned} &J(t)J''(t) - (1 + \alpha)(J'(t))^2 \\ &\geq 2J(G'(t, u), u) - 4(1 + \alpha) \int_0^t \|u(\eta)\|^2 \, d\eta \int_0^t \|u'(\eta)\|^2 \, d\eta \end{aligned}$$

$$
\begin{aligned}
&- 2\varepsilon(1+\alpha)\|u_0\|^2 \int_0^t \|u(\eta)\|^2\,d\eta - 2(1+\alpha)\frac{\|u_0\|^2}{\varepsilon}\int_0^t \|u'(\eta)\|^2\,d\eta \\
&- (1+\alpha)\|u_0\|^4. \qquad (6.8)
\end{aligned}
$$

Using notation (6.5), we transform (6.8) into the form

$$
\begin{aligned}
&J(t)J''(t) - (1+\alpha)(J'(t))^2 \\
&\geq 2J(G'(t,u),u) - 4(1+\alpha)J(t)\int_0^t \|u'(\eta)\|^2\,d\eta \\
&\quad + 4(1+\alpha)c_0^2 \int_0^t \|u'(\eta)\|^2\,d\eta - 2\varepsilon(1+\alpha)\|u_0\|^2 J(t) + 2(1+\alpha)\varepsilon c_0^2\|u_0\|^2 \\
&\quad - 2\frac{1+\alpha}{\varepsilon}\|u_0\|^2 \int_0^t \|u'(\eta)\|^2\,d\eta - (1+\alpha)\|u_0\|^4 \\
&= 2J\left[(G'(t,u),u) - 2(1+\alpha)\int_0^t \|u'(\eta)\|^2\,d\eta - \varepsilon(1+\alpha)\|u_0\|^2\right] \\
&\quad + 2(1+\alpha)\left(2c_0^2 - \frac{\|u_0\|^2}{\varepsilon}\right)\int_0^t \|u'(\eta)\|^2\,d\eta \\
&\quad + (1+\alpha)\|u_0\|^2(2\varepsilon c_0^2 - \|u_0\|^2). \qquad (6.9)
\end{aligned}
$$

Using equation (6.2), we obtain the equality

$$
\begin{aligned}
\int_0^t \|u'(\eta)\|^2\,d\eta &= \int_0^t (u'(\eta),u'(\eta))\,d\eta = \int_0^t (G'(\eta,u),u'(\eta))\,d\eta \\
&= \int_0^t \left(\frac{dG}{d\eta} - \frac{\partial G}{\partial \eta}\right)d\eta = G(t,u(t)) - \int_0^t G_\eta(\eta,u(\eta))\,d\eta - G(0,u_0). \qquad (6.10)
\end{aligned}
$$

Substituting this into (6.9), we arrive at the inequality

$$
\begin{aligned}
J(t)J''(t) - (1+\alpha)(J'(t))^2 &\geq 2M(u(t))J(t) \\
&\quad + 2(1+\alpha)\left(2c_0^2 - \frac{\|u_0\|^2}{\varepsilon}\right)\int_0^t \|u'(\eta)\|^2\,d\eta \\
&\quad + (1+\alpha)\|u_0\|^2(2\varepsilon c_0^2 - \|u_0\|^2), \qquad (6.11)
\end{aligned}
$$

where

$$M(u(t)) = (G'(t,u),u) - 2(1+\alpha)G(t,u) + 2(1+\alpha)G(0,u_0)$$

$$+ 2(1+\alpha)\int_0^t G_\eta(\eta, u)\, d\eta - \varepsilon(1+\alpha)\|u_0\|^2. \tag{6.12}$$

Choosing a positive constant

$$\varepsilon = \varepsilon_0 = \frac{\|u_0\|^2}{2c_0^2},$$

from (6.11) we obtain the following Levine-type inequality:

$$J(t)J''(t) - (1+\alpha)(J'(t))^2 \geq 2M_0(u(t))J(t), \tag{6.13}$$

where $M_0(u(t))$ means the value of $M(u(t))$ corresponding to the chosen value of ε.

Step II. Analysis of the Levine-type inequality

1. First, we consider the case of the weak inequality $M_0 \geq 0$. In this case, we have the following ordinary differential inequality:

$$J(t)J''(t) - (1+\alpha)(J'(t))^2 \geq 0. \tag{6.14}$$

Since, by definition, $J(t) \geq c_0^2 > 0$, for $\alpha > 0$, we can perform the substitution $Y(t) = J^{-\alpha}(t)$. Then from expression (6.14) we obtain the upper estimate

$$Y''(t) = -\alpha J^{-(\alpha+2)}(t)\big(J(t)J''(t) - (1+\alpha)(J'(t))^2\big) \leq 0. \tag{6.15}$$

Twice integrating by time from 0 to t, we obtain the inequality

$$Y(t) \leq Y(0) + Y'(0)t \tag{6.16}$$

or, returning to the notation $J(t)$,

$$J^{-\alpha}(t) \leq J^{-\alpha}(0) - \alpha J^{-\alpha-1}(0)J'(0)t = \frac{c_0^2 - \alpha\|u_0\|^2 t}{c_0^{2(\alpha+1)}}. \tag{6.17}$$

Due to the positiveness of the functional $J(t)$, for the Cauchy problem (6.2)–(6.3) we obtain the following result.

Theorem 6.1. *Let the condition*

$$M_0(u(t)) \geq 0 \tag{6.18}$$

be fulfilled for some $\alpha > 0$ and $c_0^2 > 0$. Then the Cauchy problem (6.2)–(6.3) *has no solutions from the class considered for $t \geq T_0$, where*

$$T_0 = \frac{c_0^2}{\alpha}\frac{1}{\|u_0\|^2}. \tag{6.19}$$

Note that the constants $\alpha > 0$ and $c_0^2 > 0$ can be chosen arbitrarily. For fixed α, we consider the optimal (with respect to c_0^2) estimate of the time of blow-up (6.19). Find the minimal positive value of c_0^2 under the condition

$$K(u) = M_0(u(t)) + (1+\alpha)\frac{\|u_0\|^4}{2c_0^2} = (G'(t,u),u) - 2(1+\alpha)G(t,u) \\ + 2(1+\alpha)\int_0^t G_\eta(\eta,u)\,d\eta + 2(1+\alpha)G(0,u_0) > 0. \tag{6.20}$$

Formula (6.19) immediately implies that the Cauchy problem has no solutions for $t \geq T_0^*$, where

$$T_0^* = \frac{1+\alpha}{2\alpha}\frac{\|u_0\|^2}{K(u)}. \tag{6.21}$$

Thus, we have the following statement.

Corollary 6.1. *If the right-hand side of equation* (6.2) *satisfies the inequalities*

$$(G'(t,u),u) \geq 2(1+\alpha)G(t,u), \quad G_t(t,u) \geq 0$$

and if the initial condition $u_0(x)$ satisfies the inequality

$$G(0,u_0) > 0,$$

then a solution of problem (6.2)–(6.3) *blows up in finite time $t < T_0^*$, where*

$$T_0^* = \frac{1}{4\alpha}\frac{\|u_0\|^2}{G(0,u_0)}. \tag{6.22}$$

Moreover, estimate (6.22) is unimprovable up to an insignificant constant with respect to the initial data u_0 in the class considered. This follows from the following simple example:

$$\frac{du}{dt} = |u|^{2\alpha}u, \quad u|_{t=0} = u_0 \neq 0. \tag{6.23}$$

Formula (6.22) implies that $G(u) = (2+2\alpha)^{-1}|u|^{2(1+\alpha)}$, and we arrive at the following estimate of the blow-up time:

$$T_0^* = \frac{1+\alpha}{2\alpha}\frac{1}{|u_0|^{2\alpha}}. \tag{6.24}$$

The exact blow-up time is

$$T_0^* = \frac{1}{2\alpha}\frac{1}{|u_0|^{2\alpha}}. \tag{6.25}$$

2. Consider the case of the strong inequality $M_0(u) \geq m_0 > 0$. In this case, inequality (6.13) is of the form

$$J(t)J''(t) - (1+\alpha)(J'(t))^2 \geq 2m_0 J(t), \quad m_0 = \text{const} > 0. \tag{6.26}$$

To analyze (6.26), we apply the method of nonlinear capacity (see Chapter 1). This method is based on the choice of an optimal nonnegative test function $\varphi \in \mathbb{C}^{(2)}(\mathbb{R}_+)$,

$$\varphi(t) = \begin{cases} 1 & \text{if } 0 \leq t \leq T, \\ 0 & \text{if } t \geq 2T. \end{cases}$$

Multiplying inequality (6.26) by the function $J^\beta(t)\varphi(t)$, $\beta \in \mathbb{R}$, and integrating by parts, we obtain

$$J^{\beta+1}(0)J'(0) - (\alpha+\beta+2)\int_0^{2T} J^\beta(t)J'(t)^2\varphi(t)\,dt + \frac{1}{\beta+2}\int_0^{2T} J^{\beta+2}(t)\varphi''(t)\,dt \geq 2m_0\int_0^{2T} J^{\beta+1}(t)\varphi(t)\,dt. \tag{6.27}$$

Choosing β such that $\beta+\alpha+2=0$, we rewrite (6.27) in the form

$$\int_0^{2T} J^{\beta+1}(t)\varphi(t)\,dt \leq c_1\int_0^{2T} J^{\beta+2}(t)\varphi''(t)\,dt - \frac{1}{2m_0}J^{\beta+1}(0)J'(0), \tag{6.28}$$

where $c_1 = 1/(2m_0(\beta+2))$. Replacing all β by α, we rewrite (6.28) for positive $\alpha > 0$ as follows:

$$\int_0^{2T} J^{-\alpha-1}(t)\varphi(t)\,dt \leq -\frac{1}{2m_0\alpha}\int_0^{2T} J^{-\alpha}(t)\varphi''(t)\,dt - \frac{1}{2m_0}J^{-\alpha-1}(0)J'(0). \tag{6.29}$$

By the Young inequality for $\eta > 0$ from (6.29) we obtain the estimate

$$\int_0^{2T} J^{-\alpha-1}(t)\varphi(t)\,dt \leq \eta\int_0^{2T} J^{-\alpha-1}(t)\varphi(t)\,dt + \frac{c(m_0,\alpha)}{2m_0}\frac{1}{\eta^\alpha}\int_0^{2T}\frac{|\varphi''(t)|^{\alpha+1}}{\varphi^\alpha(t)}\,dt - \frac{1}{2m_0}J^{-\alpha-1}(0)J'(0), \tag{6.30}$$

where the constant

$$c(m_0,\alpha) = \frac{1}{\alpha(\alpha+1)^{\alpha+1}}\frac{1}{2^\alpha m_0^\alpha}.$$

Setting $\eta = 1$, we obtain the inequality

$$0 \leq c(m_0, \alpha) \int_0^{2T} \frac{|\varphi''(t)|^{\alpha+1}}{\varphi^{\alpha}(t)} \, dt - J^{-\alpha-1}(0) J'(0). \tag{6.31}$$

To estimate the right-hand side of inequality (6.31), we rescale the time variable:

$$\varphi(t) = \varphi_0\left(\frac{t}{T}\right) = \varphi_0(\tau) = \begin{cases} 1 & \text{if } 0 \leq \tau \leq T, \\ 0 & \text{if } \tau \geq 2, \end{cases} \qquad \tau = \frac{t}{T}.$$

Then (6.31) implies the inequality

$$0 \leq c(m_0, \alpha) d_\alpha T^{-2\alpha-1} - J^{-\alpha-1}(0) J'(0), \tag{6.32}$$

where

$$d_\alpha = \int_0^2 \frac{|\varphi_0''(\tau)|^{\alpha+1}}{\varphi_0^{\alpha}(\tau)} \, d\tau < \infty.$$

Thus, the lifetime T of a possible solution of the Cauchy problem (6.2)–(6.3) satisfies the estimate

$$T \leq \left[d_\alpha c(m_0, \alpha) c_0^{2\alpha+2} \|u_0\|^{-2}\right]^{1/(2\alpha+1)}. \tag{6.33}$$

We find the optimal constant $c_0^2 > 0$ under the condition

$$M_0(u) \geq K(u) > m_0 > 0.$$

As before, we can easily verify that

$$c_0^2 = \frac{1+\alpha}{2} \frac{\|u_0\|^4}{K(u) - m_0}, \tag{6.34}$$

and the best estimate of the blow-up time is of the form

$$T \leq T^* = \frac{1}{2} \|u_0\|^2 \left(\frac{d_\alpha}{\alpha} \frac{1}{m_0^\alpha}\right)^{\frac{1}{2\alpha+1}} (K(u) - m_0)^{-\frac{\alpha+1}{2\alpha+1}}. \tag{6.35}$$

Now we state the final result.

Theorem 6.2. *For some $\alpha > 0$, let the condition $K(u) > m_0$ hold. Then the Cauchy problem (6.2)–(6.3) has no solutions for $t \geq T^*$.*

Note that, with respect to m_0, the estimates of the blow-up time (6.21) and (6.35) have the same character. Indeed, after the norming

$$K(u) = m_0 \bar{K}(u) > 0,$$

from estimates (6.21) and (6.35) we obtain

$$T_0^* = \frac{1+\alpha}{2\alpha} \frac{\|u_0\|^2}{\bar{K}(u)} \frac{1}{m_0} \quad \text{and} \quad T^* = \frac{1}{2}\left(\frac{d_\alpha}{\alpha}\right)^{\frac{1}{2\alpha+1}} \frac{\|u_0\|^2}{(K(u)-1)^{\frac{\alpha+1}{2\alpha+1}}} \frac{1}{m_0}, \tag{6.36}$$

respectively.

6.1.2 Examples

Example 6.1. Consider the following problem for the model parabolic equation with the p-Laplacian

$$\Delta_p u = \operatorname{div}(|\nabla u|^{p-2} \nabla u)$$

for a certain constant $\alpha > 0$:

$$\frac{\partial u}{\partial t} = \Delta_p u + |u|^{\sigma-2} u, \quad (x,t) \in \mathbb{R}_+^{N+1}, \tag{6.37}$$

$$u|_{t=0} = u_0, \quad \sigma \geq 2(\alpha+1) \geq p > 1. \tag{6.38}$$

Solutions of this problem are considered in the natural class of functions $u(x,t)$ such that

$$u(t,\cdot),\ u_t(t,\cdot) \in \mathbb{X}(\mathbb{R}^N) = \mathbb{W}^{1,p}(\mathbb{R}^N) \cap \mathbb{L}^\sigma(\mathbb{R}^N) \cap \mathbb{L}^2(\mathbb{R}^N),$$
$$\|u(t,\cdot)\|_{\mathbb{L}^2},\ \|u_t(t,\cdot)\|_{\mathbb{L}^2} \in \mathbb{L}^2_{\text{loc}}(\mathbb{R}_+).$$

The corresponding functional is of the form

$$G(u) = -\frac{1}{p} \int_{\mathbb{R}^N} |\nabla u|^p \, dx + \frac{1}{\sigma} \int_{\mathbb{R}^N} |u|^\sigma \, dx, \quad G_t \equiv 0. \tag{6.39}$$

Condition (6.20) holds due to the condition for σ and p since it has the form

$$\begin{aligned} (G'(u), u) - 2(1+\alpha)G(u) &= \left(-1 + \frac{2+2\alpha}{p}\right) \int_{\mathbb{R}^N} |\nabla u|^p \, dx \\ &\quad + \left(1 - \frac{2+2\alpha}{\sigma}\right) \int_{\mathbb{R}^N} |u|^\sigma \, dx \geq 0. \end{aligned} \tag{6.40}$$

Thus, the blow-up result can be stated as follows.

Theorem 6.3. *Let $u_0 \in \mathbb{X}(\mathbb{R}^N)$ and $G(u_0) > 0$. Then the Cauchy problem* (6.37)–(6.38) *has no solutions in the functional class considered for*

$$t \geq T_0^* = \frac{1}{4\alpha} \frac{\|u_0\|_{\mathbb{L}^2}^2}{G(u_0)}.$$

Example 6.2. Consider the following problem for the inhomogeneous parabolic equation and some constant $\alpha > 0$ in the same functional class as in (6.37)–(6.38):

$$\frac{\partial u}{\partial t} = \Delta_p u + |u|^{\sigma-2}u + h(x), \quad h(x) \in \mathbb{L}^2(\mathbb{R}^N) \cap \mathbb{L}^{\sigma'}(\mathbb{R}^N), \tag{6.41}$$

$$u|_{t=0} = u_0 \in \mathbb{X}(\mathbb{R}^N), \quad \sigma' = \frac{\sigma}{\sigma - 1}, \quad \sigma \geq 2(\alpha + 1) \geq p > 1. \tag{6.42}$$

The functional $G(u)$ for this problem is of the form

$$G(u) = -\frac{1}{p} \int_{\mathbb{R}^N} |\nabla u|^p \, dx + \frac{1}{\sigma} \int_{\mathbb{R}^N} |u|^\sigma \, dx + \int_{\mathbb{R}^N} hu \, dx. \tag{6.43}$$

We can apply Theorem 6.1. First, we calculate $M_0(u(t))$ by formula (6.12):

$$\begin{aligned} M_0(u(t)) = {} & \left(\frac{2(1+\alpha)}{p} - 1\right) \int_{\mathbb{R}^N} |\nabla u|^p \, dx + \left(1 - \frac{2(1+\alpha)}{\sigma}\right) \int_{\mathbb{R}^N} |u|^\sigma \, dx \\ & - (1 + 2\alpha) \int_{\mathbb{R}^N} hu \, dx + 2(1+\alpha)G(u_0) - (1+\alpha)\frac{\|u_0\|_{\mathbb{L}^2}^4}{2c_0^2}. \end{aligned} \tag{6.44}$$

Applying the Young inequality with parameter $\varepsilon > 0$, we rewrite (6.44) in the form

$$\begin{aligned} M_0(u(t)) \geq {} & \left(\frac{2(1+\alpha)}{p} - 1\right) \int_{\mathbb{R}^N} |\nabla u|^p \, dx + \left(1 - \frac{2(1+\alpha)}{\sigma} - \varepsilon\right) \int_{\mathbb{R}^N} |u|^\sigma \, dx \\ & - \frac{1}{\sigma'} \frac{(1+2\alpha)^{\sigma'}}{(\varepsilon\sigma)^{\sigma'-1}} \int_{\mathbb{R}^N} |h|^{\sigma'} \, dx + 2(1+\alpha)G(u_0) - (1+\alpha)\frac{\|u_0\|_{\mathbb{L}^2}^4}{2c_0^2}. \end{aligned} \tag{6.45}$$

Introduce the notation

$$c_{\alpha,\sigma} = \frac{1}{2(1+\alpha)\sigma'} \frac{(1+2\alpha)^{\sigma'}}{(\varepsilon\sigma)^{\sigma'-1}};$$

then, under the condition

$$G(u_0) > c_{\alpha,\sigma} \int_{\mathbb{R}^N} |h|^{\sigma'} \, dx, \tag{6.46}$$

we can find c_0^2 for which condition (6.18) holds:

$$c_0^2 = \frac{\|u_0\|_{\mathbb{L}^2}^4}{4(G(u_0) - c_{\alpha,\delta} \int_{\mathbb{R}^N} |h|^{\sigma'} dx)}. \tag{6.47}$$

Then Theorem 6.1 for problem (6.41)–(6.42) can be restated as follows.

Theorem 6.4. *Let* $u_0 \in \mathbb{X}(\mathbb{R}^N)$, $h \in \mathbb{L}^2(\mathbb{R}^N) \cap \mathbb{L}^{\sigma'}(\mathbb{R}^N)$, *and, moreover, let inequality* (6.46) *be fulfilled. Then the Cauchy problem* (6.41)–(6.42) *has no solutions in the functional class considered for*

$$t \geq T_0 = \frac{1}{4\alpha} \frac{\|u_0\|^2_{\mathbb{L}^2}}{G(u_0) - c_{\alpha,\sigma}\|h\|^{\sigma'}_{\mathbb{L}^{\sigma'}}}.$$

These results show that the energy method of Levine can be easily generalized to parabolic problems in unbounded domains. In the next section, we discuss blow-up results in unbounded domains for hyperbolic problems.

6.2 Blow-up for hyperbolic equations

The special importance of the energy method of Pokhozhaev is the ability of studying the blow-up of solutions to the Cauchy problem for the abstract quasilinear hyperbolic equation in a real Hilbert space $\mathbb{H}$:

$$\begin{gathered} \frac{d^2u}{dt^2} = f'(u), \quad t > 0, \\ u(0) = u_0 \in \mathbb{D}, \quad u_t(0) = u_1 \in \mathbb{H}, \end{gathered} \tag{6.48}$$

where $f : \mathbb{H} \to \mathbb{R}$ is a Fréchet-differentiable $\mathbb{C}^1$-function. We denote by $f' : \mathbb{D} \to \mathbb{H}$ the Fréchet derivative, where $\mathbb{D} \subset \mathbb{H}$, $(\cdot, \cdot)$ is the duality bracket, and $\|\cdot\|$ is the norm in $\mathbb{H}$.

As particular cases, this abstract problem includes the classical semilinear equation

$$u_{tt} = \Delta u + |u|^{p-1}u, \quad p > 1, \tag{6.49}$$

the higher-order equation

$$u_{tt} = -(-\Delta)^m u + |u|^{p-1}u, \quad p > 1,\ m \geq 1, \tag{6.50}$$

the quasilinear second-order equation with the p-Laplacian and its higher-order analog

$$u_{tt} = \operatorname{div}(|\nabla u|^{q-1}\nabla u) + |u|^{p-1}u, \quad p > 1,\ q > 0, \tag{6.51}$$

and also the Kirchhoff-type equation with nonlocal nonlinearity

$$u_{tt} = -\Big(\int_{\mathbb{R}^N} |\nabla^m u|^2\, dx \Big)^q (-\Delta)^m u + |u|^{p-1}u, \quad p > 1,\ q > 0,\ m \geq 1, \tag{6.52}$$

where we have used the following standard notation:

$$|\nabla^m u|^2 = |\Delta^{m/2} u|^2 \quad \text{for } m = 2k,$$

$$|\nabla^m u|^2 = \left|\nabla(\nabla^{(m-1)/2} u)\right|^2 \quad \text{for } m = 2k+1.$$

We assume the local solvability of problem (6.48). We obtain sufficient conditions of the blow-up by the method proposed by Galaktionon and Pokhozhaev [27]; this method differs from classical Levine's method (see [58]) by its applicability to Cauchy problems.

6.2.1 Abstract results

Introduce the functionals

$$E(t) = \frac{1}{2}\|u_t(t)\|^2 - f(u(t)), \quad G(t) = \|u(t)\|^2. \tag{6.53}$$

Differentiating by time, from equation (6.48) we obtain the equality

$$E'(t) = 0 \quad \Rightarrow \quad \frac{1}{2}\|u_t\|^2 - f(u) \equiv E(0) = E_0. \tag{6.54}$$

On the other hand, differentiating $G(t)$ by time and using (6.48), we obtain

$$G'(t) = 2(u, u_t), \tag{6.55}$$

$$G''(t) = 2\|u_t\|^2 + 2(u, u_{tt}) = 2\|u_t\|^2 + 2(u, f'(u)). \tag{6.56}$$

We assume that, for all $u \in \mathbb{D}$, there exists a real constant $\lambda > 2$ such that

$$(u, f'(u)) - \lambda f(u) \geq 0 \tag{6.57}$$

and the initial data satisfy the inequality

$$E_0 = E(0) \leq 0.$$

Then from (6.54), (6.56), and (6.57) we obtain the following relations for all t:

$$G'' \geq 2\|u_t\|^2 + \lambda(\|u_t\|^2 - 2E_0) = (2+\lambda)\|u_t\|^2 - 2\lambda E_0 \geq (2+\lambda)\|u_t\|^2. \tag{6.58}$$

Using the Cauchy–Bunyakovsky inequality, from (6.55) we obtain the estimate

$$(G')^2 \leq 4\|u\|^2\|u_t\|^2 = 4G\|u_t\|^2 \quad \Rightarrow \quad \|u_t\|^2 \geq \frac{(G')^2}{4G}. \tag{6.59}$$

Substituting (6.59) into (6.58), we obtain the following ordinary differential inequality for the function $G(t)$:

$$G'' \geq (2+\lambda)\frac{(G')^2(t)}{4G(t)}, \quad t > 0. \tag{6.60}$$

Applying the theory of ordinary differential inequalities, we can obtain some results on the solvability and blow-up.

Lemma 6.1. *Assume that condition* (6.57) *holds and the inequalities*

$$E_0 \le 0, \quad G'(0) = 2(u(0), u_t(0)) \equiv 2(u_0, u_1) < 0$$

are valid. Then, for all $t > 0$, *the norm* $\|u(t)\|$ *monotonically decreases and is uniformly bounded*: $\|u(t)\| < \|u(0)\|$.

Proof. Indeed, integrating inequality (6.60) written in the form

$$\frac{G''}{(G')^2} \ge \frac{2+\lambda}{4G(t)}, \tag{6.61}$$

we obtain

$$\frac{1}{G'(t)} \le \frac{1}{G'(0)} - \frac{2+\lambda}{4} \int_0^t \frac{d\tau}{G(\tau)} < 0.$$

Therefore, for all $t > 0$, we have

$$G'(t) \equiv 2(u(t), u_t(t)) < 0.$$

Rewrite (6.60) in the form

$$\frac{G''}{G'} \le \frac{(2+\lambda)G'(t)}{4G(t)}.$$

Integrating by time from 0 to t, we arrive at the inequality

$$\frac{G'(t)}{G'(0)} \le \left[\frac{G(t)}{G(0)}\right]^{\frac{2+\lambda}{4}}.$$

Another integration yields the estimate

$$\|u(t)\| \le \|u_0\| \left[1 + (\lambda - 2)\frac{|F_0| t}{4}\right]^{-2/(\lambda-2)}, \quad F_0 = 2\frac{(u_0, u_1)}{\|u_0\|^2} > 0, \tag{6.62}$$

which implies the lemma. □

On the other hand, setting $G'(0) > 0$, from (6.60) we conclude that $G'(t) > 0$ for all $t > 0$. Twice integrating by time, we obtain the inequality

$$G(t) \ge G(0)[1 - (\lambda - 2)|F_0| t/4]^{-4/(\lambda-2)}, \tag{6.63}$$

which implies the blow-up result and an estimate of the blow-up time.

Lemma 6.2. *Assume that there exists a solution of the Cauchy problem* (6.48) *under condition* (6.57) *and the inequalities*

$$E_0 \le 0, \quad G'(0) = 2(u(0), u_t(0)) \equiv 2(u_0, u_1) > 0$$

are fulfilled. Then the blow-up occurs in the following sense: the norm $\|u(t)\|^2$ *unboundedly increases on the finite time interval* $(0, T)$, *where*

$$T = \frac{4}{(\lambda - 4)F_0}. \tag{6.64}$$

6.2.2 Example

Now we present applications of Lemma 6.2 to specific problems. We do not dwell on the choice of initial data guaranteeing the blow-up (usually, the choice of such initial data is a sufficiently easy task) but verify the sufficient condition (6.57).

Example 6.3. First, we consider the semilinear higher-order hyperbolic equation

$$u_{tt} = f'(u) \equiv -(-\Delta)^m u + |u|^{p-1}u, \quad m \geq 1,\ p > 1, \tag{6.65}$$

in the domain

$$\mathbb{R}^{N+1}_+ = \mathbb{R}^N \times \mathbb{R}_+.$$

Restoring the right-hand side by the Fréchet derivative, we obtain

$$f(u) = -\frac{1}{2}\int\limits_{\mathbb{R}^N} |\nabla^m u|^2\,dx + \frac{1}{p+1}\int\limits_{\mathbb{R}^N} |u|^{p+1}\,dx,$$

$$(u, f'(u)) = -\int\limits_{\mathbb{R}^N} |\nabla^m u|^2\,dx + \int\limits_{\mathbb{R}^N} |u|^{p+1}\,dx.$$

Then condition (6.57) takes the form

$$(u, f'(u)) - \lambda f(u) = \left(\frac{\lambda}{2} - 1\right)\int\limits_{\mathbb{R}^N} |\nabla^m u|^2\,dx + \left(1 - \frac{\lambda}{p+1}\right)\int\limits_{\mathbb{R}^N} |u|^{p+1}\,dx \geq 0. \tag{6.66}$$

We choose $\lambda = p + 1$; then inequality (6.66) can be written as the condition

$$(u, f'(u)) - (p+1)f(u) = \left(\frac{p+1}{2} - 1\right)\int\limits_{\mathbb{R}^N} |\nabla^m u|^2\,dx \geq 0, \tag{6.67}$$

which is known to be fulfilled, and therefore if the initial data satisfy the conditions of Lemma 6.2, then the blow-up in finite time occurs.

Example 6.4. In the same domain, we consider the quasilinear equation with the p-Laplacian:

$$u_{tt} = f'(u) \equiv \operatorname{div}(|\nabla u|^{q-1}\nabla u) + |u|^{p-1}u, \quad q > 0,\ p > 1, \tag{6.68}$$

where the function $f(u)$ is of the form

$$f(u) = -\frac{1}{q+1}\int\limits_{\mathbb{R}^N} |\nabla u|^{q+1}\,dx + \frac{1}{p+1}\int\limits_{\mathbb{R}^N} |u|^{p+1}\,dx,$$

$$(u, f'(u)) = -\int\limits_{\mathbb{R}^N} |\nabla u|^{q+1}\,dx + \int\limits_{\mathbb{R}^N} |u|^{p+1}\,dx.$$

Condition (6.57) for $\lambda > 2$ is of the form

$$(u,f'(u)) - \lambda f(u) = \left(\frac{\lambda}{q+1} - 1\right) \int_{\mathbb{R}^N} |\nabla u|^{q+1}\,dx + \left(1 - \frac{\lambda}{p+1}\right) \int_{\mathbb{R}^N} |u|^{p+1}\,dx \geq 0; \quad (6.69)$$

therefore Lemma 6.2 can be applied if $\lambda > 2$ satisfies the inequalities $\lambda \geq q+1$ and $\lambda \leq p+1$, that is, a solution blows up in finite time if $p \geq q$.

Example 6.5. A natural generalization of the previous two examples is the following quasilinear higher-order hyperbolic equation:

$$\begin{aligned} u_{tt} = &- \sum_{|\alpha|\leq m} (-1)^{|\alpha|} D^\alpha (a_\alpha(x)|D^\alpha u|^{q-1} D^\alpha u) \\ &+ \sum_{|\beta|\leq k} (-1)^{|\beta|} D^\beta (b_\beta(x)|D^\beta u|^{p-1} D^\beta u), \end{aligned} \quad (6.70)$$

where $p > 1$ and $q > 0$ are parameters, and α and β are multi-indices with modules $|\alpha| = \alpha_1 + \cdots + \alpha_N$ and $|\beta| = \beta_1 + \cdots + \beta_N$. For equation (6.70), the following equalities hold:

$$f(u) = -\frac{1}{q+1} \int_{\mathbb{R}^N} \sum_{|\alpha|\leq m} a_\alpha(x)|D^\alpha u|^{q+1}\,dx + \frac{1}{p+1} \int_{\mathbb{R}^N} \sum_{|\beta|\leq k} b_\beta(x)|D^\beta u|^{p+1}\,dx,$$

$$(u,f'(u)) = -\int_{\mathbb{R}^N} \sum_{|\alpha|\leq m} a_\alpha(x)|D^\alpha u|^{q+1}\,dx + \int_{\mathbb{R}^N} \sum_{|\beta|\leq k} b_\beta(x)|D^\beta u|^{p+1}\,dx.$$

We rewrite condition (6.57) of Lemma 6.2 in the form

$$\begin{aligned} (u,f'(u)) - \lambda f(u) = &\left(\frac{\lambda}{q+1} - 1\right) \int_{\mathbb{R}^N} \sum_{|\alpha|\leq m} a_\alpha(x)|D^\alpha u|^{q+1}\,dx \\ &+ \left(1 - \frac{\lambda}{p+1}\right) \int_{\mathbb{R}^N} \sum_{|\beta|\leq k} b_\beta(x)|D^\beta u|^{p+1}\,dx \geq 0. \end{aligned} \quad (6.71)$$

Assume that the integrand differential forms are nonnegative,

$$\sum_{|\alpha|\leq m} a_\alpha(x)|\xi_\alpha|^{q+1} \geq 0, \quad \sum_{|\beta|\leq k} b_\beta(x)|\xi_\beta|^{p+1} \geq 0,$$

and the parameter λ satisfies the conditions

$$\lambda \geq q+1, \quad \lambda \geq p+1,$$

or, equivalently, $p \geq q$. Then, using Lemma 6.2, we can find sufficient conditions for the initial data under which the blow-up of solutions occurs. Note that the blow-up in finite time does not require any special assumptions on the positiveness of the first operator in the right-hand side of (6.70) if we set $\lambda = q+1 > 2$ and $p > q > 1$. If $p = q > 1$ and $\lambda = q+1$, then the blow-up result can be proved without any restrictions for both operators in (6.70).

Example 6.6. Now we consider the problem for the second-order Kirchhoff-type equation with nonlocal nonlinearity:

$$u_{tt} = a\Big(\int_{\mathbb{R}^N} |\nabla u|^2\,dx\Big)\Delta u + h(u), \quad (x,t) \in \mathbb{R}_+^{N+1}, \tag{6.72}$$

with smooth real-valued functions $a(s)$ and $h(s)$. Restoring the function $f(u)$ in the right-hand side by the Fréchet derivative, we obtain

$$f(u) = -\frac{1}{2}A(u)\Big(\int_{\mathbb{R}^N} |\nabla u|^2\,dx\Big) + \int_{\mathbb{R}^N} H(u)\,dx,$$

where

$$A(s) = \int_0^s a(\tau)\,d\tau, \quad H(s) = \int_0^s h(\tau)\,d\tau.$$

In this case, condition (6.57) can be written in the form

$$\begin{aligned}(u,f'(u)) - \lambda f(u) &= -a\Big(\int_{\mathbb{R}^N} |\nabla u|^2\,dx\Big)\int_{\mathbb{R}^N} |\nabla u|^2\,dx + \int_{\mathbb{R}^N} h(u)u\,dx \\ &\quad + \frac{\lambda}{2}A\Big(\int_{\mathbb{R}^N} |\nabla u|^2\,dx\Big) - \lambda\int_{\mathbb{R}^N} H(u)\,dx \\ &= \Big[\frac{\lambda}{2}A\Big(\int_{\mathbb{R}^N} |\nabla u|^2\,dx\Big) - a\Big(\int_{\mathbb{R}^N} |\nabla u|^2\,dx\Big)\int_{\mathbb{R}^N} |\nabla u|^2\,dx\Big] \\ &\quad + \int_{\mathbb{R}^N} (h(u)u - \lambda H(u))\,dx \ge 0.\end{aligned} \tag{6.73}$$

Thus, if for certain $\lambda > 2$, inequality (6.73) is fulfilled, then we can find sufficient initial conditions under which a solution blows up in finite time due to Lemma 6.2.

Example 6.7. We can naturally generalize this result to the case of nonlocal higher-order hyperbolic equations

$$u_{tt} = -\Big(\int_{\mathbb{R}^N} |\nabla^m u|^2\Big)^q (-\Delta)^m u + |u|^{p-1}u, \quad m \ge 1,\ q > 0,\ p > 1. \tag{6.74}$$

Comparing this problem with the previous one, we conclude that here

$$a(s) = s^q, \quad A(s) = \frac{s^{q+1}}{q+1}, \quad s = \int_{\mathbb{R}^N} |\nabla^m u|^2\,dx, \quad H(u) = \frac{|u|^{p+1}}{p+1},$$

$$f(u) = -\frac{1}{2(q+1)}\Big(\int_{\mathbb{R}^N} |\nabla^m u|^2\,dx\Big)^{q+1} + \frac{1}{p+1}\int_{\mathbb{R}^N} |u|^{p+1}\,dx.$$

Therefore condition (6.73) takes the form

$$(u, f'(u)) - \lambda f(u) = \Big[\frac{\lambda}{2(q+1)} - 1\Big]\Big(\int_{\mathbb{R}^N} |\nabla^m u|^2\,dx\Big)^{q+1} + \Big(1 - \frac{\lambda}{p+1}\Big)\int_{\mathbb{R}^N} |u|^{p+1}\,dx \ge 0. \tag{6.75}$$

Under additional conditions for λ, namely $\lambda > 2$, $\lambda \ge 2(q+1)$, and $\lambda \le p+1$, Lemma 6.2 can be applied to equation (6.74) in the case $p \ge 2q+1$.

6.3 Critical exponents of semilinear equations

Discussing applications of the energy method to the study of collapsing solutions of nonlinear hyperbolic equations, we must mention results of Pokhozhaev concerning critical exponents for nonlinearities of the form $|u|^p$. We briefly review these results for the three-dimensional case (the general case is described in [27]).

We study the global insolvability of the Cauchy problem for the semilinear hyperbolic equation

$$\begin{gathered} u_{tt} = \Delta u + b(x,t)|u|^p, \quad p > 1, \\ u(0,x) = u_0(x) \in \mathbb{C}_0^{(2)}(\mathbb{R}^N), \quad u_t(0,x) = u_1(x) \in \mathbb{C}_0^{(1)}(\mathbb{R}^N). \end{gathered} \tag{6.76}$$

We assume that there exists a local-in-time solution of the class $\mathbb{C}^{(2)}_{\mathrm{loc}}(\mathbb{R}^{N+1}_+)$ satisfying the initial data

$$\int_{\mathbb{R}^N} u_1(x)\,dx > 0, \quad \operatorname{supp} u_0, \operatorname{supp} u_1 \subset \{|x| < R\} \tag{6.77}$$

in the ball of radius R. We also assume that the function $b(x,t) \in \mathbb{L}^\infty_{\mathrm{loc}}(\mathbb{R}^{N+1}_+)$ satisfies the inequalities

$$\limsup_{t\to\infty} \frac{B(t)}{t^\alpha} < \infty, \quad \limsup_{t\to\infty} \frac{B_1(t)}{t^\beta} < \infty, \tag{6.78}$$

where $\alpha, \beta \in \mathbb{R}$, p, and p' are such that $(p)^{-1} + (p')^{-1} = 1$, and

$$B(t) = \Big|\int_{|x|\le R+t} b^{1-p'}\,dx\Big|^{p-1}, \quad B_1(t) = \Big|\int_{t-R<|x|\le R+t} b^{1-p'}\,dx\Big|^{p-1}. \tag{6.79}$$

The basic result on critical exponents can be stated as follows.

Theorem 6.5. *In the three-dimensional case ($N = 3$), under the condition*

$$\alpha < (p+2-\beta)(p-1)+2, \tag{6.80}$$

there are no global solutions of problem (6.76).

Proof. For a solution $u(t)$, $\operatorname{supp} u(t) \subset \{|x| \le R+t\}$ of problem (6.76), introduce the energy functional

$$E(t) = \int_{\mathbb{R}^N} u(x,t)\,dx = \int_{|x|\le R+t} u(x,t)\,dx.$$

Taking into account the dependence of the limits of the integral on time, we twice differentiate this functional by time:

$$E'' = \int_{|x|\le R+t} b(x,t)|u|^p\,dx. \tag{6.81}$$

By the Hölder inequality we have the following estimate:

$$\left|\int_{|x|\le R+t} u(x,t)\,dx\right| \le \left|\int_{|x|\le R+t} b(x,t)|u|^p\,dx\right|^{1/p} \left|\int_{|x|\le R+t} b^{1-p'}(x,t)\,dx\right|^{1/p'}. \tag{6.82}$$

Substituting this estimate into (6.81), we obtain

$$E'' = \int_{|x|\le R+t} b(x,t)|u|^p\,dx \ge B^{-1}(t)\left|\int_{|x|\le R+t} u(x,t)\,dx\right|^p = B^{-1}(t)E^p(t) \tag{6.83}$$

for all $t > 0$ and $E'(0) > 0$. In particular, for $b \equiv 1$, from the last inequality we obtain the critical Kato exponent $p_k = (N+1)/(N-1)$ (see [39]), for which $E(t)$ blows up in finite time for all $1 < p < p_k$.

However, to prove the blow-up in a more general case, we must construct an additional lower estimate for $E(t)$. For this purpose, we consider the linear Cauchy problem in the three-dimensional space with the same initial data as for (6.76):

$$v_{tt} = \Delta v, \quad v(0,x) = u_0(x), \quad v_t(0,x) = u_1(x). \tag{6.84}$$

Denoting by δ the Dirac distribution, we can represent the fundamental solution in the form

$$F_3(x,t) = \frac{1}{2\pi}\delta(t^2 - |x|^2). \tag{6.85}$$

By the properties of the fundamental solution in $\mathbb{R}^4_+$, $u(x,t) \ge v(x,t)$, and

$$E(t) = \int_{|x|\le R+t} u(x,t)\,dx \ge \int_{|x|\le R+t} v(x,t)\,dx. \tag{6.86}$$

On the other hand, by the Huygens principle, for $t > R$, we have

$$\int\limits_{|x|\le t+R} v(x,t)\,dx = \int\limits_{t-R\le|x|\le t+R} v(x,t)\,dx. \tag{6.87}$$

Then (6.86) implies the following inequalities:

$$\begin{aligned}\int\limits_{t-R\le|x|\le t+R} v(x,t)\,dx &\le \int\limits_{t-R\le|x|\le t+R} u(x,t)\,dx\\ &\le \left|\int\limits_{t-R\le|x|\le t+R} b|u|^p\,dx\right|^{1/p}\left|\int\limits_{t-R\le|x|\le t+R} b^{1-p'}\,dx\right|^{1/p'}\\ &\le \left|\int\limits_{|x|\le t+R} b|u|^p\,dx\right|^{1/p}\left|\int\limits_{t-R\le|x|\le t+R} b^{1-p'}\,dx\right|^{1/p'}.\end{aligned} \tag{6.88}$$

Inequalities (6.88) imply the estimate

$$\int\limits_{|x|\le t+R} b\,|u|^p\,dx \ge \left|\int\limits_{t-R\le|x|\le t+R} v(x,t)\,dx\right|^p B_1^{-1}(t). \tag{6.89}$$

To obtain a lower estimate for v, we integrate (6.84) over the domain $\mathbb{R}^3$ and conclude that the function

$$E_0(t) = \int\limits_{\mathbb{R}^3} v(x,t)\,dx = \int\limits_{|x|\le t+R} v\,dx \equiv \int\limits_{t-R\le|x|\le t+R} v(x,t)\,dx \tag{6.90}$$

for $t \ge 0$ satisfies the following problem:

$$E_0'' = 0, \quad E_0(0) = \int\limits_{\mathbb{R}^3} u_0\,dx \equiv U_0, \quad E_0'(0) = \int\limits_{\mathbb{R}^3} u_1\,dx \equiv U_1. \tag{6.91}$$

We can easily obtain a solution of (6.91): $E_0(t) = U_0 + U_1 t$. By (6.89), for $U_1 > 0$, we have the inequality

$$\int\limits_{|x|\le t+R} b\,|u|^p\,dx \ge (U_0 + U_1 t)^p B_1^{-1}(t). \tag{6.92}$$

Integration of (6.81) by time from 0 to t yields the required lower estimate:

$$E(t) \ge U_0 + U_1 t + H_0(t), \quad H_0(t) = \int\limits_0^t (t-\tau)(U_0 + U_1\tau)^p B_1^{-1}(\tau)\,d\tau. \tag{6.93}$$

For large time, combining (6.83) and (6.93), we obtain the following ordinary differential inequality with additional lower restriction:

$$E''(t) \geq B^{-1}(t)E^p, \quad E(t) \geq H_0(t), \quad t \gg R. \tag{6.94}$$

Using the additional conditions (6.78), we rewrite this system of inequalities in the form

$$E''(t) \geq Ct^{-\alpha}E^p(t), \quad E(t) \geq Ct^{-\beta+p+2}, \quad t \gg 1. \tag{6.95}$$

Integrating (6.95) and taking into account condition (6.80), we immediately obtain the blow-up result. □

Consider several examples. First, let

$$b(x,t) = Ct^l, \quad C, l > 0, t \gg 1,$$

be a time-dependent operator. Calculating the functions (6.79), we obtain the values of α and β:

$$B(t) = \left| \int_{|x|\leq R+t} (Ct^l)^{1-p'} \, dx \right|^{p-1}$$

$$= C_1(R+t)^{3(p-1)}t^{-l} \quad \Rightarrow \quad \alpha = 3(p-1) - l, \tag{6.96}$$

$$B_1(t) \leq C_2(R+t)^{2(p-1)-l} \quad \Rightarrow \quad \beta = 2(p-1) - l. \tag{6.97}$$

Inequality (6.80) for critical exponents takes the form

$$3(p-1) - l < (p + 2 + l - 2p + 2)(p-1) + 2,$$

which is equivalent to the condition

$$p^2 - (2+l)p - 1 < 0.$$

Applying the theorem, we conclude that the solution blows up in finite time in the subcritical interval

$$1 < p < p_c = 1 + \frac{l}{2} + \sqrt{1 + \left(1 + \frac{l}{2}\right)^2}. \tag{6.98}$$

Now we consider equation (6.76) with the operator depending only on spatial variables:

$$b(x,t) = c|x|^k, \quad k < N(p-1) = 3(p-1), \quad c > 0.$$

The functions $B(t)$ and $B'(t)$ are of the form

$$B(t) = c_1 \left| \int\limits_{|x| \le R+t} |x|^{k(1-p')}\, dx \right|^{p-1} = c_2 (R+t)^{\alpha} \quad \Rightarrow \quad \alpha = 3(p-1) - k, \tag{6.99}$$

$$\begin{aligned} B_1(t) &= c_3 \left| \int\limits_{t-R \le |x| \le t+R} |x|^{k(1-p')}\, dx \right|^{p-1} \\ &= c_4 \big((t+R)^{3-k(1-p')} - (t-R)^{3-k(p'-1)} \big)^{p-1} \\ &\le c_5 (R+t)^{\beta} \quad \Rightarrow \quad \beta = 2(p-1) - k. \end{aligned} \tag{6.100}$$

Condition (6.80) is equivalent to the quadratic inequality

$$p^2 - (2+k)p - 1 < 0;$$

therefore the solution blows up in finite time on the subcritical interval

$$1 < p < p_c = 1 + \frac{k}{2} + \sqrt{1 + \left(1 + \frac{k}{2}\right)^2}, \quad k < 3(p-1). \tag{6.101}$$

These results exactly coincide with the results of [39], where the values of critical exponents were obtained in the particular case $b(x,t) \equiv 1$.

6.4 Bibliographical notes

A detailed presentation of the modification of the energy method proposed by Pokhozhaev can be found in [85] for parabolic problems and in [27] for nonlinear hyperbolic equations. Moreover, in [27] the study of self-similar solutions and critical exponents was also performed.

Note that Pokhozhaev's method of nonlinear capacity and all modifications of Levine's energy method can be considered as versions of the method of test functions. Indeed, the method of test functions consists of multiplication by an arbitrary nonnegative test function $\varphi(x,t)$ satisfying certain condition and integration over the space $\mathbb{R}^N \times \mathbb{R}_+$, whereas the energy method consists of multiplication by a solution $u(x,t)$ or its derivative $u_t(x,t)$ and integration over the space $\mathbb{R}^N$. The basic class of the method of test functions consists of nonlinear equations with a source, for example, of the form $|u|^q$, $q > 1$, whereas the basic class of Levine's energy method consists of nonlinear equations with a source of the form $|u|^{q-1}u$, $q > 1$. Note that Pokhozhaev's modification of Levine's energy method, in contrast to classical Levine's method, allows us to examine the nonlinearity of the form $|u|^q$, $q > 1$. The following question arises: Can equations with a source of the form $|u|^{q-1}u$, $q > 1$, be examined by the method of nonlinear capacity? A partial answer was proposed in [86], where a mixed approach was used: the equation was multiplied by the function $\varphi(x,t)u$ or by its derivative $\psi(x,t)u_t$ and then integrated over the half-space $\mathbb{R}_+^{N+1}$.

7 Energy method of V. K. Kalantarov and O. A. Ladyzhenskaya

In previous chapters, we described a simple but effective method of proving the blow-up of solutions proposed by H. A. Levine for a wide class of parabolic and hyperbolic abstract equations

$$Pu_t = -Au + F(u), \quad Pu_{tt} = -Au + F(u).$$

Note that in Levine's reasonings the important role is played by the symmetry of the operators P and A in Hilbert spaces and their nonnegativeness: $P > 0$, $A \geq 0$. Based on the same idea, Kalantarov and Ladyzhenskaya [37] proposed a method that does not require the symmetry and nonnegativeness of the operator A.

7.1 Basic lemma

Lemma 7.1. *Let a positive twice-differentiable function $\Phi(t)$ satisfy for $t \geq 0$ the inequality*

$$\Phi''\Phi' - (1+\alpha)(\Phi')^2 \geq -2c_1\Phi\Phi' - c_2\Phi^2, \tag{7.1}$$

where $\alpha > 0$, $c_1 \geq 0$, and $c_2 \geq 0$ are constants. If

$$\Phi(0) > 0, \quad \Phi'(0) > -\gamma_2\alpha^{-1}\Phi(0), \quad c_1 + c_2 > 0,$$

then $\Phi(t)$ tends to infinity as $t \to T_1 \leq T_2$, where

$$T_2 = \frac{1}{2\sqrt{c_1^2 + c_2^2}} \ln \frac{\gamma_1\Phi(0) + \alpha\Phi'(0)}{\gamma_2\Phi(0) + \alpha\Phi'(0)}, \quad \gamma_{1,2} = c_1 \pm \sqrt{c_1^2 + \alpha c_2}. \tag{7.2}$$

If

$$\Phi(0) > 0, \quad \Phi'(0) > 0, \quad \text{and} \quad c_1 = c_2 = 0,$$

then $\Phi(t)$ tends to infinity as

$$t \to T_1 \leq T_2 = \Phi(0)[\alpha\Phi'(0)]^{-1}$$

if there exists a nonextendable in time solution in the corresponding class on the interval $[0, T_1)$.

Levine's energy method is based on the second assertion of the lemma. The first, stronger assertion is the base of the method proposed by Kalantarov and Ladyzhenskaya.

https://doi.org/10.1515/9783110602074-007

Proof. Let $\Psi = \Phi^{-\alpha}$. Then we have the following expressions:

$$\Psi'(t) = -\alpha\frac{\Phi'(t)}{\Phi^{1+\alpha}(t)}, \quad \Psi''(t) = -\alpha\frac{\Phi''(t)\Phi'(t) - (\alpha+1)(\Phi'(t))^2}{\Phi^{2+\alpha}(t)}.$$

By these expressions we can rewrite (7.1) in the form

$$\Psi''(t) + 2c_1\Psi'(t) - c_2\Psi(t) = f(t) \le 0. \tag{7.3}$$

Integrating this inequality in the case where $c_1 + c_2 > 0$, we have

$$\begin{aligned}\Psi(t) &= \beta_1 e^{\gamma_1 t} + \beta_2 e^{\gamma_2 t} + \frac{1}{\gamma_1 - \gamma_2}\int_0^t f(\tau)[e^{\gamma_1(t-\tau)} - e^{\gamma_2(t-\tau)}]d\tau \\ &\le \beta_1 e^{\gamma_1 t} + \beta_2 e^{\gamma_2 t}.\end{aligned} \tag{7.4}$$

The constants β_1 and β_2 are defined by the system of two equations

$$\beta_1 + \beta_2 = \Psi(0), \quad \beta_1\gamma_1 + \beta_2\gamma_2 = \Psi'(0).$$

Omitting simple calculations, we obtain

$$\beta_1 = \frac{\Psi'(0) - \gamma_2\Psi(0)}{\gamma_1 - \gamma_2} = -\frac{\alpha\Psi'(0) + \gamma_2\Psi(0)}{(\gamma_1 - \gamma_2)\Psi^{\alpha+1}(0)} < 0,$$

$$\beta_2 = \frac{\alpha\Psi'(0) + \gamma_1\Psi(0)}{(\gamma_1 - \gamma_2)\Psi^{\alpha+1}(0)} > 0.$$

This and the assumption of the lemma imply that $\Psi(t)$ vanishes in finite time, and therefore $\Phi(t) \to +\infty$ as $t \to T_0$. The case $c_1 = c_2 = 0$ immediately follows from (7.3). The lemma is proved. □

Other energy methods (in particular, a modified method discussed in Chapter 8) are based on analogs of this lemma. These analogs are proved in the Appendix.

7.2 Blow-up for parabolic equations

Consider the Cauchy problem for the equation

$$Pu_t = -Au + Bu + F(t, u), \quad u(0) = u_0, \tag{7.5}$$

in a Hilbert space $\mathbb{H}$, where P, A, and B are linear and symmetric operators acting from $\mathbb{H}$ to $\mathbb{H}$, P is positive, A is nonnegative, and $F(t, u)$ is a nonlinear operator, which, for some fixed time moment, is the Fréchet derivative of a certain nonlinear functional $G(t, u)$:

$$\frac{d\,G(t, u(\tau))}{d\tau} = (F(t, u(\tau)), u_\tau(\tau)).$$

Moreover, the functional $G(t,u)$ smoothly depends on t, so that the functions $u(t)$ that smoothly depend on time satisfy the relation

$$\frac{dG(t,u(t))}{dt} = (F(t,u(t)),\ u_t(t)) + G_t(t,u(t)). \tag{7.6}$$

All operators may by unbounded on an arbitrary linear set $D \subset \mathbb{H}$.

We examine the behavior of a solution $u(t)$ as time increases. For simplicity, we assume that the solution is "almost classical": all terms of (7.5), the solution $u(\cdot)$, and its derivative $u_t(\cdot)$ are elements of the space $\mathbb{L}_2(0,T;\mathbb{H})$, and all terms in (7.6) are summable on $0 < T < T_0$, where $[0,T_0)$ is the maximal interval of the existence of a nonextendable solution.

Assume that the operator B (which may be nonlinear) is subordinated to the operators $A^{1/2}$ and $P^{1/2}$:

$$\|P^{-\frac{1}{2}}Bu\| \le m_1\|A^{1/2}u\| + m_2\|P^{1/2}u\|. \tag{7.7}$$

We assume that the nonlinearity satisfies the inequality

$$(F(t,u),u) \ge 2(1+\alpha_1)G(t,u), \quad \alpha_1 > 0. \tag{7.8}$$

Moreover, on the first stage, we assume that, for some $\beta \in (0,\alpha_1)$ and $\varepsilon \in (0,1)$, the following estimate holds:

$$G_t(t,u) \ge m_3(F(t,u),u), \quad m_3 = \frac{1+\alpha_1}{\alpha_1-\beta}\frac{1+\varepsilon}{1-\varepsilon}\frac{m_1^2}{4}. \tag{7.9}$$

Then the blow-up result can be stated as follows.

Theorem 7.1. *Let the initial data satisfy the inequalities*

$$\begin{gathered} b_0 = \|P^{1/2}u_0\| > 0, \quad a_0 > \frac{(1+\alpha)^2}{4\alpha(\alpha_1+1)}b_0, \\ a_0 = -\frac{1}{2}\|A^{1/2}u_0\|^2 - \frac{1}{2}m_3b_0 + G(0,u_0), \quad \alpha = -1+\sqrt{1+\beta}. \end{gathered} \tag{7.10}$$

Then the solution of the Cauchy problem for equation (7.5) *blows up in finite time*:

$$\|P^{1/2}u\| \to \infty \quad \text{as } t \to T_0 \le T_1, \tag{7.11}$$

where

$$\begin{gathered} T_1 = \frac{1}{2\sqrt{c_1^2+\alpha c_2}}\frac{4\alpha^2(\alpha_1+1)a_0+\gamma_1(1+\alpha)^2b_0}{4\alpha^2(\alpha_1+1)a_0+\gamma_2(1+\alpha)^2b_0}, \\ c_1 = \frac{m_1^2}{4\alpha_1} + m_2, \quad c_2 = 4(1+\alpha_1)m_4, \\ m_4 = m_3\left(\frac{m_1^2}{4\varepsilon}+m_2\right) + \frac{m_2^2}{4}\frac{1+\alpha_1}{\alpha_1-\beta}\frac{1+\varepsilon}{\varepsilon}. \end{gathered}$$

Proof. The proof is based on two energy equalities. The first is obtained by the scalar multiplication of equation (7.5) by u:

$$(Pu_t, u) = \frac{1}{2}\frac{d}{dt}\|P^{1/2}u\|^2 = -\|A^{1/2}u\|^2 + (Bu, u) + (F(t, u), u). \tag{7.12}$$

The second is obtained by the scalar multiplication of equation (7.5) by u_t:

$$\begin{aligned}\|P^{1/2}u_t\|^2 &= -(Au, u_t) + (Bu, u_t) + (F(t, u), u_t)\\ &= -\frac{1}{2}\frac{d}{dt}\|A^{1/2}u\|^2 + (Bu, u_t) + \frac{d}{dt}G(t, u) - G_t(t, u).\end{aligned} \tag{7.13}$$

For convenience, we introduce the following notation:

$$J(t) = -\frac{1}{2}\|A^{1/2}u\|^2 + G(t, u). \tag{7.14}$$

From the first energy equality (7.12), substituting estimates (7.6) and (7.7), we obtain the following chain of inequalities:

$$\begin{aligned}\frac{1}{2}\frac{d}{dt}\|P^{1/2}u\|^2 &\geq -\|A^{1/2}u\|^2 - \|P^{-\frac{1}{2}}Bu\|\,\|P^{1/2}u\| + 2(\alpha_1 + 1)G(t, u)\\ &\geq 2(\alpha_1 + 1)J(t) + \alpha_1\|A^{1/2}u\|^2 - m_1\|A^{1/2}u\|\,\|P^{1/2}u\| - m_2\|P^{1/2}u\|^2\\ &\geq 2(\alpha_1 + 1)J(t) + \alpha_1\|A^{1/2}u\|^2\\ &\quad - \frac{\alpha_1}{2}\|A^{1/2}u\|^2 - \frac{m_1^2}{2\alpha_1}\|P^{1/2}u\|^2 - m_2\|P^{1/2}u\|^2\\ &\geq 2(\alpha_1 + 1)J(t) - c_1\|P^{1/2}u\|^2.\end{aligned} \tag{7.15}$$

The second energy equality (7.13) for any constant $\varepsilon_{1,2} > 0$ yields the estimate

$$\begin{aligned}\frac{dJ(t)}{dt} &\geq \|P^{1/2}u_t\|^2 - \|P^{-\frac{1}{2}}Bu\|\,\|P^{1/2}u_t\| + G_t(t, u)\\ &\geq (1 - \varepsilon_1)\|P^{1/2}u_t\|^2 - \frac{1}{4\varepsilon_1}\|P^{-\frac{1}{2}}Bu\|^2 + G_t(t, u) \geq (1 - \varepsilon_1)\|P^{1/2}u_t\|^2\\ &\quad - \frac{m_1^2}{4\varepsilon_1}(1 + \varepsilon_2)\|A^{1/2}u\|^2 - \frac{m_2^2}{4\varepsilon_1}\left(1 + \frac{1}{\varepsilon_2}\right)\|P^{1/2}u\|^2 + G_t(t, u).\end{aligned} \tag{7.16}$$

Integrating (7.12) by time and applying (7.7), for any constant $\varepsilon_3 > 0$, we obtain the energy estimate

$$\begin{aligned}\frac{1}{2}\|P^{1/2}u(t)\|^2 + \int_0^t \|A^{1/2}u\|^2\,d\tau &\leq \frac{1}{2}\|P^{1/2}u_0\|^2 + \int_0^t (Bu, u)\,d\tau + \int_0^t (F, u)\,d\tau\\ &\leq \frac{1}{2}\|P^{1/2}u_0\|^2 + \int_0^t \|P^{-\frac{1}{2}}Bu\|\,\|P^{1/2}u\|\,d\tau + \int_0^t (F, u)\,d\tau\end{aligned}$$

$$\leq \frac{1}{2}\|P^{1/2}u_0\|^2 + \varepsilon_3 \int_0^t \|A^{1/2}u\|^2\, d\tau$$

$$+ \frac{m_1^2}{4\varepsilon_3} \int_0^t \|P^{1/2}u\|^2\, d\tau + m_2 \int_0^t \|P^{1/2}u\|^2\, d\tau + \int_0^t (F, u)\, d\tau.$$

Omitting the clearly nonnegative term, we obtain the inequality

$$(1-\varepsilon_3) \int_0^t \|A^{1/2}u\|^2\, d\tau \leq \frac{1}{2}\|P^{1/2}u_0\|^2$$

$$+ \left(\frac{m_1^2}{4\varepsilon_3} + m_2\right) \int_0^t \|P^{1/2}u\|^2\, d\tau + \int_0^t (F, u)\, d\tau. \tag{7.17}$$

Integrating estimate (7.16) by time from 0 to t and applying (7.17), for all $\varepsilon_3 \in (0,1)$, we obtain the expression

$$J(t) \geq J(0) + (1-\varepsilon_1) \int_0^t \|P^{1/2}u_\tau\|^2\, d\tau - \frac{m_1^2}{4\varepsilon_1}(1+\varepsilon_2) \int_0^t \|A^{1/2}u\|^2\, d\tau$$

$$- \frac{m_2^2}{4\varepsilon_1}\left(1 + \frac{1}{\varepsilon_2}\right) \int_0^t \|P^{1/2}u\|^2\, d\tau + \int_0^t G_\tau(\tau, u)\, d\tau$$

$$\geq J(0) + (1-\varepsilon_1) \int_0^t \|P^{1/2}u_\tau\|^2\, d\tau - \frac{m_1^2(1+\varepsilon_2)}{8\varepsilon_1(1-\varepsilon_3)}\|P^{1/2}u_0\|^2$$

$$- \left[\frac{m_1^2(1+\varepsilon_2)}{4\varepsilon_1(1-\varepsilon_3)}\left(\frac{m_1^2}{4\varepsilon_3} + m_2\right) + \frac{m_2^2}{4\varepsilon_1}\left(1 + \frac{1}{\varepsilon_2}\right)\right] \int_0^t \|P^{1/2}u\|^2\, d\tau$$

$$+ \int_0^t \left(G_\tau(\tau, u) - \frac{m_1^2(1+\varepsilon_2)}{4\varepsilon_1(1-\varepsilon_3)}(F, u)\right) d\tau. \tag{7.18}$$

Taking into account condition (7.9), we choose

$$\varepsilon_1 = \frac{\alpha_1 - \beta}{1+\alpha_1}, \quad \varepsilon_2 = \varepsilon_3 = \varepsilon;$$

then we can rewrite (7.18) in the form

$$J(t) \geq \frac{1+\beta}{1+\alpha_1} \int_0^t \|P^{1/2}u_\tau\|^2\, d\tau - m_4 \int_0^t \|P^{1/2}u\|^2\, d\tau + a_0, \tag{7.19}$$

where

$$m_4 = m_3\left(\frac{m_1^2}{4\varepsilon} + m_2\right) + \frac{m_2^2}{4}\frac{1+\alpha_1}{\alpha_1 - \beta}\frac{1+\varepsilon}{\varepsilon}, \quad a_0 = J(0) - \frac{m_3}{2}\|P^{1/2}u_0\|^2.$$

Combining inequalities (7.15) and (7.19), we obtain the estimate

$$\frac{d}{dt}\|P^{1/2}u\|^2 \geq 4(1+\beta)\int_0^t \|P^{1/2}u_\tau\|^2\,d\tau + 4(\alpha_1+1)a_0 - 4(\alpha_1+1)m_4\int_0^t \|P^{1/2}u\|^2\,d\tau - 2c_1\|P^{1/2}u\|^2. \tag{7.20}$$

Using inequality (7.20), we show that the function

$$\Phi(t) = \int_0^t \|P^{1/2}u\|^2\,d\tau + c_3$$

satisfies the conditions of the lemma and tends to infinity in finite time under an appropriate choice of the constant c_3 and the initial value u_0.

Introducing the notation

$$\int_0^t \|P^{1/2}u\|^2\,d\tau = a_1, \qquad \int_0^t \|P^{1/2}u_\tau\|^2\,d\tau = a_2$$

and applying the Newton–Leibnitz formula and the Cauchy inequality, we can easily prove the following inequalities for any $\varepsilon_4 > 0$:

$$\begin{aligned}(\Phi'(t))^2 = \|P^{1/2}u(t)\|^4 &= \left[\|P^{1/2}u_0\|^2 + 2\int_0^t (Pu, u_\tau)\,d\tau\right]^2 \\ &\leq \left[\|P^{1/2}u_0\|^2 + 2\sqrt{a_1a_2}\right]^2 \leq \left(1+\frac{1}{\varepsilon_4}\right)\|P^{1/2}u_0\|^4 + 4(1+\varepsilon_4)a_1a_2.\end{aligned} \tag{7.21}$$

We rewrite inequality (7.20):

$$\Phi'' \geq 4(1+\beta)a_2 + 4(\alpha_1+1)a_0 + 4(\alpha_1+1)m_4(c_3-\Phi) - 2c_1\Phi'.$$

Using (7.21), we prove the following result:

$$\begin{aligned}\Phi''(t)\Phi(t) - (1+\alpha)(\Phi'(t))^2 \geq{}& 4(1+\beta)a_2(a_1+c_3) - 4(1+\alpha)(1+\varepsilon_4)a_1a_2 \\ &+ (4(\alpha_1+1)(a_0 - m_4\Phi + m_4c_3) - 2c_1\Phi')\Phi \\ &- (1+\alpha)\left(1+\frac{1}{\varepsilon_4}\right)\|P^{1/2}u_0\|^4.\end{aligned} \tag{7.22}$$

Choose $\alpha = \varepsilon_4$ such that $(1+\alpha)^2 = 1+\beta$; then in the right-hand side of (7.22), the terms with a_1a_2 are cancelled. Omitting the nonnegative terms in the right-hand side, from (7.22) we obtain the inequality

$$\Phi''(t)\Phi(t) - (1+\alpha)(\Phi'(t))^2 \geq -2c_1\Phi'(t)\Phi(t) - 4(\alpha_1+1)m_4\Phi^2(t)$$
$$+ \left[4(\alpha_1+1)(a_0+m_4c_3)c_3 - \frac{(1+\alpha)^2}{\alpha}\|P^{1/2}u_0\|^4\right]. \quad (7.23)$$

We can apply Lemma 7.1 if the following conditions hold:

$$4(\alpha_1+1)(a_0+m_4c_3)c_3 - \frac{(1+\alpha)^2}{\alpha}\|P^{1/2}u_0\|^4 \geq 0, \quad (7.24)$$

$$\Phi'(0) > \delta\Phi(0), \quad \delta = -\alpha^{-1}\gamma_2 \geq 0. \quad (7.25)$$

The first condition (7.24) is fulfilled if

$$\Phi'(0) = \|P^{1/2}u_0\|^2 > 0, \quad a_0 > 0;$$

then we must take

$$c_3 = \frac{(1+\alpha)^2}{4\alpha(\alpha_1+1)}\frac{\|P^{1/2}u_0\|^4}{a_0}.$$

The second condition will hold if we impose the following restriction to the initial data:

$$\delta\Phi(0) = \delta c_3 = \frac{\delta(1+\alpha)^2}{4\alpha(\alpha_1+1)}\frac{\|P^{1/2}u_0\|^4}{a_0} < \|P^{1/2}u_0\|^2 = \Phi'(0).$$

Now the blow-up result is a direct consequence of Lemma 7.1. The theorem is proved. □

Remark 7.1. Note that the operator A may depend on time, but it must be differentiable:

$$|(A_t u, u)| \leq m_1'\|A^{1/2}u\|^2 + m_2'\|P^{1/2}u\|^2. \quad (7.26)$$

Indeed, the dependence of $A(t)$ on time affects only the right-hand side of (7.13), which contains the term $(A_t u, u)/2$. In turn, this leads to the appearance of additional terms $m_1'/2$ and $m_2'/2$ in the coefficients of $\|A^{1/2}u\|^2$ and $\|P^{1/2}u\|^2$ in inequality (7.16). The operator B may also depend on time, but it does not require any additional conditions.

As for the nonlinear operator, condition (7.9) can be weakened. In particular, we can prove an assertion similar to Theorem 7.1, in which restriction (7.9) is replaced by the following two conditions:

$$G_t(t,u) \geq 0 \quad \text{and} \quad 2\alpha_1\lambda G(t,u) \geq m_3(F(t,u),u). \quad (7.27)$$

The statement and the proof of this theorem can be found in [37], but the interested reader can prove this assertion applying Theorem 7.1 to the function $v(t) = u(t)\exp(-\lambda t)$.

7.3 Blow-up for hyperbolic equations

In this section, we consider the equation of the form

$$Pu_{tt} = -Au + Bu - aPu_t + F(t,u), \tag{7.28}$$

where the operators P and A are linear and symmetric: $P > 0$, $A \geq 0$, the real number $a \geq 0$ is nonnegative, and the operator B (possibly, nonlinear) is subordinated to the operators A and P:

$$\|P^{-1/2}Bu\| \leq m_1\|A^{1/2}u\|. \tag{7.29}$$

The nonlinear term $F(t,u)$ must satisfy the same conditions as for parabolic equations. Namely, we assume that there exists a potential $G(t,u)$ such that

$$(F(t,u),u) \geq 2(1+2\alpha)G(t,u), \quad \alpha > 0. \tag{7.30}$$

Moreover, we impose the condition

$$G_t(t,u) \geq m_1 G(t,u). \tag{7.31}$$

Then the following blow-up result holds.

Theorem 7.2. *Let all conditions listed be fulfilled, and, moreover, let the initial data u_0 satisfy the inequalities*

$$-\frac{1}{2}\|A^{1/2}u(0)\|^2 - \frac{1}{2}\|P^{1/2}u_t(0)\|^2 + G(0,u(0)) \geq 0, \tag{7.32}$$

$$2(Pu(0),u_t(0)) > \frac{\alpha}{4}\left[a + \sqrt{a^2+2m_1^2}\right]\|P^{1/2}u(0)\|^2. \tag{7.33}$$

Then the solution $u(x,t)$ of the Cauchy problem for equation (7.28) *does not exist globally in time:*

$$\|P^{1/2}u(t)\| \to \infty \quad \text{as } t \to T_0 \leq T_1,$$

$$T_1 = \frac{2}{\sqrt{a^2+2m_1^2}} \ln \frac{-\frac{1}{4}(a-\sqrt{a^2+2m_1^2})\Phi(0) + \alpha\Phi'(0)}{-\frac{1}{4}(a+\sqrt{a^2+2m_1^2})\Phi(0) + \alpha\Phi'(0)}, \tag{7.34}$$

$$\Phi(0) \equiv \|P^{1/2}u(0)\|^2, \quad \Phi'(0) \equiv 2(Pu(0),u_t(0)).$$

Proof. We prove that the function $\Phi(t) = \|P^{1/2}u(t)\|^2$ satisfies all the conditions of Lemma 7.1. Twice differentiating $\Phi(t)$ by time, we obtain

$$\Phi'(t) = 2(P^{1/2}u, P^{1/2}u_t), \quad \Phi''(t) = 2\|P^{1/2}u_t\|^2 + 2(Pu_{tt},u).$$

Combining the first and second derivatives, we have

$$\Phi''\Phi - (1+\alpha)(\Phi'(t))^2 = 4(1+\alpha)[\|P^{1/2}u_t\|^2\|P^{1/2}u\|^2 - (P^{1/2}u, P^{1/2}u_t)^2] + 2\Phi[(Pu_{tt}, u) - (1+2\alpha)\|P^{1/2}u_t\|^2] \geq 2\Phi R(t), \tag{7.35}$$

where

$$R(t) \equiv (Pu_{tt}, u) - (1+2\alpha)\|P^{1/2}u_t\|^2.$$

Substituting into $R(t)$ the expression for the second derivative from (7.28) and applying conditions (7.29) and (7.30) and the equality

$$a(Pu_t, u) = \frac{a}{2}\frac{d}{dt}\|P^{1/2}u\|^2,$$

we estimate $R(t)$ from below:

$$\begin{aligned} R(t) &= -\|A^{1/2}u\|^2 + (P^{-\frac{1}{2}}Bu, P^{1/2}u) - a(Pu_t, u) + (F(t,u), u) \\ &\quad - (1+2\alpha)\|P^{1/2}u_t\|^2 \geq 2(1+2\alpha)\left[-\frac{1}{2}\|A^{1/2}u\|^2 - \frac{1}{2}\|P^{1/2}u_t\|^2 + G(t,u)\right] \\ &\quad + 2\alpha\|A^{1/2}u\|^2 - m_1\|A^{1/2}u\|\,\|P^{1/2}u\| - \frac{a}{2}\frac{d}{dt}\|P^{1/2}u\|^2. \end{aligned} \tag{7.36}$$

Denoting

$$J(t) = -\frac{1}{2}\|A^{1/2}u\|^2 - \frac{1}{2}\|P^{1/2}u_t\|^2 + G(t,u),$$

we continue the estimate:

$$R(t) \geq 2(1+2\alpha)J(t) + 2\alpha\|A^{1/2}u\|^2 - \varepsilon_1\|A^{1/2}u\|^2 - \frac{m_1^2}{4\varepsilon_1}\|P^{1/2}u\|^2 - \frac{a}{2}\frac{d}{dt}\|P^{1/2}u\|^2.$$

If we take $\varepsilon_1 = 2\alpha$, then the lower estimate takes the form

$$R(t) \geq 2(1+2\alpha)J(t) - \frac{m_1^2}{8\alpha}\|P^{1/2}u\|^2 - \frac{a}{2}\frac{d}{dt}\|P^{1/2}u\|^2. \tag{7.37}$$

Multiplying equation (7.28) by u_t and integrating over the whole space, we have

$$\begin{aligned} &\frac{d}{dt}\left(-\frac{1}{2}\|P^{1/2}u\|^2 - \frac{1}{2}\|A^{1/2}u\|^2 + G(t,u)\right) \\ &\quad = -(P^{-\frac{1}{2}}Bu, P^{1/2}u_t) + a\|P^{1/2}u_t\|^2 + G_t(t,u) \\ &\quad \geq -m_1\|A^{1/2}u\|\,\|P^{1/2}u_t\| + G_t(t,u) \\ &\quad \geq -\frac{m_1}{2}\|A^{1/2}u\|^2 - \frac{m_1}{2}\|P^{1/2}u_t\|^2 + G_t(t,u), \end{aligned} \tag{7.38}$$

where we have used estimate (7.6). Therefore, by the definition of $J(t)$, from condition (7.31) we see that

$$\frac{dJ(t)}{dt} \geq m_1 J(t) + G_t(t,u) - m_1 G(t,u) \geq m_1 J(t). \tag{7.39}$$

Since, by the condition of the theorem $J(0) > 0$, we obtain $J(t) > 0$. Then from (7.37) we conclude that

$$R(t) \geq -\frac{m_1^2}{8\alpha}\Phi(t) - \frac{a}{2}\Phi'(t).$$

Therefore, we can continue the chain (7.35):

$$\Phi''\Phi - (1+\alpha)(\Phi')^2 \geq -\frac{m_1^2}{8\alpha}\Phi^2 - \frac{a}{2}\Phi'\Phi. \tag{7.40}$$

Thus, to complete the proof of the theorem, it suffices to apply Lemma 7.1, which is possible due to assumption (7.33). The theorem is proved. □

Remark 7.2. As in the parabolic case, Theorem 7.2 admits a generalization. For example, the operators A and B and the coefficient a may depend on time. In this case, $A(t)$ must be differentiable with respect to t and satisfy the inequality

$$|(A_t u, u)| \leq m_1^* \|A^{1/2}u\|^2,$$

and the coefficient $a(t)$ must be nonnegative for all t and bounded from above by a constant $a = \text{const} < \infty$. As for condition (7.31) for the nonlinear operator, it can also be replaced by the weaker inequality $G_t(t,u) \geq 0$; this can be easily proved by considering the corresponding problem for the function $v(t) = u(t)\exp(-\lambda t)$.

7.4 Examples

One of the most important advantages of the energy method proposed by Kalantarov and Ladyzhenskaya is that this method allows one to prove the blow-up for equations and systems that were considered “completely integrable.” In particular, the blow-up of solutions was proved for equations that possess Lax L-M pairs and multisoliton solutions and Cauchy problems for which they are expected to be globally solvable.

Example 7.1. Zakharov [111] considered the equation of nonlinear string

$$u_{tt} = u_{xx} - c_1 u_{xxxx} + c_2(u^2)_{xx}, \quad c_1 \geq 0, \tag{7.41}$$

and proved that equation (7.41) has a Lax L-M pair and can be examined as the operator equation $L_t = [L, M]$, which, however, does not guarantee the existence of a global-in-time solution. If $c_1 > 0$, then the corresponding problem for the hyperbolic equation is well-posed: it has a solution for small times. We show that this conclusion is invalid "in the whole."

Consider the initial-boundary-value problem for equation (7.41) on the segment $x \in [0,1]$ with the following conditions:

$$u|_{t=0} = u_0(x), \quad u_t|_{t=0} = u_1(x), \tag{7.42}$$

$$u|_{x=0} = u|_{x=1} = u_{xx}|_{x=0} = u_{xx}|_{x=1} = 0, \tag{7.43}$$

where m is an integer number, and the constant c_2 is negative for odd m and positive otherwise.

It is well known that, for the boundary conditions (7.43), the operator $-\partial^2_{xx}$ is invertible and positive definite. We denote by P the inverse operator: $P(u_{xx}) = -u$, $P(u_{xxxx}) = -u_{xx}$. Then problem (7.41)–(7.43) can be written in the following equivalent form:

$$Pu_{tt} = -u + c_1 u_{xx} - c_2 u^m. \tag{7.44}$$

Theorem 7.2 can be immediately applied to equation (7.44), where

$$A = -c_1\partial^2_{xx} + I, \quad B = 0, \quad a = 0, \quad F(t,u) = -c_2 u^m;$$

it guarantees the blow-up of a solution in finite time. Moreover, we can prove the blow-up of the solution of the Cauchy problem, in which the operator $-\partial^2_{xx}$ is positive but not positive definite.

Example 7.2. Another interesting example is the following nonlinear partial differential equation:

$$u_{tt} = \pm(u_x^m)_x. \tag{7.45}$$

For any m, it has an infinite number of conservation laws. This can be easily verified by using the integral

$$\int (u_{tt} \mp (u_x)_x^m) u_t^k \, dx = 0, \quad k = 1, 2, \ldots.$$

We present an explicit form of the first two integrals:

$$I_1 = \int \left[\frac{1}{2}u_t^2 \pm \frac{1}{m+1} u_x^{m+1}\right] dx, \quad I_2 = \int \left[\frac{1}{3}u_t^3 \pm \frac{2}{m+1} u_t u_x^{m+1}\right] dx.$$

Despite an infinite number of conservation laws, the blow-up of solutions occurs for $m > 2$ in the Cauchy problem and in initial-boundary-value problems.

To prove the blow-up, it suffices to note that Theorems 7.1 and 7.2 remain valid for the nonlinearity of the form

$$F^*(t,u) = D^* F(t, Du),$$

where D is a linear operator, D^* is its conjugate, and $F(t,v)$ satisfies the same conditions as $F(t,v)$ in Theorems 7.1 and 7.2. Indeed, the conditions for P, A, and B and the conditions for F are related only by the constant m_3, whereas the conditions for F contain only the expressions $(F(t,u),u)$ and $G(t,u)$. For the nonlinearity of the form $F^*(t,u)$, the following equalities hold:

$$(F^*(t,u),u) = (F(t,v),v), \quad v = Du, \quad G^*(t,u) = G(t,Du) = G(t,v),$$

since

$$\begin{aligned}\frac{d}{d\tau}G^*(t,u(\tau)) &= \frac{d}{d\tau}G(t,Du(\tau)) = (F(t,Du(\tau)),Du_\tau(\tau))\\ &= (D^*F(t,Du(\tau)),u_\tau(\tau)) = (F^*(t,u(\tau)),u_\tau(\tau)).\end{aligned}$$

This implies that $F^*(t,u)$ and $G^*(t,u)$ satisfy the conditions of Theorems 7.1 and 7.2 if $F(t,v)$ and $G(t,v)$ satisfy these conditions for $v = Du(t)$.

Thus, the occurrence of the blow-up is a consequence of the fact that (7.45) is a particular case of the equation

$$Pu_{tt} = -Au - Bu - aPu_t + F^*(t,u),$$

where $P = I$, $A = B = 0$, $D = \partial_x$, $D^* = -\partial_x$, and $F(t,v) = \mp v^m$.

Example 7.3. Another particular case of this equation is, for example, the Zabusky equation

$$u_{tt} = -u_{xxxx} - u_{xx} \mp (u_x^2)_x. \tag{7.46}$$

It satisfies all the conditions of the blow-up theorem for

$$P = I, \quad A = I + \partial^4_{xxxx}, \quad B = I - \partial^2_{xx}, \quad D = \partial_x, \quad D^* = -\partial_x.$$

Example 7.4. The blow-up of solutions of the Boussinesq equation

$$u_{tt} = -u_{xxxx} - u_{xx} - u + u^3 \tag{7.47}$$

occurs since

$$P = I, \quad A = I + \partial^4_{xxxx}, \quad B = -\partial^2_{xx}, \quad F(t,u) = u^3.$$

For the continual analog of the Fermi–Pasta–Ulam chain

$$\xi_n'' = \xi_{n+1} - 2\xi_n + \xi_{n-1} + (\xi_{n+1} - \xi_n)^2 - (\xi_n - \xi_{n-1})^2,$$

that is, for the equation

$$u_{tt} = u_{xx} + (u_x^2)_x$$

with an infinite number of conservation laws, the blow-up of solutions in finite time also occurs since, in this case,

$$A = -\partial^2_{xx}, \quad B = 0, \quad F^* = D^*F(Du), \quad F(t,v) = -v^2, \quad D = \partial_x.$$

7.5 Bibliographical notes

Other problems that admit the study of the blow-up by the method of Kalantarov and Ladyzhenskaya and certain interesting examples on the global insolvability can be found in [37].

One of the advantages of the energy method of Kalantarov and Ladyzhenskaya is that this method, in contrast to the classical method of Levine, allows us to take into account subordinated nonpotential terms in an equation, for example, for the following equation that appears in the theory of quasi-stationary processes in semiconductors (see, e. g., [1]):

$$\frac{\partial}{\partial t}(\Delta u - u) + \Delta u + c\frac{\partial |u|^p}{\partial x} + |u|^{q-1}u = 0, \quad q > 1,\ p > 0,\ c \neq 0.$$

Certainly, this method is applicable to a wide class of problems for equations that appear in the theory of nonlinear waves in dispersive media (see, e. g., [1]):

$$u_{tt} + \Delta^2 u - \Delta u + \Delta f(x,u) = 0, \quad u|_{\partial\Omega} = \Delta u|_{\partial\Omega} = 0,$$

where the function $f(x,u)$ satisfies, for some $\theta > 2$, the condition

$$\int_\Omega f(x,v)v\,dx \geq \theta \int_\Omega \int_0^v f(x,s)\,ds\,dx$$

for all $v \in \mathbb{H}^2(\Omega) \cap \mathbb{H}_0^1(\Omega)$, and the smooth domain Ω is bounded in $\mathbb{R}^N$.

8 Energy method of M. O. Korpusov and A. G. Sveshnikov

In this chapter, we consider a modification of the classical energy method of H. A. Levine, developed by M. O. Korpusov and A. G. Sveshnikov in [44–48] (see also [1]). As compared with Levine's method, the method proposed has a specific range of applicability.

8.1 Parabolic equations with double nonlinearities

8.1.1 Introduction

In this section, we consider initial-boundary-value problems of the following form:

$$\frac{\partial \varphi(x,u)}{\partial t} + A(u) = 0, \quad u|_{\partial\Omega} = 0, \quad u(x,0) = u_0(x), \tag{8.1}$$

where $\Omega \subset \mathbb{R}^N$ is a bounded domain with sufficiently smooth boundary. The function $\varphi(x,u)$ is of the form

$$\varphi(x,u) = u + \sum_{k=1}^{n} a_k(x)|u|^{p_k-2}u \tag{8.2}$$

and satisfies the following conditions.

Conditions for $\varphi(x,s)$

(i)$_1$ $a_k(x) \geq 0$ for almost all $x \in \Omega$; moreover, $a_k(x) \in \mathbb{L}^\infty(\Omega)$;
(ii)$_1$ $p_k > 2$ for all $k = \overline{1,n}$.

The operator $A(u)$ is a nonlinear elliptic operator of the form to be specified later. Note that equations of the form (8.1) and (8.2) are called parabolic equations with double nonlinearities.

8.1.2 Problem of the combustion theory

Consider one problem of the combustion theory (see [94]), namely the problem with double nonlinearity of the following form:

$$\frac{\partial \varphi(x,u)}{\partial t} - \Delta(|u|^{q_1}u) = |u|^{q_2}u, \tag{8.3}$$

https://doi.org/10.1515/9783110602074-008

$$\varphi(x,u) = u + \sum_{l=1}^{n} a_l(x)|u|^{p_l-2}u, \tag{8.4}$$

$$u|_{\partial\Omega} = 0, \quad u(x,0) = u_0(x), \tag{8.5}$$

where $\Omega \subset \mathbb{R}^N$ is a bounded domain with smooth boundary $\partial\Omega \in \mathbb{C}^{2,\delta}$, where $\delta \in (0,1]$ and $q_1, q_2 > 0$.

We will treat solutions of problem (8.3)–(8.5) in the classical sense

$$u(x)(t) \in \mathbb{C}^{(1)}([0,T_0); \mathbb{C}^{(2)}(\overline{\Omega})) \quad \text{for some } T_0 > 0.$$

First, we assume that there exists some maximal $T_0 > 0$ for which there exists a nonextendable classical solution of problem (8.3)–(8.5).

Introduce the following notation:

$$\begin{aligned}\Phi(t) \equiv{}& \frac{1}{q_1+2}\int_0^t \|u\|_{q_1+2}^{q_1+2}\,ds + \sum_{l=1}^{n}\frac{p_l-1}{q_1+p_l}\int_0^t\int_\Omega a_l(x)|u|^{p_l+q_1}\,dx\,ds \\ &+ \frac{1}{(q_1+2)(q_1+p_{k_0})}\|u_0\|_{q_1+2}^{q_1+2} \\ &+ \sum_{l=1}^{n}\frac{p_l-1}{(q_1+p_l)(q_1+p_{k_0})}\int_\Omega a_l(x)|u_0|^{p_l+q_1}\,dx,\end{aligned} \tag{8.6}$$

$$\begin{aligned}\mathrm{J}(t) \equiv{}& \int_0^t\int_\Omega |u|^{q_1}(u')^2\,dx\,ds + \sum_{l=1}^{n}(p_l-1)\int_0^t\int_\Omega a_l(x)|u|^{q_1+p_l-2}(u')^2\,dx\,ds \\ &+ \frac{1}{q_1+2}\|u_0\|_{q_1+2}^{q_1+2} + \sum_{l=1}^{n}\frac{p_l-1}{q_1+p_l}\int_\Omega a_l(x)|u_0|^{p_l+q_1}\,dx.\end{aligned} \tag{8.7}$$

Lemma 8.1. *For all $t \in [0,T_0)$, we have the inequality*

$$(\Phi')^2 \le (q_1+p_{k_0})\Phi\mathrm{J}, \quad p_{k_0} = \max_{k=\overline{1,n}} p_k. \tag{8.8}$$

Proof. We have

$$\begin{aligned}\Phi' ={}& \frac{1}{q_1+2}\|u\|_{q_1+2}^{q_1+2} + \sum_{l=1}^{n}\frac{p_l-1}{q_1+p_l}\int_\Omega a_l(x)|u|^{p_l+q_1}\,dx \\ ={}& \frac{1}{q_1+2}\int_0^t \frac{d}{ds}\|u\|_{q_1+2}^{q_1+2}(s)\,ds + \sum_{l=1}^{n}\frac{p_l-1}{q_1+p_l}\int_0^t\frac{d}{ds}\int_\Omega a_l(x)|u|^{p_l+q_1}\,dx\,ds \\ &+ \frac{1}{q_1+2}\|u_0\|_{q_1+2}^{q_1+2} + \sum_{l=1}^{n}\frac{p_l-1}{q_1+p_l}\int_\Omega a_l(x)|u_0|^{q_1+p_l}\,dx.\end{aligned} \tag{8.9}$$

We have the following estimates:

$$\left|\frac{1}{q_1+2}\int_0^t \frac{d}{ds}\|u\|_{q_1+2}^{q_1+2}(s)\,ds\right| \le \int_0^t\int_\Omega |u|^{q_1+1}|u'|\,dx\,ds$$
$$\le \int_0^t \|u\|_{q_1+2}^{(q_1+2)/2}\Big(\int_\Omega |u|^{q_1}|u'|^2\,dx\Big)^{1/2} ds$$
$$\le \Big(\int_0^t \|u\|_{q_1+2}^{q_1+2}\,ds\Big)^{1/2}\Big(\int_0^t\int_\Omega |u|^{q_1}|u'|^2\,dx\,ds\Big)^{1/2}, \tag{8.10}$$

$$\left|\frac{p_l-1}{q_1+p_l}\int_0^t \frac{d}{ds}\int_\Omega a_l(x)|u|^{q_1+p_l}\,dx\,ds\right|$$
$$\le (p_l-1)\int_0^t\int_\Omega a_l(x)|u|^{q_1+p_l-2}|u||u'|\,dx\,ds$$
$$\le (p_l-1)\int_0^t\Big(\int_\Omega a_l(x)|u|^{q_1+p_l}\,dx\Big)^{1/2}\Big(\int_\Omega a_l(x)|u|^{q_1+p_l-2}|u'|^2\,dx\Big)^{1/2} ds$$
$$\le (p_l-1)\Big(\int_0^t\int_\Omega a_l(x)|u|^{q_1+p_l-2}|u'|^2\,dx\,ds\Big)^{1/2}\Big(\int_0^t\int_\Omega a_l(x)|u|^{q_1+p_l}\,dx\,ds\Big)^{1/2}. \tag{8.11}$$

Thus, from equation (8.9), in view of inequalities (8.10) and (8.11), we obtain the estimate

$$(\Phi')^2 \le \Bigg[\Big(\int_0^t \|u\|_{q_1+2}^{q_1+2}\,ds\Big)^{1/2}\Big(\int_0^t\int_\Omega |u|^{q_1}|u'|^2\,dx\,ds\Big)^{1/2}$$
$$+\sum_{l=1}^n (p_l-1)\Big(\int_0^t\int_\Omega a_l(x)|u|^{q_1+p_l-2}|u'|^2\,dx\,ds\Big)^{1/2}$$
$$\times\Big(\int_0^t\int_\Omega a_l(x)|u|^{q_1+p_l}\,dx\,ds\Big)^{1/2}$$
$$+\frac{1}{q_1+2}\|u_0\|_{q_1+2}^{q_1+2}+\sum_{l=1}^n\frac{p_l-1}{q_1+p_l}\int_\Omega a_l(x)|u_0|^{q_1+p_l}\,dx\Bigg]^2. \tag{8.12}$$

Squaring the right-hand side, grouping factors appropriately, and using the inequality $2ab \le a^2+b^2$, we obtain the inequality

$$(\Phi')^2 \le \Big(\int_0^t \|u\|_{q_1+2}^{q_1+2}\,ds + \sum_{l=1}^n (p_l-1)\int_0^t\int_\Omega a_l(x)|u|^{q_1+p_l}\,dx\,ds$$

$$+\frac{1}{q_1+2}\|u_0\|_{q_1+2}^{q_1+2}+\sum_{l=1}^{n}\frac{p_l-1}{q_1+p_l}\int_{\Omega}a_l(x)|u_0|^{q_1+p_l}\,dx\Bigg)$$

$$\times\Bigg(\int_0^t\int_{\Omega}|u|^{q_1}|u'|^2\,dx\,ds+\sum_{l=1}^{n}(p_l-1)\int_0^t\int_{\Omega}a_l(x)|u|^{q_1+p_l-2}|u'|^2\,dx\,ds$$

$$+\frac{1}{q_1+2}\|u_0\|_{q_1+2}^{q_1+2}+\sum_{l=1}^{n}\frac{p_l-1}{q_1+p_l}\int_{\Omega}a_l(x)|u_0|^{q_1+p_l}\,dx\Bigg). \tag{8.13}$$

Using the notation p_{k_0} introduced in the lemma and taking into account (8.6) and (8.7), from inequality (8.13) we obtain

$$\begin{aligned}(\Phi')^2 &\le (q_1+p_{k_0})\Bigg(\frac{1}{q_1+p_{k_0}}\int_0^t\|u\|_{q_1+2}^{q_1+2}\,ds+\sum_{l=1}^{n}\frac{p_l-1}{q_1+p_{k_0}}\int_0^t\int_{\Omega}a_l(x)|u|^{q_1+p_l}\,dx\,ds\\
&\quad+\frac{1}{(q_1+2)(q_1+p_{k_0})}\|u_0\|_{q_1+2}^{q_1+2}\\
&\quad+\sum_{l=1}^{n}\frac{p_l-1}{(q_1+p_{k_0})(q_1+p_l)}\int_{\Omega}a_l(x)|u_0|^{q_1+p_l}\,dx\Bigg)\mathrm{J}\\
&\le (q_1+p_{k_0})\Bigg(\frac{1}{q_1+2}\int_0^t\|u\|_{q_1+2}^{q_1+2}\,ds+\sum_{l=1}^{n}\frac{p_l-1}{q_1+p_l}\int_0^t\int_{\Omega}a_l(x)|u|^{q_1+p_l}\,dx\,ds\\
&\quad+\frac{1}{(q_1+2)(q_1+p_{k_0})}\|u_0\|_{q_1+2}^{q_1+2}\\
&\quad+\sum_{l=1}^{n}\frac{p_l-1}{(q_1+p_{k_0})(q_1+p_l)}\int_{\Omega}a_l(x)|u_0|^{q_1+p_l}\,dx\Bigg)\mathrm{J}\\
&=(q_1+p_{k_0})\Phi\mathrm{J}.\end{aligned} \tag{8.14}$$

The proof is complete. □

Now we obtain the energy relations. First, we multiply both sides of equation (8.3) by $|u|^{q_1}u$ and integrate over the domain Ω. Integrating by parts, we obtain the first energy equality

$$\frac{d^2\Phi}{dt^2}+\int_{\Omega}\left|\nabla(|u|^{q_1}u)\right|^2dx=\|u\|_{q_1+q_2+2}^{q_1+q_2+2}. \tag{8.15}$$

To obtain the second energy relation, we multiply both sides of equation (8.3) by $(|u|^{q_1}u)'$ and again integrate over the domain Ω. Integrating by parts, we obtain

$$(q_1+1)\frac{d\mathrm{J}}{dt}+\frac{1}{2}\frac{d}{dt}\int_{\Omega}\left|\nabla(|u|^{q_1}u)\right|^2dx=\frac{q_1+1}{q_1+q_2+2}\frac{d}{dt}\|u\|_{q_1+q_2+2}^{q_1+q_2+2}, \tag{8.16}$$

which implies

$$(q_1+q_2+2)\frac{dJ}{dt}+\frac{q_1+q_2+2}{2q_1+2}\frac{d}{dt}\int\limits_{\Omega}\left|\nabla(|u|^{q_1}u)\right|^2dx=\frac{d}{dt}\|u\|_{q_1+q_2+2}^{q_1+q_2+2}. \tag{8.17}$$

Integrating both sides of (8.17) by time, we arrive at the equality

$$(q_1+q_2+2)J+\frac{q_1+q_2+2}{2q_1+2}\int\limits_{\Omega}\left|\nabla(|u|^{q_1}u)\right|^2dx-E(0)=\|u\|_{q_1+q_2+2}^{q_1+q_2+2}, \tag{8.18}$$

where the integration constant is

$$\begin{aligned}E(0)&\equiv(q_1+q_2+2)\left(\frac{1}{q_1+2}\|u_0\|_{q_1+2}^{q_1+2}+\sum_{l=1}^{n}\frac{p_l-1}{q_1+p_l}\int\limits_{\Omega}a_l(x)|u_0|^{p_l+q_1}\,dx\right)\\&\quad+\frac{q_1+q_2+2}{2q_1+2}\int\limits_{\Omega}\left|\nabla(|u_0|^{q_1}u_0)\right|^2dx-\|u_0\|_{q_1+q_2+2}^{q_1+q_2+2}.\end{aligned} \tag{8.19}$$

Using (8.18), from (8.15) we finally obtain the relation

$$\frac{d^2\Phi}{dt^2}+E(0)=(q_1+q_2+2)J+\left(\frac{q_1+q_2+2}{2q_1+2}-1\right)\int\limits_{\Omega}\left|\nabla(|u|^{q_1}u)\right|^2dx. \tag{8.20}$$

Now we assume that $q_2\geq q_1$. Then (8.20) can be simplified:

$$\frac{d^2\Phi}{dt^2}+E(0)\geq(q_1+q_2+2)J. \tag{8.21}$$

Combining (8.8) and (8.21), we obtain the required differential inequality

$$\Phi\Phi''-\alpha(\Phi')^2+\beta\Phi\geq 0, \tag{8.22}$$

where

$$\alpha=\frac{q_1+q_2+2}{q_1+p_{k_0}},\quad \beta=E(0). \tag{8.23}$$

This differential inequality if a fundamental tool in the energy method proposed: it is discussed in detail in the Appendix (see Section A.1). We use Theorem A.1 (see p. 299), which states sufficient conditions of blow-up and estimates of the lifetime of a solution.

Assume that

$$q_2+2>p_{k_0}\quad\Rightarrow\quad\alpha>1.$$

We must consider the cases $E(0)>0$ and $E(0)\leq 0$ separately.

Let $E(0) > 0$. Applying Theorem A.1 (see the Appendix, Section A.1, $\gamma = 0$), we obtain that under the conditions

$$\Phi(0) > 0, \quad \Phi'(0) > \left(\frac{2\beta}{2\alpha - 1}\Phi(0)\right)^{1/2} > 0, \tag{8.24}$$

blow-up in finite time occurs. Obviously, the first inequality in (8.24) is fulfilled for the nontrivial solutions. The second inequality can be transformed to the form

$$\begin{aligned} \|u_0\|_{q_1+q_2+2}^{q_1+q_2+2} > \frac{q_1+q_2+2}{2q_1+2} \int_\Omega |\nabla(|u_0|^{q_1}u_0)|^2\,dx \\ + \frac{q_1+p_{k_0}}{2}\left(\frac{1}{q_1+2}\|u_0\|_{q_1+2}^{q_1+2} + \sum_{l=1}^{n}\frac{p_l-1}{q_1+p_l}\int_\Omega a_l(x)|u_0|^{p_l+q_1}\,dx\right). \end{aligned} \tag{8.25}$$

The case $E(0) \le 0$ is much easier: the blow-up result follows from the same formulas of the Appendix, in which we must set $\beta = 0$.

The final result can be stated as the following:

Theorem 8.1. *Assume that the conditions for the function $\varphi(x, s)$ are fulfilled. If the initial function satisfies the condition*

$$\begin{aligned} \|u_0\|_{q_1+q_2+2}^{q_1+q_2+2} > \frac{q_1+q_2+2}{2q_1+2} \int_\Omega |\nabla(|u_0|^{q_1}u_0)|^2\,dx \\ + \frac{q_1+p_{k_0}}{2}\left(\frac{1}{q_1+2}\|u_0\|_{q_1+2}^{q_1+2} + \sum_{l=1}^{n}\frac{p_l-1}{q_1+p_l}\int_\Omega a_l(x)|u_0|^{p_l+q_1}\,dx\right) \end{aligned}$$

and the conditions

$$q_2 \ge q_1, \quad q_2 + 2 > p_{k_0},$$

then $T_0 < +\infty$; more precisely,

$$T \le T_0 \le T_\infty = \Phi^{1-\alpha}(0)A^{-1},$$

and the following limit relation holds:

$$\lim_{t \uparrow T_0} \Phi(t) = +\infty,$$

where

$$A^2 = (\alpha-1)^2\Phi^{-2\alpha}(0)\left[(\Phi'(0))^2 - \frac{2\beta}{2\alpha-1}\Phi(0)\right],$$

$$\alpha = \frac{q_1+q_2+2}{q_1+p_{k_0}}, \quad \beta = \begin{cases} E(0) & \text{if } E(0) > 0, \\ 0 & \text{if } E(0) \le 0, \end{cases}$$

$$E(0) \equiv (q_1 + q_2 + 2)\left(\frac{1}{q_1 + 2}\|u_0\|_{q_1+2}^{q_1+2} + \sum_{l=1}^{n} \frac{p_l - 1}{q_1 + p_l} \int_\Omega a_l(x)|u_0|^{p_l+q_1}\, dx \right)$$
$$+ \frac{q_1 + q_2 + 2}{2q_1 + 2} \int_\Omega |\nabla(|u_0|^{q_1}u_0)|^2\, dx - \|u_0\|_{q_1+q_2+2}^{q_1+q_2+2}.$$

Thus, we have obtained an explicit estimate of the lifetime of a classical solution of the problem considered. Now we introduce the notion of a weak generalized solution and verify that the energy method of Korpusov and Sveshnikov allows us to prove the blow-up for weak solutions.

8.1.3 Second-order problem of the general form

Consider the question on sufficient conditions of blow-up of a weak solution of the following initial-boundary-value problem:

$$\frac{\partial \varphi(x,u)}{\partial t} - \operatorname{div}(h(x,|\nabla u|)\nabla u) + g(x,u) = f(x,u), \tag{8.26}$$

$$\varphi(x,u) = u + \sum_{k=1}^{n} a_k(x)|u|^{p_k-2}u, \tag{8.27}$$

$$u|_{\partial\Omega} = 0, \quad u(x,0) = u_0(x), \tag{8.28}$$

where $x \in \Omega \subset \mathbb{R}^N$, and Ω is a bounded domain with smooth boundary $\partial\Omega \in \mathbb{C}^{2,\delta}$, $\delta \in (0,1]$. We introduce the conditions for the functions $\varphi(x,s)$, $h(x,s)$, $g(x,s)$, and $f(x,s)$.

Conditions for the function $\varphi(x,u)$

$(\text{i})_1$ $a_k(x) \geq 0$ for almost all $x \in \Omega$; moreover, $a_k(x) \in \mathbb{L}^\infty(\Omega)$;
$(\text{ii})_1$ $p_{k_0} = \max_{k\in\overline{1,n}} p_k < p^*$, and, in addition, $a_{k_0}(x) \geq a_0 > 0$;
$(\text{iii})_1$ $p_k > 2$ for all $k = \overline{1,n}$.

Conditions for the function $h(x,s)$

$(\text{i})_2$ $h(x,s) : \Omega \times \mathbb{R}^1_+ \to \mathbb{R}^1$ is a Carathéodori function;
$(\text{ii})_2$ for almost all $x \in \Omega$,

$$c_2 s^{p-2} \leq h(x,s) \leq c_1 + c_2 s^{p-2}, \quad p \geq 2; \tag{8.29}$$

$(\text{iii})_2$ for any $v(x) \in \mathbb{W}^{1,p}_0(\Omega)$,

$$0 \leq \int_\Omega h(x,|\nabla v(x)|)|\nabla v(x)|^2\, dx \leq \theta_1 \int_\Omega \mathcal{H}(x,|\nabla v(x)|)\, dx \tag{8.30}$$

for some positive constant $\theta_1 > 0$, where

$$\mathcal{H}(x,s) = \int_0^s h(x,\sigma)\sigma\, d\sigma.$$

Conditions for the function $g(x,s)$

(i)$_3$ $g(x,s) : \Omega \times \mathbb{R}^1 \to \mathbb{R}^1$ is a Carathéodori function;
(ii)$_3$ for almost all $x \in \Omega$,

$$|g(x,s)| \le c_4 + c_5|s|^{q_1+1}, \quad q_1 \in [0, p^* - 2); \tag{8.31}$$

(iii)$_3$ for any $v(x) \in \mathbb{L}^{q_1+2}(\Omega)$,

$$0 \le \int_\Omega v(x) g(x, v(x))\, dx \le \theta_2 \int_\Omega \mathcal{G}(x, v(x))\, dx \tag{8.32}$$

for some positive constant $\theta_2 > 0$, where

$$\mathcal{G}(x,s) = \int_0^s g(x,\sigma)\, d\sigma.$$

Conditions for the function $f(x,s)$

(i)$_4$ $f(x,s) : \Omega \times \mathbb{R}^1 \to \mathbb{R}^1$ is a Carathéodori function;
(ii)$_4$ for almost all $x \in \Omega$,

$$|f(x,s)| \le c_6 + c_7|s|^{q_2+1}, \quad q_2 \in (0, p^* - 2); \tag{8.33}$$

(iii)$_4$ for any $v(x) \in \mathbb{L}^{q_2+2}(\Omega)$,

$$\int_\Omega v(x) f(x, v(x))\, dx \ge \theta_3 \int_\Omega \mathcal{F}(x, v(x))\, dx \tag{8.34}$$

for some positive constant $\theta_3 > 0$, where

$$\mathcal{F}(x,s) = \int_0^s f(x,\sigma)\, d\sigma.$$

Here and further we use the traditional notation

$$p^* = \begin{cases} \frac{Np}{N-p} & \text{if } N > p, \\ +\infty & \text{if } N \le p; \end{cases}$$

obviously, $p^* > 2$ for $p \ge 2$.

Now we give the definition of a weak generalized solution.

Definition 8.1. A *weak generalized solution* of problem (8.26)–(8.28) is a function

$$u(x)(t) \in \mathbb{L}^\infty(0,T;\mathbb{W}_0^{1,p}(\Omega)), \quad u'(x)(t) \in \mathbb{L}^2(0,T;\mathbb{L}^2(\Omega)),$$
$$(|u|^{p_k/2})' \in \mathbb{L}^2(Q_T), \quad Q_T = (0,T)\times\Omega,$$
$$u(x)(0) = u_0(x) \in \mathbb{W}_0^{1,p}(\Omega),$$

satisfying the equality

$$\begin{aligned}&\int_\Omega \left[u' + \sum_{k=1}^n (p_k - 1)\frac{2}{p_k} a_k(x)|u|^{p_k/2-2}u(|u|^{p_k/2})'\right] v\,dx \\ &+ \int_\Omega h(x,|\nabla u|)(\nabla u, \nabla v)\,dx \\ &+ \int_\Omega [g(x,u) - f(x,u)]v(x)(t)\,dx = 0 \end{aligned} \tag{8.35}$$

for all $v(x)(t) \in \mathbb{L}^\infty(0,T;\mathbb{W}_0^{1,p}(\Omega))$ and almost all $t \in [0,T]$ for some $T > 0$.

Remark. Note that after a possible redefinition on a subset of zero Lebesgue measure, the weak generalized solution in the sense of Definition 8.1 belongs to the class

$$u(x)(t) \in \mathbb{C}([0,T];\mathbb{L}^2(\Omega)),$$

and hence the initial condition makes sense.

Assume that problem (8.26)–(8.28) has a weak generalized solution of the class

$$u(x)(t) \in \mathbb{C}^{(1)}([0,T_0);\mathbb{W}_0^{1,p}(\Omega)),$$

where $T_0 > 0$ is the maximal time of existence of a nonextendable in time solution.

Similarly to problem (8.3)–(8.5), introduce the following notation:

$$\begin{aligned}\Phi(t) \equiv &\frac{1}{2}\int_0^t \|u\|_2^2\,ds + \sum_{l=1}^n \frac{p_l - 1}{p_l}\int_0^t\int_\Omega a_l(x)|u|^{p_l}\,dx\,ds \\ &+ \frac{1}{2p_{k_0}}\|u_0\|_2^2 + \sum_{l=1}^n \frac{p_l - 1}{p_{k_0}p_l}\int_\Omega a_l(x)|u_0|^{p_l}\,dx,\end{aligned} \tag{8.36}$$

$$\begin{aligned}\mathrm{J}(t) \equiv &\int_0^t \|u'\|_2^2(s)\,ds + \sum_{l=1}^n (p_l - 1)\int_0^t\int_\Omega a_l(x)|u'|^2|u|^{p_l-2}\,dx\,ds \\ &+ \frac{1}{2}\|u_0\|_2^2 + \sum_{l=1}^n \frac{p_l - 1}{p_l}\int_\Omega a_l(x)|u_0|^{p_l}\,dx.\end{aligned} \tag{8.37}$$

Lemma 8.2. *We have the differential inequality*

$$(\Phi')^2 \leq p_{k_0}\Phi J \quad \textit{for all } t \in [0, T_0), \quad p_{k_0} = \max_{k\in\overline{1,n}} p_k. \tag{8.38}$$

Proof. First, we note the following equality:

$$\begin{aligned}\Phi' &= \frac{1}{2}\|u\|_2^2 + \sum_{l=1}^{n} \frac{p_l - 1}{p_l} \int_\Omega a_l(x)|u|^{p_l}\,dx \\ &= \frac{1}{2}\int_0^t \frac{d}{ds}\|u\|_2^2(s)\,ds + \sum_{l=1}^{n} \frac{p_l-1}{p_l}\int_0^t \frac{d}{ds}\int_\Omega a_l(x)|u|^{p_l}\,dx\,ds \\ &+ \frac{1}{2}\|u_0\|_2^2 + \sum_{l=1}^{n} \frac{p_l-1}{p_l}\int_\Omega a_l(x)|u_0|^{p_l}\,dx.\end{aligned} \tag{8.39}$$

We estimate the terms containing derivatives under the integral sign:

$$\begin{aligned}\left|\frac{1}{2}\int_0^t \frac{d}{ds}\|u\|_2^2(s)\,ds\right| &\leq \int_0^t\int_\Omega |u||u'|\,dx\,ds \\ &\leq \int_0^t \|u\|_2\|u'\|_2\,ds \leq \left(\int_0^t \|u\|_2^2\,ds\right)^{1/2}\left(\int_0^t \|u'\|_2^2\,ds\right)^{1/2},\end{aligned} \tag{8.40}$$

$$\begin{aligned}&\left|\frac{p_l-1}{p_l}\int_0^t \frac{d}{ds}\int_\Omega a_l(x)|u|^{p_l}\,dx\,ds\right| \\ &\leq (p_l-1)\int_0^t\int_\Omega a_l(x)|u|^{p_l-2}|u||u'|\,dx\,ds \\ &\leq (p_l-1)\int_0^t\left(\int_\Omega a_l(x)|u|^{p_l}\,dx\right)^{1/2}\left(\int_\Omega a_l(x)|u|^{p_l-2}|u'|^2\,dx\right)^{1/2} ds \\ &\leq (p_l-1)\left(\int_0^t\int_\Omega a_l(x)|u|^{p_l-2}|u'|^2\,dx\,ds\right)^{1/2}\left(\int_0^t\int_\Omega a_l(x)|u|^{p_l}\,dx\,ds\right)^{1/2}.\end{aligned} \tag{8.41}$$

Thus, taking into account inequalities (8.40) and (8.41), from equation (8.39) we obtain the estimate

$$\begin{aligned}(\Phi')^2 \leq \Bigg[&\left(\int_0^t \|u\|_2^2\,ds\right)^{1/2}\left(\int_0^t \|u'\|_2^2\,ds\right)^{1/2} \\ &+ \sum_{l=1}^{n}(p_l-1)\left(\int_0^t\int_\Omega a_l(x)|u|^{p_l-2}|u'|^2\,dx\,ds\right)^{1/2}\left(\int_0^t\int_\Omega a_l(x)|u|^{p_l}\,dx\,ds\right)^{1/2}\end{aligned}$$

$$+\frac{1}{2}\|u_0\|_2^2+\sum_{l=1}^{n}\frac{p_l-1}{p_l}\int_\Omega a_l(x)|u_0|^{p_l}\,dx\Bigg]^2. \tag{8.42}$$

Squaring the right-hand side of (8.42), grouping terms appropriately, and using the inequality $2ab \le a^2+b^2$, we obtain the following important inequality:

$$\begin{aligned}(\Phi')^2 \le {}&\Bigg(\int_0^t \|u\|_2^2\,ds+\sum_{l=1}^{n}(p_l-1)\int_0^t\int_\Omega a_l(x)|u|^{p_l}\,dx\,ds\\&+\frac{1}{2}\|u_0\|_2^2+\sum_{l=1}^{n}\frac{p_l-1}{p_l}\int_\Omega a_l(x)|u_0|^{p_l}\,dx\Bigg)\\&\times\Bigg(\int_0^t \|u'\|_2^2\,ds+\sum_{l=1}^{n}(p_l-1)\int_0^t\int_\Omega a_l(x)|u|^{p_l-2}|u'|^2\,dx\,ds\\&+\frac{1}{2}\|u_0\|_2^2+\sum_{l=1}^{n}\frac{p_l-1}{p_l}\int_\Omega a_l(x)|u_0|^{p_l}\,dx\Bigg).\end{aligned} \tag{8.43}$$

Using the definition of p_{k_0} and taking into account notation (8.36)–(8.37), from inequality (8.43) we obtain the inequalities

$$\begin{aligned}(\Phi')^2 &\le p_{k_0}\Bigg(\frac{1}{p_{k_0}}\int_0^t \|u\|_2^2\,ds+\sum_{l=1}^{n}\frac{p_l-1}{p_{k_0}}\int_0^t\int_\Omega a_l(x)|u|^{p_l}\,dx\,ds\\&\quad+\frac{1}{2p_{k_0}}\|u_0\|_2^2+\sum_{l=1}^{n}\frac{p_l-1}{p_{k_0}p_l}\int_\Omega a_l(x)|u_0|^{p_l}\,dx\Bigg)\mathrm{J}\\&\le p_{k_0}\Bigg(\frac{1}{2}\int_0^t \|u\|_2^2\,ds+\sum_{l=1}^{n}\frac{p_l-1}{p_l}\int_0^t\int_\Omega a_l(x)|u|^{p_l}\,dx\,ds\\&\quad+\frac{1}{2p_{k_0}}\|u_0\|_2^2+\sum_{l=1}^{n}\frac{p_l-1}{p_{k_0}p_l}\int_\Omega a_l(x)|u_0|^{p_l}\,dx\Bigg)\mathrm{J}\\&= p_{k_0}\Phi\mathrm{J}.\end{aligned} \tag{8.44}$$

The proof is complete. □

Now we set $v = u(x)(t)$ in the definition of a weak generalized solution (8.35); then, after integration by parts, we obtain the first energy equality

$$\frac{d^2\Phi(t)}{dt^2}+\int_\Omega h(x,|\nabla u|)|\nabla u|^2\,dx+\int_\Omega g(x,u)u\,dx=\int_\Omega f(x,u)u\,dx. \tag{8.45}$$

To obtain the second energy equality, we set $v = u'(x)(t)$ in (8.35) and, integrating by parts, obtain

$$\frac{dJ}{dt} + \frac{d}{dt}\int_\Omega \mathcal{H}(x, |\nabla u|)\,dx + \frac{d}{dt}\int_\Omega \mathcal{G}(x,u)\,dx = \frac{d}{dt}\int_\Omega \mathcal{F}(x,u)\,dx. \tag{8.46}$$

Integrating (8.46) by time, we arrive at the equality

$$J + \int_\Omega \mathcal{H}(x, |\nabla u|)\,dx + \int_\Omega \mathcal{G}(x,u)\,dx - E(0) = \int_\Omega \mathcal{F}(x,u)\,dx, \tag{8.47}$$

where the integration constant is

$$\begin{aligned} E(0) \equiv \frac{1}{2}\|u_0\|_2^2 + \sum_{l=1}^{n} \frac{p_l - 1}{p_l}\int_\Omega a_l(x)|u_0|^{p_l}\,dx \\ + \int_\Omega \mathcal{H}(x, |\nabla u_0|)\,dx + \int_\Omega \mathcal{G}(x,u_0)\,dx - \int_\Omega \mathcal{F}(x,u_0)\,dx. \end{aligned} \tag{8.48}$$

From inequality (8.34), multiplying (8.47) by the constant θ_3, we obtain the estimate

$$\begin{aligned} \theta_3 J + \theta_3\int_\Omega \mathcal{H}(x, |\nabla u|)\,dx + \theta_3\int_\Omega \mathcal{G}(x,u)\,dx \\ - \theta_3 E(0) \le \int_\Omega f(x,u)u\,dx. \end{aligned} \tag{8.49}$$

From (8.45) and (8.49) we obtain the inequality

$$\begin{aligned} \frac{d^2\Phi(t)}{dt^2} + \int_\Omega h(x, |\nabla u|)|\nabla u|^2\,dx \\ + \int_\Omega g(x,u)u\,dx \ge \theta_3 J + \theta_3\int_\Omega \mathcal{H}(x, |\nabla u|)\,dx + \theta_3\int_\Omega \mathcal{G}(x,u)\,dx - \theta_3 E(0). \end{aligned} \tag{8.50}$$

Applying properties (8.30) and (8.32), we obtain the estimate

$$\begin{aligned} \frac{d^2\Phi(t)}{dt^2} + \theta_3 E(0) \ge (\theta_3 - \theta_1)\int_\Omega \mathcal{H}(x, |\nabla u|)\,dx \\ + (\theta_3 - \theta_2)\int_\Omega \mathcal{G}(x,u)\,dx + \theta_3 J. \end{aligned} \tag{8.51}$$

Impose the additional conditions

$$\theta_3 \ge \theta_1, \quad \theta_3 \ge \theta_2,$$

which allow us to eliminate two terms in the right-hand side of the inequality (8.51):

$$\frac{d^2\Phi(t)}{dt^2} + \theta_3 E(0) \geq \theta_3 J. \tag{8.52}$$

By Lemma 8.2, from (8.52) we obtain the basic differential inequality

$$\Phi\Phi'' - \alpha(\Phi')^2 + \beta\Phi \geq 0, \tag{8.53}$$

with parameters

$$\alpha = \frac{\theta_3}{p_{k_0}} \quad \text{and} \quad \beta = \theta_3 E(0). \tag{8.54}$$

As before, we apply Theorem A.1 (see the Appendix, p. 299) and obtain the blow-up result. For this, we impose the condition

$$\theta_3 > p_{k_0} = \max_{l=\overline{1,n}} p_l. \tag{8.55}$$

We separately consider the cases $\beta > 0$ and $\beta \leq 0$.

Consider the first, a more complicated case. Theorem A.1 implies that blow-up occurs if

$$(\Phi'(0))^2 > \frac{2\beta}{2\alpha - 1}\Phi(0).$$

In our case, this condition can be represented in the form

$$\frac{1}{2}\|u_0\|_2^2 + \sum_{l=1}^{n} \frac{p_l - 1}{p_l} \int_\Omega a_l(x)|u_0|^{p_l}\, dx > \frac{2\theta_3}{2\theta_3 - p_{k_0}} E(0) > 0. \tag{8.56}$$

If this condition holds, then the following lower estimate is fulfilled:

$$\Phi(t) \geq \frac{1}{[\Phi^{1-\alpha}(0) - At]^{1/(\alpha-1)}}, \tag{8.57}$$

where A is a positive constant:

$$A^2 = (\alpha - 1)^2 \Phi^{-2\alpha}(0)\left[(\Phi'(0))^2 - \frac{2\beta}{2\alpha - 1}\Phi(0)\right] > 0. \tag{8.58}$$

Using the expression of $E(0)$ in terms of the initial function, we rewrite (8.56) in the form

$$\int_\Omega \mathcal{F}(x, u_0)\, dx > \int_\Omega \mathcal{H}(x, |\nabla u_0|)\, dx + \int_\Omega \mathcal{G}(x, u_0)\, dx$$

$$+\frac{p_{k_0}}{2\theta_3}\left(\frac{1}{2}\|u_0\|_2^2+\sum_{l=1}^{n}\frac{p_l-1}{p_l}\int_\Omega a_l(x)|u_0|^{p_l}\,dx\right),\quad p_{k_0}<\theta_3. \tag{8.59}$$

We complement this condition with the condition $\beta > 0$:

$$\frac{1}{2}\|u_0\|_2^2+\sum_{l=1}^{n}\frac{p_l-1}{p_l}\int_\Omega a_l(x)|u_0|^{p_l}\,dx$$
$$+\int_\Omega \mathcal{H}(x,|\nabla u_0|)\,dx+\int_\Omega \mathcal{G}(x,u_0)\,dx>\int_\Omega \mathcal{F}(x,u_0)\,dx. \tag{8.60}$$

Now we consider the simpler case $\beta \le 0$:

$$\int_\Omega \mathcal{F}(x,u_0)\,dx\ge\frac{1}{2}\|u_0\|_2^2+\sum_{l=1}^{n}\frac{p_l-1}{p_l}\int_\Omega a_l(x)|u_0|^{p_l}\,dx$$
$$+\int_\Omega \mathcal{H}(x,|\nabla u_0|)\,dx+\int_\Omega \mathcal{G}(x,u_0)\,dx. \tag{8.61}$$

In this case, the solution blows up, and inequality (8.57) holds with $\beta = 0$.

Thus, the following blow-up result is proved.

Theorem 8.2. *Let all conditions for the function $\varphi(x,s)$, $h(x,s)$, $g(x,s)$, and $f(x,s)$ be fulfilled. Then, under the conditions*

$$\int_\Omega \mathcal{F}(x,u_0)\,dx>\int_\Omega \mathcal{H}(x,|\nabla u_0|)\,dx+\int_\Omega \mathcal{G}(x,u_0)\,dx$$
$$+\frac{p_{k_0}}{2\theta_3}\left(\frac{1}{2}\|u_0\|_2^2+\sum_{l=1}^{n}\frac{p_l-1}{p_l}\int_\Omega a_l(x)|u_0|^{p_l}\,dx\right), \tag{8.62}$$

$$p_{k_0}<\theta_3,\quad \theta_3\ge\theta_1,\quad \theta_3\ge\theta_2, \tag{8.63}$$

there are no nontrivial generalized weak global-in-time solutions of problem (8.26)–(8.28), *and the limit relation*

$$\limsup_{t\uparrow T_0}\Phi(t)=+\infty$$

and the upper estimate

$$T_0\le T_\infty=\Phi^{1-\alpha}(0)A^{-1}$$

hold, where the constant $A > 0$ is defined by formula (8.58), *and*

$$\beta=\begin{cases}\theta_3 E(0) & \text{if } E(0)>0,\\ 0 & \text{if } E(0)\le 0,\end{cases}$$

where

$$E(0) \equiv \frac{1}{2}\|u_0\|_2^2 + \sum_{l=1}^{n} \frac{p_l - 1}{p_l} \int_{\Omega} a_l(x)|u_0|^{p_l}\, dx$$
$$+ \int_{\Omega} \mathcal{H}(x, |\nabla u_0|)\, dx + \int_{\Omega} \mathcal{G}(x, u_0)\, dx - \int_{\Omega} \mathcal{F}(x, u_0)\, dx.$$

Applying the energy equalities, we can obtain sufficient conditions for blow-up of weak generalized solutions for the class of nonlinear parabolic equations. Speaking about the class, we emphasize that the right-hand side and nonlinear terms were not strictly defined: their choice was restricted only by several conditions. However, the equations themselves are local, which is not a necessary condition for the application of the modified energy method.

8.1.4 Nonlocal equation

Consider the following initial-boundary-value problem for the nonlocal equation::

$$\frac{\partial \varphi(x,u)}{\partial t} - \chi(\|\nabla u\|_2^2)\Delta u = f(x,u), \tag{8.64}$$

$$\varphi(x,u) = u + \sum_{k=1}^{n} a_k(x)|u|^{p_k-2}u, \tag{8.65}$$

$$u|_{\partial\Omega} = 0, \quad u(x,0) = u_0(x), \tag{8.66}$$

where $\Omega \subset \mathbb{R}^3$ is a bounded domain with smooth boundary $\partial\Omega \in \mathbb{C}^{2,\delta}$, $\delta \in (0,1]$.

For the function $\varphi(x,u)$, we take the same conditions as in Section 8.1.3; the conditions for the functions $\chi(s)$ and $f(x,s)$ are as follows.

Conditions for the function $\chi(s)$

(i)$_2$ $\chi(s) \in \mathbb{C}^{(1)}([0,+\infty))$;

(ii)$_2$ for all $s \in \mathbb{R}^1_+$, the following growth conditions are fulfilled:

$$c_1 \le \chi(s) \le c_1 + c_2 s^{q_1}; \tag{8.67}$$

(iii)$_2$ for all $s \in \mathbb{R}^1_+$,

$$\chi(s)s \le \frac{\theta_1}{2}\int_0^s \chi(\sigma)\, d\sigma, \tag{8.68}$$

where $\theta_1 > 0$.

Conditions for the function $f(x, s)$

(i)$_3$ $f(x,s) : \Omega \times \mathbb{R}^1 \to \mathbb{R}^1$ is a Carathéodori function such that $f(x,0) = 0$ for almost all $x \in \partial\Omega$;

(ii)$_3$ we have the upper estimate

$$|f(x,s)| \leq c_3 + c_4|s|^{q_2+1} \tag{8.69}$$

for almost all $x \in \Omega$;

(iii)$_3$ for all $v(x) \in \mathbb{L}^{q_2+2}(\Omega)$,

$$\int_\Omega f(x,v(x))v(x) \geq \theta_2 \int_\Omega \mathcal{F}(x,v(x))\,dx, \tag{8.70}$$

where $\theta_2 > 0$ and

$$\mathcal{F}(x,s) = \int_0^s d\sigma f(x,\sigma).$$

We formulate the definition of a generalized solution of problem (8.64)–(8.66).

Definition 8.2. A *weak generalized solution* of the problem (8.64)–(8.66) is a function $u(x)(t)$ that belongs to the functional classes

$$u(x)(t) \in \mathbb{L}^\infty(0,T;\mathbb{H}_0^1(\Omega) \cap \mathbb{H}^2(\Omega)), \quad u'(x)(t) \in \mathbb{L}^2(0,T;\mathbb{L}^2(\Omega)),$$

and satisfies the equality

$$\int_0^T \int_\Omega [u' - \chi(\|\nabla u\|_2^2)\Delta u - f(x,u)]v(x)(t)\,dx\,dt = 0 \tag{8.71}$$

for all

$$v(x)(t) \in \mathbb{L}^{q_2+2}(0,T;\mathbb{H}_0^1(\Omega) \cap \mathbb{H}^2(\Omega)),$$
$$u(x)(0) = u_0(x) \in \mathbb{H}_0^1(\Omega) \cap \mathbb{H}^2(\Omega).$$

Remark. Definition 8.2 implies that after a possible redefinition on a subset of zero Lebesgue measure, the weak generalized solution $u(x)(t)$ belongs to the class

$$u(x)(t) \in \mathbb{C}([0,T];\mathbb{L}^2(\Omega)).$$

We assume that a weak generalized solution of the class

$$u(x)(t) \in \mathbb{C}^{(1)}([0,T_0);\mathbb{H}_0^1(\Omega) \cap \mathbb{H}^2(\Omega))$$

exists and satisfies equation (8.71), where the constant $T_0 > 0$ is the maximal time moment at which the nonextendable solution exists.

Introduce the following notation:

$$\Phi(t) \equiv \frac{1}{2}\int_0^t \|u\|_2^2\,ds + \sum_{l=1}^n \frac{p_l - 1}{p_l}\int_0^t\int_\Omega a_l(x)|u|^{p_l}\,dx\,ds$$
$$+ \frac{1}{2p_{k_0}}\|u_0\|_2^2 + \sum_{l=1}^n \frac{p_l - 1}{p_{k_0}p_l}\int_\Omega a_l(x)|u_0|^{p_l}\,dx, \tag{8.72}$$

$$\mathrm{J}(t) \equiv \int_0^t \|u'\|_2^2(s)\,ds + \sum_{l=1}^n (p_l - 1)\int_0^t\int_\Omega a_l(x)|u'|^2|u|^{p_l-2}\,dx\,ds$$
$$+ \frac{1}{2}\|u_0\|_2^2 + \sum_{l=1}^n \frac{p_l - 1}{p_l}\int_\Omega a_l(x)|u_0|^{p_l}\,dx. \tag{8.73}$$

Due to Lemma 8.2,

$$(\Phi')^2 \le p_{k_0}\Phi\mathrm{J} \tag{8.74}$$

for all $t \in [0, T_0)$, where

$$p_{k_0} = \max_{k\in\overline{1,n}} p_k.$$

First, we obtain the energy equalities. In the definition of a weak solution (8.71), we set

$$v(x)(t) = \begin{cases} u(x)(s) & \text{if } s \in [0,t], \\ 0 & \text{if } s \in (t,T]. \end{cases}$$

Integrating by parts and differentiating by time in the class considered, we obtain the first energy equality

$$\frac{d^2\Phi}{dt^2} + \chi(\|\nabla u\|_2^2)\|\nabla u\|_2^2 = \int_\Omega f(x,u)u\,dx. \tag{8.75}$$

Setting in (8.71)

$$v(x)(t) = \begin{cases} u'(x)(s) & \text{if } s \in [0,t], \\ 0 & \text{if } s \in (t,T], \end{cases}$$

and integrating by parts, we obtain the second equality

$$\mathrm{J} + \frac{1}{2}\int_0^{\|\nabla u\|_2^2} \chi(s)\,ds - E(0) = \int_\Omega \mathcal{F}(x,u)\,dx, \tag{8.76}$$

where $E(0)$ is the constant determined by the initial conditions:

$$E(0) = \frac{1}{2}\|u_0\|_2^2 + \sum_{l=1}^{n} \frac{p_l - 1}{p_l} \int_{\Omega} a_l(x)|u_0|^{p_l}\,dx + \frac{1}{2} \int_{0}^{\|\nabla u_0\|_2^2} \chi(s)\,ds - \int_{\Omega} \mathcal{F}(x, u_0)\,dx. \tag{8.77}$$

Using condition (8.70) for the function $f(x,u)$, we obtain the inequality

$$\theta_2 J + \frac{\theta_2}{2} \int_{0}^{\|\nabla u\|_2^2} \chi(s)\,ds - \theta_2 E(0) \le \int_{\Omega} f(x,u)u\,dx, \tag{8.78}$$

which by (8.75) implies

$$\frac{d^2\Phi}{dt^2} + \chi(\|\nabla u\|_2^2)\|\nabla u\|_2^2 \ge \theta_2 J + \frac{\theta_2}{2} \int_{0}^{\|\nabla u\|_2^2} \chi(s)\,ds - \theta_2 E(0). \tag{8.79}$$

Finally, using condition (8.68) for the function $\chi(s)$, we arrive at the inequality

$$\frac{d^2\Phi}{dt^2} + \theta_2 E(0) \ge \theta_2 J + \frac{\theta_2 - \theta_1}{2} \int_{0}^{\|\nabla u\|_2^2} \chi(s)\,ds. \tag{8.80}$$

If $\theta_2 \ge \theta_1$, then we can eliminate the last term in the right-hand side and obtain

$$\frac{d^2\Phi}{dt^2} + \theta_2 E(0) \ge \theta_2 J. \tag{8.81}$$

This inequality and (8.74) imply the ordinary differential inequality

$$\Phi\Phi'' - \alpha(\Phi')^2 + \beta\Phi \ge 0, \tag{8.82}$$

$$\alpha = \frac{\theta_2}{p_{k_0}}, \quad \beta = \theta_2 E(0). \tag{8.83}$$

To apply Theorem A.1 (see Appendix, p. 299), we impose the condition

$$\theta_2 > p_{k_0}, \quad p_{k_0} = \max_{l=\overline{1,n}} p_l. \tag{8.84}$$

Similarly to theorems proved before, we obtain the following blow-up result.

Theorem 8.3. *Let all conditions for the functions $\varphi(x,s)$, $\chi(s)$, and $f(x,s)$ be fulfilled. Then under the conditions*

$$\int_{\Omega} \mathcal{F}(x, u_0)\,dx > \frac{1}{2} \int_{0}^{\|\nabla u_0\|_2^2} \chi(s)\,ds$$

$$+\frac{p_{k_0}}{2\theta_2}\left(\frac{1}{2}\|u_0\|_2^2+\sum_{l=1}^{n}\frac{p_l-1}{p_l}\int_\Omega a_l(x)|u_0|^{p_l}\,dx\right), \tag{8.85}$$

$$p_{k_0}<\theta_2, \quad \theta_2\ge\theta_1, \tag{8.86}$$

there are no global-in-time weak generalized solutions of problem (8.64)–(8.66). *Moreover, we have the limit relation*

$$\limsup_{t\uparrow T_0}\Phi(t)=+\infty$$

and the upper estimate

$$T_0\le T_\infty=\Phi^{1-\alpha}(0)A^{-1},$$

where

$$A^2=(\alpha-1)^2\Phi^{-2\alpha}(0)\left[(\Phi'(0))^2-\frac{2\beta}{2\alpha-1}\Phi(0)\right]>0,$$

$$\alpha=\frac{\theta_2}{p_{k_0}}, \quad \beta=\begin{cases}\theta_2E(0) & \textit{if } E(0)>0,\\ 0 & \textit{if } E(0)\le 0,\end{cases}$$

and

$$E(0)=\frac{1}{2}\|u_0\|_2^2+\sum_{l=1}^{n}\frac{p_l-1}{p_l}\int_\Omega a_l(x)|u_0|^{p_l}\,dx$$
$$+\frac{1}{2}\int_0^{\|\nabla u_0\|_2^2}\chi(s)\,ds-\int_\Omega\mathcal{F}(x,u_0)\,dx.$$

Thus, the modified energy method allows us to obtain sufficient conditions of blow-up for nonlocal problems. Applications of this method to nonlocal problems of mathematical physics are presented in detail in [45] (see also the references therein).

8.2 Hyperbolic equations with positive energy

8.2.1 Nonlinear wave equations of nonlinear mechanics

We consider several initial-boundary-value problems for hyperbolic nonlinear equations that appear in various branches of mathematical physics. Despite the differences in these equations, the modified energy method allows us to reduce them to the ordinary differential inequality

$$\Phi\Phi''-\alpha(\Phi')^2+\beta\Phi\ge 0$$

and apply Theorem A.1 (see Appendix, p. 299). All the subtlety is only a competent choice of the function Φ, which is defined by the problem studied.

First, we consider the homogeneous Dirichlet problem for the following equation of the Klein–Gordon type:

$$\begin{gathered}\frac{\partial^2 u}{\partial t^2} - \Delta u + u = u^2 + u^3,\\ u|_{\partial\Omega} = 0,\\ u(x,0) = u_0(x), \quad u'(x,0) = u_1(x).\end{gathered} \tag{8.87}$$

All constant are further assumed to be positive. Moreover, consider the homogeneous Dirichlet problem for the classical system of wave equation, which describes longitudinal bending oscillations of the plate:

$$\frac{\partial^2 u}{\partial t^2} - c_0^2 \frac{\partial^2 u}{\partial x^2} + c_0^2 \frac{\partial w}{\partial x}\frac{\partial^2 w}{\partial x^2} = 0, \quad u|_{x=0,l} = 0, \quad l > 0, \tag{8.88}$$

$$\frac{\partial^2 w}{\partial t^2} + c_1^2 \frac{\partial^4 w}{\partial x^4} + c_0^2 \frac{\partial}{\partial x}\left(\frac{\partial w}{\partial x}\frac{\partial u}{\partial x}\right) = 0, \quad w|_{x=0,l} = w_x|_{x=0,l} = 0, \tag{8.89}$$

$$u(x,0) = u_0(x), \quad w(x,0) = w_0(x), \quad u'(x,0) = u_1(x), \quad w'(x,0) = w_1(x).$$

Consider also the initial-boundary-value problem

$$\begin{gathered}\frac{\partial^2 u}{\partial t^2} + \alpha \frac{\partial^4 u}{\partial x^4} = \alpha^2 \lambda \left[\frac{\partial^4 u}{\partial x^4}\left(\frac{\partial^2 u}{\partial x^2}\right)^2 + 2\left(\frac{\partial^3 u}{\partial x^3}\right)^2 \frac{\partial^2 u}{\partial x^2}\right],\\ u|_{x=0,l} = u_x|_{x=0,l} = 0, \quad u(x,0) = u_0(x).\end{gathered} \tag{8.90}$$

Finally, we consider the problem for a higher-order wave equation with unit coefficients:

$$\begin{gathered}\frac{\partial^2}{\partial t^2}\left(-\frac{\partial^2 u}{\partial x^2} + u\right) + \frac{\partial^4 u}{\partial x^4} - \frac{\partial^2 u}{\partial x^2} = -\frac{\partial}{\partial x}\left(\left(\frac{\partial u}{\partial x}\right)^2\right),\\ u|_{x=0,l} = u_x|_{x=0,l} = 0, \quad u(x,0) = u_0(x).\end{gathered} \tag{8.91}$$

Equation (8.87) describes mechanics of crystals (see [64]). System (8.88)–(8.89) describes longitudinal bending oscillations of the plate (see [17]). Equation (8.90) describes bending oscillations of a physically nonlinear rod (see [40]). Equation (8.91) describes bending waves in a thin stretched rod (see [17]).

By the classical Levine method, more or less the same results were obtained for problems mentioned. Namely, in [64], the blow-up of solutions of equation (8.87) for negative and moderately positive values of the initial energy was proved. The fact that the energy can take a negative or sufficiently large positive value satisfying three conditions (see [106]) was proved in other papers. However, these results have significant disadvantages: first, it is difficult to verify the compatibility of blow-up conditions, and, second, for fixed u_0 and sufficiently large u_1, these conditions are incompatible.

First, we consider discuss problem (8.87) in a bounded domain $\Omega \subset \mathbb{R}^3$ with smooth boundary $\partial\Omega \in \mathbb{C}^{(2,\gamma)}$, $\gamma \in (0,1]$. Solutions of problem (8.87) will be treated in the following classical sense.

Definition 8.3. A classical solution of problem (8.87) is a function $u(x,t)$ of the class

$$\mathbb{C}^{(2)}([0,T_0);\mathbb{C}^{(2)}(\overline{\Omega})),$$

where $T_0 > 0$ is the maximal time of existence of a nonextendable solution satisfying equation (8.87) pointwise.

Assume that there exists $T_0 > 0$ such that a classical solution of problem (8.87) exists. Multiplying both parts of equation (8.87) by u and integrating by parts, we obtain the first energy equality

$$\frac{1}{2}\Phi'' - \mathrm{J} + \|\nabla u\|_2^2 + \|u\|_2^2 - \|u\|_4^4 = \int_\Omega u^3\,dx, \tag{8.92}$$

where

$$\Phi = \int_\Omega |u|^2\,dx \quad \text{and} \quad \mathrm{J} = \int_\Omega |u'|^2\,dx. \tag{8.93}$$

To obtain the second energy equality, we multiply both sides of (8.87) by u', integrate by parts, and then integrate by time; we obtain the following equation:

$$\frac{1}{2}\mathrm{J} + \frac{1}{2}\|\nabla u\|_2^2 + \frac{1}{2}\|u\|_2^2 - \frac{1}{4}\|u\|_4^4 - E(0) = \frac{1}{3}\int_\Omega u^3\,dx, \tag{8.94}$$

where

$$E(0) = \frac{1}{2}\mathrm{J}(0) + \frac{1}{2}\|\nabla u_0\|_2^2 + \frac{1}{2}\|u_0\|_2^2 - \frac{1}{4}\|u_0\|_4^4 - \frac{1}{3}\int_\Omega u_0^3\,dx \tag{8.95}$$

is the integration constant. From (8.94) we obtain

$$\int_\Omega u^3\,dx = \frac{3}{2}\mathrm{J} + \frac{3}{2}\|\nabla u\|_2^2 + \frac{3}{2}\|u\|_2^2 - \frac{3}{4}\|u\|_4^4 - 3E(0). \tag{8.96}$$

Substituting the expression for the integral

$$\int_\Omega u^3\,dx$$

from equation (8.96) into (8.92), we obtain

$$\frac{1}{2}\Phi'' - \mathrm{J} + \|\nabla u\|_2^2 + \|u\|_2^2 - \|u\|_4^4$$

$$= \frac{3}{2}J + \frac{3}{2}\|\nabla u\|_2^2 + \frac{3}{2}\|u\|_2^2 - \frac{3}{4}\|u\|_4^4 - 3E(0), \tag{8.97}$$

which, in its turn, implies the equality

$$\frac{1}{2}\Phi'' - \frac{5}{2}J - \frac{1}{2}[\|\nabla u\|_2^2 + \|u\|_2^2] - \frac{1}{4}\|u\|_4^4 + 3E(0) = 0. \tag{8.98}$$

Since norms are nonnegative, we arrive at the inequality

$$\frac{1}{2}\Phi'' + 3E(0) \geq \frac{5}{2}J. \tag{8.99}$$

Using the Cauchy–Bunyakovsky inequality, we deduce the estimate

$$(\Phi')^2 \leq 4J\Phi; \tag{8.100}$$

combining (8.100) with inequality (8.99), we arrive at the ordinary differential inequality

$$\Phi\Phi'' - \frac{5}{4}(\Phi')^2 + 6E(0)\Phi \geq 0. \tag{8.101}$$

Comparing this differential inequality with the inequality from Theorem A.1 (see Appendix, p. 299), we obtain

$$\alpha = \frac{5}{4}, \quad \beta = 6E(0), \quad \gamma = 0.$$

We have

$$\frac{2\beta}{2\alpha - 1} = \frac{12E(0)}{5/2 - 1} = 8E(0).$$

By Theorem 8.1 we conclude that, under the conditions

$$\Phi'(0) > (8E(0)\Phi(0))^{1/2} > 0, \quad E(0) > 0, \tag{8.102}$$

where

$$\Phi(0) = \int_\Omega |u_0|^2\, dx, \quad \Phi'(0) = 2\int_\Omega u_1 u_0\, dx > 0,$$

the lifetime $T > 0$ of the solution $u(x)(t)$ of problem (8.87) is bounded from above. Namely, we have the following inequality:

$$\Phi(t) \geq (\Phi^{-1/4}(0) - At)^{-1/4}, \tag{8.103}$$

where

$$A^2 = \frac{1}{16}\Phi^{-5/2}(0)[(\Phi'(0))^2 - 8E(0)\Phi(0)]. \tag{8.104}$$

Thus inequality (8.103) implies that, under conditions (8.102), the classical solution of the problem considered blows up in finite time.

Remark. The most important point in the proof of sufficient conditions (8.102) is the verification of their compatibility. Choose an initial function $u_0(x) \in \mathbb{C}^{(2)}(\overline{\Omega})$ so large that

$$\frac{1}{2}\|\nabla u_0\|_2^2 + \frac{1}{2}\|u_0\|_2^2 < \frac{1}{4}\|u_0\|_4^4 + \frac{1}{3}\int_\Omega u_0^3\,dx. \tag{8.105}$$

Such functions $u_0(x)$ exist if they are sufficiently "large." We fix such a function $u_0(x)$, set $u_1(x) = \lambda u_0(x)$, and choose $\lambda > 0$ so large that

$$E(0) = \frac{\lambda^2}{2}\int_\Omega |u_0|^2\,dx + \frac{1}{2}\|\nabla u_0\|_2^2 + \frac{1}{2}\|u_0\|_2^2 - \frac{1}{4}\|u_0\|_4^4 - \frac{1}{3}\int_\Omega u_0^3\,dx > 0. \tag{8.106}$$

Note that, for this initial function, we have the relation

$$(\Phi'(0))^2 = 4\Big(\int_\Omega u_0 u_1\,dx\Big)^2 = 4\lambda^2\Big(\int_\Omega |u_0|^2\,dx\Big)^2; \tag{8.107}$$

therefore condition (8.102) takes the form

$$\begin{aligned}(\Phi'(0))^2 = 4\lambda^2\Big(\int_\Omega |u_0|^2\,dx\Big)^2 > 8E(0)\Phi(0) = 4\lambda^2\Big(\int_\Omega |u_0|^2\,dx\Big)^2 \\ + 8\Phi(0)\Big[\frac{1}{2}\|\nabla u_0\|_2^2 + \frac{1}{2}\|u_0\|_2^2 - \frac{1}{4}\|u_0\|_4^4 - \frac{1}{3}\int_\Omega u_0^3\,dx\Big].\end{aligned} \tag{8.108}$$

Obviously, this inequality holds by (8.105). Thus conditions (8.102) are compatible.

Now we examine the system of equations (8.88)–(8.89).

Definition 8.4. A *classical solution* of system (8.88)–(8.89) is a pair of functions $u(x)(t)$ and $w(x)(t)$ of the class

$$u(x)(t) \in \mathbb{C}^{(2)}([0,T_0);\mathbb{C}(\overline{\Omega})) \cap \mathbb{C}([0,T];\mathbb{C}^{(2)}(\overline{\Omega})),$$
$$w(x)(t) \in \mathbb{C}^{(2)}([0,T_0);\mathbb{C}(\overline{\Omega})) \cap \mathbb{C}([0,T];\mathbb{C}^{(4)}(\overline{\Omega})),$$

satisfying the initial and boundary condition treated pointwise, where $T_0 > 0$ is the maximal time of existence of a nonextendable solution.

Assume that there exists $T_0 > 0$ for which a classical solution of problem (8.88)–(8.89) exists and prove that, under certain sufficient conditions, T_0 cannot be equal to $+\infty$; for this, we obtain an upper estimate for T_0.

Introduce the functionals

$$\Phi(t) \equiv \int_\Omega [|u|^2 + |w|^2]\,dx, \quad \mathrm{J}(t) \equiv \int_\Omega [|u'|^2 + |w'|^2]\,dx. \tag{8.109}$$

It is easy to prove that by the Cauchy–Bunyakovsky inequality the following differential inequality is fulfilled:

$$(\Phi')^2 \leq 4J\Phi. \tag{8.110}$$

Multiplying equation (8.88) by $u(x)(t)$ and equation (8.89) by $w(x)(t)$, integrating over the domain Ω, and adding the relations obtained, we arrive at the first energy equality

$$\frac{1}{2}\frac{d^2}{dt^2}\Phi(t) - J(t) + c_0^2\|u_x\|_2^2 + c_1^2\|w_{xx}\|_2^2 = \frac{3c_0^2}{2}\int_\Omega (w_x)^2 u_x\,dx. \tag{8.111}$$

Multiplying equation (8.88) by $u'(x)(t)$ and equation (8.89) by $w'(x)(t)$, integrating over the domain Ω, and adding, we get the second energy equality

$$\frac{d}{dt}\left(\frac{1}{2}J + \frac{c_0^2}{2}\|u_x\|_2^2 + \frac{c_1^2}{2}\|w_{xx}\|_2^2\right) = \frac{c_0^2}{2}\frac{d}{dt}\int_\Omega (w_x)^2 u_x\,dx. \tag{8.112}$$

Integrating both sides of equation (8.112) by time, we obtain the equation

$$\frac{1}{2}J + \frac{c_0^2}{2}\|u_x\|_2^2 + \frac{c_1^2}{2}\|w_{xx}\|_2^2 - E(0) = \frac{c_0^2}{2}\int_\Omega (w_x)^2 u_x\,dx, \tag{8.113}$$

where

$$E(0) = \frac{1}{2}J(0) + \frac{c_0^2}{2}\|u_{0x}\|_2^2 + \frac{c_1^2}{2}\|w_{0xx}\|_2^2 - \frac{c_0^2}{2}\int_\Omega (w_{0x})^2 u_{0x}\,dx \tag{8.114}$$

is the integration constant. Substituting the expression for

$$\frac{c_0^2}{2}\int_\Omega (w_x)^2 u_x\,dx$$

from (8.113) into (8.111), we obtain the equality

$$\begin{aligned}\frac{1}{2}\frac{d^2}{dt^2}\Phi(t) - J(t) + c_0^2\|u_x\|_2^2 + c_1^2\|w_{xx}\|_2^2 \\ = \frac{3}{2}J + \frac{3c_0^2}{2}\|u_x\|_2^2 + \frac{3c_1^2}{2}\|w_{xx}\|_2^2 - 3E(0),\end{aligned} \tag{8.115}$$

which immediately implies the inequality

$$\frac{1}{2}\Phi'' - \frac{5}{2}J + 3E(0) \geq 0. \tag{8.116}$$

Due to estimate (8.110), from inequality (8.116) we obtain the ordinary differential inequality

$$\Phi\Phi'' - \frac{5}{4}(\Phi')^2 + 6E(0)\Phi \geq 0. \tag{8.117}$$

Comparing it with the inequality from Theorem A.1 (see Appendix, p. 299), we conclude that

$$\alpha = \frac{5}{4}, \quad \beta = 6E(0), \quad \gamma = 0, \quad \frac{2\beta}{2\alpha - 1} = 8E(0).$$

Thus by Theorem 8.1 we obtain the following blow-up result.

Theorem 8.4. *Assume that there exists a maximal $T_0 > 0$ for which a classical nonextendable solution of the problem (8.88)–(8.89) exists. Then, under the conditions*

$$(\Phi'(0))^2 > 8E(0)\Phi(0) > 0, \quad E(0) > 0, \quad \Phi'(0) > 0, \tag{8.118}$$

we have the following lower estimate:

$$\Phi(t) \geq (\Phi^{-1/4}(0) - At)^{-1/4}.$$

Therefore,

$$T_0 \leq \Phi^{-1/4}(0)A^{-1},$$

where

$$\Phi(t) = \int_\Omega [|u|^2 + |w|^2]\, dx, \quad \Phi'(0) = 2\int_\Omega [u_0 u_1 + w_0 w_1]\, dx,$$

$$A^2 = \frac{1}{16}\Phi^{-5/2}(0)[(\Phi'(0))^2 - 8E(0)\Phi(0)].$$

Remark. As usual, we must verify the compatibility of the sufficient conditions (8.118). First, we choose initial functions $u_0(x) \in \mathbb{C}^{(2)}(\overline{\Omega})$ and $w_0(x) \in \mathbb{C}^{(4)}(\overline{\Omega})$ sufficiently "large" so that

$$\frac{c_0^2}{2}\|u_{0x}\|_2^2 + \frac{c_1^2}{2}\|w_{0xx}\|_2^2 < \frac{c_0^2}{2}\int_\Omega (w_{0x})^2 u_{0x}\, dx. \tag{8.119}$$

Such functions obviously exist. Now, for fixed functions u_0 and w_0, we consider the functions

$$u_1 = \lambda u_0 \quad \text{and} \quad w_1 = \lambda w_0, \tag{8.120}$$

where $\lambda > 0$ is so large that

$$\begin{aligned} E(0) = {} & \frac{\lambda^2}{2}\int_\Omega [|u_0|^2 + |w_0|^2]\, dx + \frac{c_0^2}{2}\|u_{0x}\|_2^2 \\ & + \frac{c_1^2}{2}\|w_{0xx}\|_2^2 - \frac{c_0^2}{2}\int_\Omega (w_{0x})^2 u_{0x}\, dx > 0. \end{aligned} \tag{8.121}$$

For these initial data,

$$(\Phi'(0))^2 = 4\Big(\int_\Omega [u_0u_1 + w_0w_1]\,dx\Big)^2 = 4\lambda^2\Big(\int_\Omega [|u_0|^2 + |w_0|^2]\,dx\Big)^2. \tag{8.122}$$

Therefore, we can rewrite condition (8.118) in the form

$$\begin{aligned}(\Phi'(0))^2 &= 4\Big(\int_\Omega [u_0u_1 + w_0w_1]\,dx\Big)^2\\ &= 4\lambda^2\Big(\int_\Omega [|u_0|^2 + |w_0|^2]\,dx\Big)^2 > 8E(0)\Phi(0)\\ &= 4\lambda^2\Big(\int_\Omega [|u_0|^2 + |w_0|^2]\,dx\Big)^2\\ &\quad + 8\Phi(0)\Big(\frac{c_0^2}{2}\|u_{0x}\|_2^2 + \frac{c_1^2}{2}\|w_{0xx}\|_2^2 - \frac{c_0^2}{2}\int_\Omega (w_{0x})^2u_{0x}\,dx\Big).\end{aligned} \tag{8.123}$$

Clearly, inequality (8.123) holds if condition (8.119) does. Thus, the compatibility of conditions (8.118) and hence the theorem are proved.

Now we consider problem (8.90). First, we note that the equation can be reduced to a more convenient form. Indeed, we have the equalities

$$w_{xxxx}(w_{xx})^2 + 2w_{xx}(w_{xxx})^2 = \frac{\partial}{\partial x}(w_{xxx}(w_{xx})^2) = \frac{1}{3}((w_{xx})^3)_{xx}.$$

Therefore we consider the following more general multidimensional problem:

$$\frac{\partial^2 u}{\partial t^2} + \Delta^2 u = \Delta(\Delta u)^3, \quad u|_{\partial\Omega} = \frac{\partial u}{\partial n_x}\Big|_{\partial\Omega} = 0, \quad u(x,0) = u_0(x), \tag{8.124}$$

in a bounded domain $\Omega \subset \mathbb{R}^3$ with smooth boundary $\partial\Omega \in \mathbb{C}^{4,\delta}$.

Now we give the definition of a classical solution of problem (8.124).

Definition 8.5. A *classical solution* of problem (8.124) is a function $u(x)(t)$ of the class

$$u(x)(t) \in \mathbb{C}^{(2)}([0,T_0); \mathbb{C}(\overline{\Omega})) \cap \mathbb{C}([0,T_0); \mathbb{C}^{(4)}(\overline{\Omega})),$$

satisfying the initial and boundary conditions pointwise, where T_0 is the maximal time of existence of a nonextendable solution.

Assume that, for some $T_0 > 0$, a classical solution of problem (8.124) exists. Multiplying both sides of (8.124) by u, we obtain the first energy equality

$$\frac{1}{2}\frac{d^2}{dt^2}\Phi - J + \|\Delta u\|_2^2 = \|\Delta u\|_4^4, \tag{8.125}$$

where

$$\Phi(t) \equiv \int_{\Omega} |u|^2\,dx, \quad J \equiv \int_{\Omega} |u'|^2\,dx. \tag{8.126}$$

Multiplying both sides of equation (8.124) by u', we obtain the second energy equality

$$\frac{d}{dt}\left(\frac{1}{2}J(t) + \frac{1}{2}\|\Delta u\|_2^2\right) = \frac{1}{4}\frac{d}{dt}\|\Delta u\|_4^4. \tag{8.127}$$

Integrating both sides of equation (8.127) by t, we obtain the equation

$$\frac{1}{2}J(t) + \frac{1}{2}\|\Delta u\|_2^2 - E(0) = \frac{1}{4}\|\Delta u\|_4^4, \tag{8.128}$$

where

$$E(0) = \frac{1}{2}J(0) + \frac{1}{2}\|\Delta u_0\|_2^2 - \frac{1}{4}\|\Delta u_0\|_4^4 \tag{8.129}$$

is the integration constant. Substituting from equation (8.128) the norm $\|\Delta u\|_4^4$ of (8.125), we arrive at the equality

$$\frac{1}{2}\frac{d^2}{dt^2}\Phi - J + \|\Delta u\|_2^2 = 2J(t) + 2\|\Delta u\|_2^2 - 4E(0),$$

which immediately implies the inequality

$$\frac{1}{2}\Phi'' - 3J + 4E(0) \geq 0. \tag{8.130}$$

On the other hand, as we have seen before,

$$(\Phi')^2 \leq 4\Phi J. \tag{8.131}$$

Combining (8.130) and (8.131), we obtain the required differential inequality

$$\Phi\Phi'' - \frac{3}{2}(\Phi')^2 + 8E(0)\Phi \geq 0. \tag{8.132}$$

Comparing this inequality with the differential inequality from Theorem A.1 (Appendix, p. 299), we obtain the coefficients

$$\alpha = \frac{3}{2}, \quad \beta = 8E(0), \quad \frac{2\beta}{2\alpha - 1} = 8E(0), \quad \gamma = 0,$$

for which the following blow-up result holds.

Theorem 8.5. *Assume that, for some $T_0 > 0$, there exists a classical solution of problem (8.124). Then, under the conditions*

$$(\Phi'(0))^2 > 8E(0)\Phi(0) > 0, \quad E(0) > 0, \quad \Phi'(0) > 0, \tag{8.133}$$

we have the inequality

$$\Phi(t) \geq (\Phi^{-1/2}(0) - At)^{-1/2}$$

and the following upper estimate for T_0:

$$T_0 \leq \Phi^{-1/2}(0)A^{-1},$$

where

$$\Phi(t) = \int_\Omega |u|^2\,dx, \quad \Phi'(0) = 2\int_\Omega u_0 u_1\,dx,$$
$$A^2 = \frac{1}{4}\Phi^{-3}(0)[(\Phi'(0))^2 - 8E(0)\Phi(0)].$$

The compatibility of conditions (8.133) can be verified as before.

Finally, we consider problem (8.91).

Definition 8.6. A *classical solution* of problem (8.91) is a function $u(x)(t)$ of the class

$$u(x)(t) \in \mathbb{C}([0,T_0);\mathbb{C}^{(4)}(\overline{\Omega})) \cap \mathbb{C}^{(2)}([0,T_0);\mathbb{C}^{(2)}(\overline{\Omega}))$$

satisfying the initial and boundary conditions pointwise.

Assume that there exists a maximal time $T_0 > 0$ for which a nonextendable classical solution of problem (8.91) exists. Multiplying both sides of equation (8.91) by $u(x)(t)$, we obtain the first energy equality

$$\frac{1}{2}\frac{d^2}{dt^2}\Phi(t) - \mathrm{J}(t) + \|u_{xx}\|_2^2 + \|u_x\|_2^2 = \int_\Omega (u_x)^3\,dx, \tag{8.134}$$

where

$$\Phi(t) \equiv \int_\Omega [|u_x|^2 + |u|^2]\,dx, \quad \mathrm{J}(t) \equiv \int_\Omega [|u_x'|^2 + |u'|^2]\,dx. \tag{8.135}$$

Next, multiplying both sides of (8.91) by $u'(x)(t)$ and integrating by parts, we obtain the second energy equality

$$\frac{d}{dt}\left[\frac{1}{2}\mathrm{J}(t) + \frac{1}{2}\|u_{xx}\|_2^2 + \frac{1}{2}\|u_x\|_2^2\right] = \frac{1}{3}\frac{d}{dt}\int_\Omega (u_x)^3\,dx. \tag{8.136}$$

Integrating it by t, we get the equation

$$\frac{1}{2}\mathrm{J}(t) + \frac{1}{2}\|u_{xx}\|_2^2 + \frac{1}{2}\|u_x\|_2^2 - E(0) = \frac{1}{3}\int_\Omega (u_x)^3\,dx, \tag{8.137}$$

where

$$E(0) \equiv \frac{1}{2}\mathrm{J}(0) + \frac{1}{2}\|u_{0xx}\|_2^2 + \frac{1}{2}\|u_{0x}\|_2^2 - \frac{1}{3}\int_\Omega (u_{0x})^3\,dx \tag{8.138}$$

is the integration constant. Expressing the integral

$$\int_\Omega (u_x)^3\,dx = \frac{3}{2}\mathrm{J}(t) + \frac{3}{2}\|u_{xx}\|_2^2 + \frac{3}{2}\|u_x\|_2^2 - 3E(0) \tag{8.139}$$

from (8.137) and substituting it into equation (8.134), we obtain

$$\frac{1}{2}\frac{d^2}{dt^2}\Phi(t) - \mathrm{J}(t) + \|u_{xx}\|_2^2 + \|u_x\|_2^2 = \frac{3}{2}\mathrm{J}(t) + \frac{3}{2}\|u_{xx}\|_2^2 + \frac{3}{2}\|u_x\|_2^2 - 3E(0).$$

Eliminating the norms, we obtain the inequality

$$\frac{1}{2}\frac{d^2}{dt^2}\Phi(t) - \frac{5}{2}\mathrm{J}(t) + 3E(0) \geq 0. \tag{8.140}$$

Moreover, as usual, we have the inequality

$$(\Phi')^2 \leq 4\Phi\mathrm{J}. \tag{8.141}$$

From inequalities (8.140) and (8.141) we obtain the required inequality

$$\Phi\Phi'' - \frac{5}{4}(\Phi')^2 + 6E(0)\Phi \geq 0. \tag{8.142}$$

Applying Theorem A.1 (Appendix, p. 299), we obtain the coefficients

$$\alpha = \frac{5}{4}, \quad \beta = 6E(0), \quad \gamma = 0, \quad \frac{2\beta}{2\alpha - 1} = 8E(0).$$

Thus, we obtain the following blow-up result.

Theorem 8.6. *Assume that for certain $T_0 > 0$, there exists a classical solution of the problem* (8.91). *Then under the conditions*

$$(\Phi'(0))^2 > 8E(0)\Phi(0) > 0, \quad E(0) > 0, \quad \Phi'(0) > 0, \tag{8.143}$$

the inequality

$$\Phi(t) \geq \left(\Phi^{-1/4}(0) - At\right)^{-1/4}$$

holds and the following upper estimate is valid:

$$T_0 \le \Phi^{-1/4}(0)A^{-1},$$

where

$$\Phi(t) = \int_\Omega [|u|^2 + |u_x|^2]\,dx, \quad \Phi'(0) = 2\int_\Omega [u_0 u_1 + u_{0x}u_{1x}]\,dx,$$

$$A^2 = \frac{1}{16}\Phi^{-5/2}(0)[(\Phi'(0))^2 - 8E(0)\Phi(0)].$$

The compatibility of conditions (8.143) can be performed as before. Thus we are convinced that the modified energy method is applicable for various hyperbolic equations. In the following section, we consider a vector-valued hyperbolic problem that appears in electromagnetism and leads to an ordinary differential inequality, which is somewhat different from those encountered before.

8.2.2 System of nonlinear wave equations of the theory of electromagnetic waves

In this section, we consider the following homogeneous Dirichlet problem for the nonlinear system of hyperbolic equation (see [53]):

$$\frac{\partial^2 \mathbf{A}}{\partial t^2} + \mu\frac{\partial \mathbf{A}}{\partial t} - \Delta\mathbf{A} - h(x, |\mathbf{A}|)\mathbf{A} = \nabla\psi, \quad \mu \ge 0, \quad \operatorname{div}\mathbf{A} = 0, \tag{8.144}$$

$$\mathbf{A}(x,0) = \mathbf{A}_0(x), \quad \mathbf{A}'(x,0) = \mathbf{A}_1(x), \quad \mathbf{A}|_{\partial\Omega} = 0, \tag{8.145}$$

in which the vector-valued function $\mathbf{A}(x,t)$ has the sense of the vector potential, and the function $\psi(x,t)$ is related to the scalar potential φ by the formula

$$\psi(x,t) = -\frac{\varepsilon}{c}\frac{\partial\varphi}{\partial t}.$$

System (8.144) is obtained in the so-called Coulomb gauge (see [56])

$$\operatorname{div}\mathbf{A} = 0.$$

We consider problem (8.144)–(8.145) in a bounded domain $\Omega \subset \mathbb{R}^3$ with smooth boundary $\partial\Omega \in \mathbb{C}^{2,\delta}$, $\delta \in (0,1]$. Using the modified energy method, we find sufficient conditions of blow-up.

First, we formulate the definitions of a classical solution of the problem and the auxiliary space $\mathbf{H}(\Omega)$.

Definition 8.7. A *classical solution* $\mathbf{A}(x,t)$ of problem (8.144)–(8.145) is a function of the class

$$\mathbf{A}(x)(t) \in \mathbb{C}^{(1)}([0,T_0);\mathbf{H}(\Omega)), \quad \mathbf{H}(\Omega) = \{\mathbf{v} : \mathbf{v} \in \mathbf{H}_0^1(\Omega),\ \operatorname{div}\mathbf{v} = 0\},$$

where $T_0 > 0$ is the maximal time of existence of a nonextendable solution.

Now we obtain energy equalities. Assume that there exists $T_0 > 0$ such that a classical solution

$$\mathbf{A}(x,t) \in \mathbb{C}^{(2)}([0,T];\mathbf{H}(\Omega))$$

exists. Multiplying scalarly both sides of equation (8.144) by $\mathbf{A}(x,t)$ and integrating over the domain Ω, we obtain

$$\frac{1}{2}\frac{d^2\Phi}{dt^2} + \frac{\mu}{2}\frac{d\Phi}{dt} - \mathrm{J}(t) + ((\mathbf{A},\mathbf{A})) = \int_\Omega h(x,|\mathbf{A}|)|\mathbf{A}|^2\,dx, \tag{8.146}$$

where

$$\Phi(t) \equiv \int_\Omega |\mathbf{A}|^2\,dx, \quad \mathrm{J}(t) \equiv \int_\Omega |\mathbf{A}'|^2\,dx. \tag{8.147}$$

To obtain the second energy equality, we multiply scalarly both sides of (8.144) by $\mathbf{A}'_t(x,t)$ and integrate by parts; then we obtain the equality

$$\frac{d}{dt}\left[\frac{1}{2}\mathrm{J}(t) + \frac{1}{2}((\mathbf{A},\mathbf{A})) - \int_\Omega \mathcal{H}(x,|\mathbf{A}|)\,dx\right] + \mu\mathrm{J}(t) = 0. \tag{8.148}$$

Integrating by time, we arrive at the inequality

$$\frac{1}{2}\mathrm{J}(t) + \frac{1}{2}((\mathbf{A},\mathbf{A})) - E(0) \le \int_\Omega \mathcal{H}(x,|\mathbf{A}|)\,dx, \tag{8.149}$$

where

$$E(0) \equiv \frac{1}{2}\mathrm{J}(0) + \frac{1}{2}((\mathbf{A}_0,\mathbf{A}_0)) - \int_\Omega \mathcal{H}(x,|\mathbf{A}_0|)\,dx \tag{8.150}$$

is the integration constant. Applying inequality (8.144), we obtain the estimate

$$\frac{\theta}{2}\mathrm{J}(t) + \frac{\theta}{2}((\mathbf{A},\mathbf{A})) - \theta E(0) \le \int_\Omega h(x,|\mathbf{A}|)|\mathbf{A}|^2\,dx. \tag{8.151}$$

From (8.146) and (8.151) we have

$$\frac{1}{2}\frac{d^2\Phi}{dt^2} + \frac{\mu}{2}\frac{d\Phi}{dt} - \mathrm{J}(t) + ((\mathbf{A},\mathbf{A})) \ge \frac{\theta}{2}\mathrm{J}(t) + \frac{\theta}{2}((\mathbf{A},\mathbf{A})) - \theta E(0),$$

which immediately implies

$$\frac{1}{2}\frac{d^2\Phi}{dt^2} + \frac{\mu}{2}\frac{d\Phi}{dt} + \theta E(0) \ge \left(1 + \frac{\theta}{2}\right)\mathrm{J}(t), \tag{8.152}$$

since $\theta > 2$. Applying the Cauchy–Bunyakovsky inequality, we obtain the auxiliary inequality

$$(\Phi'(t))^2 \leq 4J(t)\Phi(t). \tag{8.153}$$

From (8.152) and (8.153) we obtain an ordinary differential inequality, to which we apply Theorem A.1 from the Appendix:

$$\Phi\Phi'' - \frac{1}{2}\left(1+\frac{\theta}{2}\right)(\Phi')^2 + \mu\Phi\Phi' + 2\theta E(0)\Phi \geq 0. \tag{8.154}$$

We rewrite it in a more convenient form

$$\Phi\Phi'' - \alpha(\Phi')^2 + \gamma\Phi\Phi' + \beta\Phi \geq 0, \tag{8.155}$$

where the coefficients are defined by the formulas

$$\alpha = \frac{1}{2}\left(1+\frac{\theta}{2}\right), \quad \beta = 2\theta E(0), \quad \gamma = \mu.$$

Moreover, by the condition $\theta > 2$ we have $\alpha > 1$. Impose the condition

$$E(0) > 0 \quad \Rightarrow \quad \beta > 0$$

and apply Theorem A.1 (see Appendix). We conclude that, under the conditions

$$\Phi'(0) > \frac{\gamma}{\alpha-1}\Phi(0), \tag{8.156}$$

$$\left(\Phi'(0) - \frac{\gamma}{\alpha-1}\Phi(0)\right)^2 > \frac{2\beta}{2\alpha-1}\Phi(0), \tag{8.157}$$

a solution $\mathbf{A}(x)(t)$ blows up in finite time due to the lower estimate

$$\Phi(t) \geq \frac{e^{\gamma t/(\alpha-1)}}{[\Phi^{1-\alpha}(0) - At]^{1/(\alpha-1)}}, \tag{8.158}$$

where

$$A^2 \equiv (\alpha-1)^2\Phi^{-2\alpha}(0)\left[\left(\Phi'(0) - \frac{\gamma}{\alpha-1}\Phi(0)\right)^2 - \frac{2\beta}{2\alpha-1}\Phi(0)\right].$$

The general result can be formulated as follows.

Theorem 8.7. *For any initial data*

$$\mathbf{A}_0 \in \mathbf{H}(\Omega), \quad \mathbf{A}_1 \in \mathbf{X}(\Omega)$$

satisfying the conditions

$$\Phi'(0) > \frac{4\mu}{\theta-2}\Phi(0) + (8E(0))^{1/2}\Phi^{1/2}(0), \quad \Phi(0) > 0, \quad E(0) > 0, \tag{8.159}$$

there are no global-in-time classical solutions of problem (8.144)–(8.145). *Moreover, the value of* $T_0 > 0$ *has the following upper estimate*:

$$T_0 \le \Phi^{(2-\theta)/4}(0)A^{-1},$$

and the solution itself satisfies the limit equality

$$\limsup_{t\uparrow T_0}[((\mathbf{A},\mathbf{A})) + \|\mathbf{A}'\|_2] = +\infty,$$

where

$$A^2 \equiv \frac{(\theta-2)^2}{16}\Phi^{-1-\theta/2}(0)\left[\left(\Phi'(0) - \frac{4\mu}{\theta-2}\Phi(0)\right)^2 - 8E(0)\Phi(0)\right],$$

$$E(0) \equiv \frac{1}{2}\int_\Omega |\mathbf{A}_1|^2\,dx + \frac{1}{2}((\mathbf{A}_0,\mathbf{A}_0)) - \int_\Omega \mathcal{H}(x,|\mathbf{A}_0|)\,dx,$$

$$\Phi(0) = \int_\Omega |\mathbf{A}_0|^2\,dx, \quad \Phi'(0) = 2\int_\Omega (\mathbf{A}_1,\mathbf{A}_0)\,dx.$$

Now we verify the compatibility of the sufficient conditions (8.159). We will see that these conditions are compatible for sufficiently small μ. First, we choose an initial function $\mathbf{A}_0 \in \mathbf{H}(\Omega)$ so large that

$$\int_\Omega \mathcal{H}(x,|\mathbf{A}_0|)\,dx > \frac{1}{2}((\mathbf{A}_0,\mathbf{A}_0)) + \frac{2}{\theta-2}\int_\Omega |\mathbf{A}_0|^2\,dx. \tag{8.160}$$

For a fixed function $\mathbf{A}_0$, we take

$$\mathbf{A}_1 = \lambda\mathbf{A}_0, \quad \lambda > \frac{2\mu}{\theta-2}, \tag{8.161}$$

where $\lambda > 0$ is so large that

$$E(0) \equiv \frac{\lambda^2}{2}\int_\Omega |\mathbf{A}_0|^2\,dx + \frac{1}{2}((\mathbf{A}_0,\mathbf{A}_0)) - \int_\Omega \mathcal{H}(x,|\mathbf{A}_0|)\,dx > 0.$$

Note that, for this λ,

$$\Phi'(0) > \frac{4\mu}{\theta-2}\Phi(0) > 0. \tag{8.162}$$

Indeed, we have:

$$\Phi'(0) - \frac{4\mu}{\theta-2}\Phi(0) = 2\int_\Omega \left(\mathbf{A}_1 - \frac{2\mu}{\theta-2}\mathbf{A}_0, \mathbf{A}_0\right)$$

$$> 2\left(\lambda - \frac{2\mu}{\theta - 2}\right) \int_\Omega |\mathbf{A}_0|^2\, dx > 0. \tag{8.163}$$

Now we consider condition (8.159), which by (8.163) is equivalent to the condition

$$\left(\Phi'(0) - \frac{4\mu}{\theta - 2}\Phi(0)\right)^2 > 8E(0)\Phi(0). \tag{8.164}$$

For the initial functions chosen, we have the following equalities for the left-hand side of (8.164):

$$\begin{aligned}\left(\Phi'(0) - \frac{4\mu}{\theta - 2}\Phi(0)\right)^2 &= 4\left(\lambda - \frac{2\mu}{\theta - 2}\right)^2 \left(\int_\Omega |\mathbf{A}_0|^2\, dx\right)^2 \\ &= \left(4\lambda^2 - \frac{16\lambda\mu}{\theta - 2} + \frac{16\mu^2}{(\theta - 2)^2}\right)\left(\int_\Omega |\mathbf{A}_0|^2\, dx\right)^2,\end{aligned} \tag{8.165}$$

whereas the right-hand side of (8.164) can be rewritten in the form

$$\begin{aligned}8E(0)\Phi(0) = \Bigg[4\lambda^2 \int_\Omega |\mathbf{A}_0|^2\, dx &+ 4((\mathbf{A}_0, \mathbf{A}_0)) \\ &- 8\int_\Omega \mathcal{H}(x, |\mathbf{A}_0|)\, dx\Bigg] \int_\Omega |\mathbf{A}_0|^2\, dx.\end{aligned} \tag{8.166}$$

Combining (8.164)–(8.166), we see that the inequality

$$\int_\Omega \mathcal{H}(x, |\mathbf{A}_0|)\, dx + \frac{2\mu^2}{(\theta - 2)^2} \int_\Omega |\mathbf{A}_0|^2\, dx > \frac{2\lambda\mu}{\theta - 2} \int_\Omega |\mathbf{A}_0|^2\, dx + \frac{1}{2}((\mathbf{A}_0, \mathbf{A}_0)) \tag{8.167}$$

is a sufficient condition. We set

$$\lambda = \frac{1}{\mu} \quad \text{for } \mu > 0.$$

For this choice of λ, condition (8.161) takes the form

$$\mu < \left(\frac{\theta - 2}{2}\right)^{1/2}.$$

Clearly, $\lambda \to +\infty$ as $\mu \to +0$. Rewrite (8.167) as the inequality

$$\int_\Omega \mathcal{H}(x, |\mathbf{A}_0|)\, dx + \frac{2\mu^2}{(\theta - 2)^2} \int_\Omega |\mathbf{A}_0|^2\, dx > \frac{2}{\theta - 2} \int_\Omega |\mathbf{A}_0|^2\, dx + \frac{1}{2}((\mathbf{A}_0, \mathbf{A}_0)),$$

which obviously holds under condition (8.160).

Thus, it remains to verify (8.160). Let $h(x,s) = s^{p-2}$, $p > 2$; then

$$\mathcal{H}(x, |\mathbf{A}|) = \frac{1}{2}|\mathbf{A}|^p, \quad \int_\Omega \mathcal{H}(x, |\mathbf{A}|) = \frac{1}{p}\int_\Omega |\mathbf{A}|^p\, dx.$$

Since $p > 2$, it is clear that, for sufficiently large $|\mathbf{A}_0|$, inequality (8.160) holds, and, moreover, $\theta = p > 2$. Thus, for sufficiently small $\mu \geq 0$, the compatibility of the conditions and hence the blow-up result are proved.

8.2.3 System of the theory of charged mesons

The last problem for a hyperbolic equation that will be presented here is the initial-boundary-value problem for the following system in a bounded domain $\Omega \subset \mathbb{R}^3$ with smooth boundary $\partial\Omega \in \mathbb{C}^{2,\delta}$, $\delta \in (0,1]$:

$$u_{tt} + \mu u_t - \Delta u + m_1^2 u = 4(u+av)^3 + 2buv^2, \tag{8.168}$$

$$v_{tt} + \mu v_t - \Delta v + m_2^2 v = 4a(u+av)^3 + 2bu^2 v, \tag{8.169}$$

$$u(x,0) = u_0(x), \quad v(x,0) = v_0(x), \tag{8.170}$$

$$u'(x,0) = u_1(x), \quad v'(x,0) = v_1(x), \tag{8.171}$$

$$u|_{\partial\Omega} = v|_{\partial\Omega} = 0, \tag{8.172}$$

where

$$\mu \geq 0, \quad a > 0, \quad b > 0, \quad m_1 > 0, \quad m_2 > 0.$$

We give a review of results obtained for problem (8.168)–(8.172) for $\mu = 0$. First of all, we note the paper [91], where system (8.168)–(8.169), which describes the interaction of two scalar fields in the kink model, was obtained. Note also the papers [98, 36, 68, 70, 71], in which the local and global weak solvability was examined. In [62, 63], sufficient conditions of blow-up and the global solvability were studied. In these papers, important blow-up results for arbitrary positive energy was obtained. In addition to the positivity condition for the initial energy, two other conditions were needed, but the compatibility of all these three conditions has not been verified. Now we complement the positivity condition for the initial energy with another condition (note that it is contained in the three conditions of [62, 63] mentioned before) and verify their compatibility.

We consider weak generalized solutions of problem (8.168)–(8.172).

Definition 8.8. A *weak generalized solution* of problem (8.168)–(8.172) is a pair of functions (u,v) of the class

$$(u,v) \in \mathbb{L}^\infty(0,T;\mathbb{H}_0^1(\Omega) \otimes \mathbb{H}_0^1(\Omega)), \quad (u',v') \in \mathbb{L}^\infty(0,T;\mathbb{L}^2(\Omega) \otimes \mathbb{L}^2(\Omega)),$$

$$(u'', v'') \in \mathbb{L}^\infty(0, T; \mathbb{H}^{-1}(\Omega) \otimes \mathbb{H}^{-1}(\Omega)),$$

satisfying the equality

$$\int_0^T dt \langle D(u, v), w \rangle = 0 \quad \text{for all } w(x)(t) \in \mathbb{L}^1(0, T; \mathbb{H}_0^1(\Omega) \otimes \mathbb{H}_0^1(\Omega)), \tag{8.173}$$

where

$$D(u, v) = (D_1(u, v), D_2(u, v)), \quad D_1(u, v) = u_{tt} + \mu u_t - \Delta u + m_1^2 u - 4(u + av)^3 - 2buv^2,$$
$$D_2(u, v) = v_{tt} + \mu v_t - \Delta v + m_2^2 v - 4a(u + av)^3 - 2bu^2 v,$$

and $\langle \cdot, \cdot \rangle$ is the duality bracket between the Hilbert spaces $\mathbb{H}_0^1(\Omega) \otimes \mathbb{H}_0^1(\Omega)$ and $\mathbb{H}^{-1}(\Omega) \otimes \mathbb{H}^{-1}(\Omega)$.

In [62], the following existence theorem was proved.

Theorem 8.8. *For any functions*

$$(u_0, v_0) \in \mathbb{H}_0^1(\Omega) \otimes \mathbb{H}_0^1(\Omega) \cap \mathbb{H}^2(\Omega) \otimes \mathbb{H}^2(\Omega), \quad (u_1, v_1) \in \mathbb{H}_0^1(\Omega) \otimes \mathbb{H}_0^1(\Omega)$$

there exists a weak generalized solution of problem (8.168)–(8.172) *in the class*

$$\mathbb{C}^{(2)}([0, T_0); \mathbb{L}^2(\Omega) \otimes \mathbb{L}^2(\Omega)), \quad \mathbb{C}^{(1)}([0, T_0); \mathbb{H}_0^1(\Omega) \otimes \mathbb{H}_0^1(\Omega)).$$

Moreover, $T_0 = T_0(u_0, v_0, u_1, v_1) > 0$, *and either* $T_0 = +\infty$ *or* $T_0 < +\infty$. *In the latter case, we have*

$$\limsup_{t \uparrow T_0} [\|\nabla u'\|_2^2 + \|\nabla v'\|_2^2 + \|\nabla u\|_2^2 + \|\nabla v\|_2^2 + \|\Delta u\|_2^2 + \|\Delta v\|_2^2] = +\infty.$$

Now we obtain sufficient conditions for blow-up in finite time of a weak generalized solution of problem (8.168)–(8.172). By the existence theorem this solution belongs to the class

$$\mathbb{C}^{(2)}([0, T_0); \mathbb{L}^2(\Omega) \otimes \mathbb{L}^2(\Omega)), \quad \mathbb{C}^{(1)}([0, T_0); \mathbb{H}_0^1(\Omega) \otimes \mathbb{H}_0^1(\Omega)).$$

To obtain the first energy equality, we choose $w = (u, v)$ in definition (8.173); integrating by parts, we obtain the formula

$$\begin{aligned} &\frac{1}{2} \frac{d^2 \Phi}{dt^2} + \frac{\mu}{2} \frac{d\Phi}{dt} - \mathrm{J} + \|\nabla u\|_2^2 + \|\nabla v\|_2^2 + m_1^2 \|u\|_2^2 + m_2^2 \|v\|_2^2 \\ &\quad = 4 \int_\Omega (u + av)^4 \, dx + 4b \int_\Omega u^2 v^2 \, dx, \end{aligned} \tag{8.174}$$

where $\Phi(t)$ and $\mathrm{J}(t)$ are of the form

$$\Phi(t) \equiv \int_\Omega [u^2 + v^2]\,dx, \quad \mathrm{J}(t) \equiv \int_\Omega [|u'|^2 + |v'|^2]\,dx. \tag{8.175}$$

Setting $w = (u', v')$ in equation (8.173) and integrating by parts, we get the second energy equality

$$\frac{d}{dt}\left[\frac{1}{2}\mathrm{J} + \frac{1}{2}\|\nabla u\|_2^2 + \frac{1}{2}\|\nabla v\|_2^2 + \frac{m_1^2}{2}\|u\|_2^2 + \frac{m_2^2}{2}\|v\|_2^2\right] + \mu \mathrm{J}$$
$$= \frac{d}{dt}\int_\Omega (u + av)^4\,dx + \frac{d}{dt} b \int_\Omega u^2v^2\,dx. \tag{8.176}$$

Integrating by time, we obtain the estimate

$$\frac{1}{2}\mathrm{J} + \frac{1}{2}\|\nabla u\|_2^2 + \frac{1}{2}\|\nabla v\|_2^2 + \frac{m_1^2}{2}\|u\|_2^2 + \frac{m_2^2}{2}\|v\|_2^2 - E(0)$$
$$\leq \int_\Omega (u + av)^4\,dx + b\int_\Omega u^2v^2\,dx, \tag{8.177}$$

where

$$E(0) \equiv \frac{1}{2}\int_\Omega [|u_1|^2 + |v_1|^2]\,dx + \frac{1}{2}\|\nabla u_0\|_2^2 + \frac{1}{2}\|\nabla v_0\|_2^2$$
$$+ \frac{m_1^2}{2}\|u_0\|_2^2 + \frac{m_2^2}{2}\|v_0\|_2^2 - \int_\Omega (u_0 + av_0)^4\,dx - b\int_\Omega u_0^2v_0^2\,dx \tag{8.178}$$

is the integration constant. Combining (8.174) and (8.177), we obtain the inequality

$$\frac{1}{2}\frac{d^2\Phi}{dt^2} + \frac{\mu}{2}\frac{d\Phi}{dt} - \mathrm{J} + \|\nabla u\|_2^2 + \|\nabla v\|_2^2 + m_1^2\|u\|_2^2 + m_2^2\|v\|_2^2$$
$$\geq 2\mathrm{J} + 2\|\nabla u\|_2^2 + 2\|\nabla v\|_2^2 + 2m_1^2\|u\|_2^2 + 2m_2^2\|v\|_2^2 - 4E(0), \tag{8.179}$$

which, in turn, implies

$$\frac{1}{2}\frac{d^2\Phi}{dt^2} + \frac{\mu}{2}\frac{d\Phi}{dt} + 4E(0) \geq 3\mathrm{J}. \tag{8.180}$$

Applying the Cauchy–Bunyakovsky inequality, we arrive at the differential inequality

$$(\Phi')^2 \leq 4\Phi\mathrm{J}. \tag{8.181}$$

Finally, we obtain the traditional differential inequality

$$\Phi\Phi'' - \frac{3}{2}(\Phi')^2 + \mu\Phi\Phi' + 8E(0)\Phi \geq 0. \tag{8.182}$$

Comparing it with the differential inequality from Theorem A.1 (see Appendix), we have

$$\alpha = \frac{3}{2}, \quad \beta = 8E(0), \quad \gamma = \mu, \quad \frac{2\beta}{2\alpha - 1} = 8E(0), \quad \frac{\gamma}{\alpha - 1} = 2\mu.$$

Thus, we have proved the following blow-up result.

Theorem 8.9. *Assume that the initial data of problem* (8.168)–(8.172) *belong to the class*

$$(u_0, v_0) \in \mathbb{H}_0^1(\Omega) \otimes \mathbb{H}_0^1(\Omega) \cap \mathbb{H}^2(\Omega) \otimes \mathbb{H}^2(\Omega), \quad (u_1, v_1) \in \mathbb{H}_0^1(\Omega) \otimes \mathbb{H}_0^1(\Omega)$$

and satisfy the conditions

$$\Phi'(0) > 2\mu\Phi(0) + (8E(0)\Phi(0))^{1/2} > 0, \quad E(0) > 0. \tag{8.183}$$

Then the lifetime of a weak generalized solution T_0 is bounded from above:

$$T_0 \leq \Phi^{-1/2}(0)A^{-1}.$$

Moreover,

$$\limsup_{t \uparrow T_0} [\|\nabla u'\|_2^2 + \|\nabla v'\|_2^2 + \|\nabla u\|_2^2 + \|\nabla v\|_2^2 + \|\Delta u\|_2^2 + \|\Delta v\|_2^2] = +\infty,$$

where

$$\Phi(0) = \int_\Omega [|u_0|^2 + |v_0|^2]\, dx, \quad \Phi'(0) = 2\int_\Omega [u_1 u_0 + v_1 v_0]\, dx,$$

$$E(0) \equiv \frac{1}{2}\int_\Omega [|u_1|^2 + |v_1|^2]\, dx + \frac{1}{2}\|\nabla u_0\|_2^2 + \frac{1}{2}\|\nabla v_0\|_2^2 + \frac{m_1^2}{2}\|u_0\|_2^2 + \frac{m_2^2}{2}\|v_0\|_2^2 - \int_\Omega (u_0 + a v_0)^4\, dx - b\int_\Omega u_0^2 v_0^2\, dx,$$

and

$$A^2 \equiv \frac{1}{4}\Phi^{-3}(0)[(\Phi'(0) - 2\mu\Phi(0))^2 - 8E(0)\Phi(0)].$$

Now we verify the compatibility of conditions (8.183) for sufficiently small $\mu \geq 0$. First, we choose

$$\mathbf{w}_0 = (u_0, v_0) \in \mathbb{H}_0^1(\Omega) \otimes \mathbb{H}_0^1(\Omega) \cap \mathbb{H}^2(\Omega) \otimes \mathbb{H}^2(\Omega)$$

so “small” that

$$\int_\Omega (u_0 + a v_0)^4\, dx + b\int_\Omega u_0^2 v_0^2\, dx$$

$$> \int_\Omega [|u_0|^2 + |v_0|^2]\,dx + \frac{1}{2}\|\nabla u_0\|_2^2 + \frac{1}{2}\|\nabla v_0\|_2^2$$
$$+ \frac{m_1^2}{2}\|u_0\|_2^2 + \frac{m_2^2}{2}\|v_0\|_2^2; \tag{8.184}$$

such functions obviously exist. Fix $\mathbf{w}_0 = (u_0, v_0)$ and take $\mathbf{w}_1 = \lambda(u_0, v_0)$, $\lambda > \mu$. Then

$$\Phi'(0) - 2\mu\Phi(0) = 2(\lambda - \mu)\Phi(0) > 0 \quad \text{for } \lambda > \mu. \tag{8.185}$$

Increasing $\lambda > \mu$, we achieve the positivity condition for $E(0)$:

$$E(0) \equiv \frac{\lambda^2}{2} \int_\Omega [|u_0|^2 + |v_0|^2]\,dx + \frac{1}{2}\|\nabla u_0\|_2^2 + \frac{1}{2}\|\nabla v_0\|_2^2$$
$$+ \frac{m_1^2}{2}\|u_0\|_2^2 + \frac{m_2^2}{2}\|v_0\|_2^2 - \int_\Omega (u_0 + av_0)^4\,dx - b\int_\Omega u_0^2 v_0^2\,dx > 0. \tag{8.186}$$

Note that, using (8.185), we can rewrite the first inequality in (8.183) in the form

$$(\Phi'(0) - 2\mu\Phi(0))^2 > 8E(0)\Phi(0), \tag{8.187}$$

where the left-hand side is equal to

$$(\Phi'(0) - 2\mu\Phi(0))^2 = 4(\lambda - \mu)^2(\Phi(0))^2 = (4\lambda^2 - 8\lambda\mu + 4\mu^2)(\Phi(0))^2, \tag{8.188}$$

whereas the right-hand side is equal to

$$8E(0)\Phi(0) = 4\lambda^2(\Phi(0))^2 + 8\Phi(0)\Big[\frac{1}{2}\|\nabla u_0\|_2^2 + \frac{1}{2}\|\nabla v_0\|_2^2$$
$$+ \frac{m_1^2}{2}\|u_0\|_2^2 + \frac{m_2^2}{2}\|v_0\|_2^2 - \int_\Omega (u_0 + av_0)^4\,dx - b\int_\Omega u_0^2 v_0^2\,dx\Big]. \tag{8.189}$$

Combining (8.188) and (8.189), we obtain the following condition:

$$(\Phi'(0) - 2\mu\Phi(0))^2 = 4(\lambda - \mu)^2(\Phi(0))^2 = (4\lambda^2 - 8\lambda\mu + 4\mu^2)(\Phi(0))^2$$
$$> 8E(0)\Phi(0) = 4\lambda^2(\Phi(0))^2 + 8\Phi(0)\Big[\frac{1}{2}\|\nabla u_0\|_2^2 + \frac{1}{2}\|\nabla v_0\|_2^2$$
$$+ \frac{m_1^2}{2}\|u_0\|_2^2 + \frac{m_2^2}{2}\|v_0\|_2^2 - \int_\Omega (u_0 + av_0)^4\,dx - b\int_\Omega u_0^2 v_0^2\,dx\Big]. \tag{8.190}$$

Further, we have

$$\int_\Omega (u_0 + av_0)^4\,dx + b\int_\Omega u_0^2 v_0^2\,dx + \frac{\mu^2}{2}\int_\Omega [|u_0|^2 + |v_0|^2]\,dx$$

$$> \lambda\mu \int_\Omega [|u_0|^2 + |v_0|^2]\,dx + \frac{1}{2}\|\nabla u_0\|_2^2 + \frac{1}{2}\|\nabla v_0\|_2^2 + \frac{m_1^2}{2}\|u_0\|_2^2 + \frac{m_2^2}{2}\|v_0\|_2^2. \tag{8.191}$$

If $\mu > 0$, then we set $\lambda = 1/\mu$; if $\mu = 0$, then we simply choose sufficiently large λ. Clearly, for sufficiently small μ, all previous requirements remains valid, and now we can rewrite (8.191) in the form

$$\int_\Omega (u_0 + av_0)^4\,dx + b\int_\Omega u_0^2 v_0^2\,dx + \frac{\mu^2}{2}\int_\Omega [|u_0|^2 + |v_0|^2]\,dx > \int_\Omega [|u_0|^2 + |v_0|^2]\,dx + \frac{1}{2}\|\nabla u_0\|_2^2 + \frac{1}{2}\|\nabla v_0\|_2^2 + \frac{m_1^2}{2}\|u_0\|_2^2 + \frac{m_2^2}{2}\|v_0\|_2^2. \tag{8.192}$$

Clearly, under condition (8.184), inequality (8.187) also holds. Thus, for sufficiently small $\mu \geq 0$, we have proved the compatibility of conditions (8.183). The theorem is proved.

8.3 Pseudo-parabolic equations with double nonlinearities

In this section, we consider initial-boundary-value problems for the abstract equation

$$\frac{\partial B(u)}{\partial t} = A(u), \tag{8.193}$$

where $B(u)$ and $A(u)$ are elliptic operators (note that A can be a zero-order operator).

8.3.1 Nonlinear equation with two p-Laplacians

Consider the following initial-boundary-value problem:

$$\frac{\partial}{\partial t}(\Delta u + \Delta_{p_1} u) + \Delta u - \Delta_{p_2} u = 0, \tag{8.194}$$

$$u|_{\partial\Omega} = 0, \quad u(x,0) = u_0(x) \in \mathbb{W}_0^{1,p}(\Omega), \tag{8.195}$$

where

$$\Delta_p u = \operatorname{div}(|\nabla u|^{p-2}\nabla u),$$

$\Omega \subset \mathbb{R}^3$ is a bounded domain with smooth boundary $\partial\Omega \in \mathbb{C}^{(2,\delta)}$, $\delta \in (0,1]$, and $p = \max\{p_1, p_2\}$ with $p_1, p_2 > 2$.

From the physical point of view, the nonlinear problem (8.194)–(8.195) describes the effect of negative differential conductivity in semiconductors in the absence of an external electric field (see [1]). We obtain sufficient conditions of blow-up by using the modified energy method. First, we introduce the notion of a strong generalized solution.

Definition 8.9. A *strong generalized solution* of problem (8.194)–(8.195) is a function $u \in \mathbb{C}^{(1)}([0,T]; \mathbb{W}_0^{1,p}(\Omega))$ satisfying the conditions

$$\left\langle \frac{\partial}{\partial t}(\Delta u + \Delta_{p_1} u) + \Delta u - \Delta_{p_2} u, w \right\rangle = 0 \tag{8.196}$$

for all $w \in \mathbb{W}_0^{1,p}(\Omega)$ and all $t \in [0,T]$,

$$p = \max\{p_1, p_2\}, \quad u(0) = u_0 \in \mathbb{W}_0^{1,p}(\Omega),$$

where $\langle\cdot,\cdot\rangle$ is the duality bracket between the Banach spaces $\mathbb{W}_0^{1,p}(\Omega)$ and $\mathbb{W}^{-1,p'}(\Omega)$.

Denote

$$\Phi(t) = \frac{1}{2}\|\nabla u\|_2^2 + \frac{p_1 - 1}{p_1}\|\nabla u\|_{p_1}^{p_1}$$

and assume that the following existence conjecture is fulfilled.

Conjecture 8.1. *For any $u_0 \in \mathbb{W}_0^{1,p}(\Omega)$, there exists a unique strong generalized solution $u \in \mathbb{C}^{(1)}([0,T_0); \mathbb{W}_0^{1,p}(\Omega))$ of problem* (8.194)–(8.195)*; moreover, either $T_0 = +\infty$ or $T_0 < +\infty$, and in the latter case,*

$$\limsup_{t\uparrow T_0} \|\nabla u\|_p = +\infty. \tag{8.197}$$

Theorem 8.10. *Let $u_0 \in \mathbb{W}_0^{1,p}(\Omega)$, and let Conjecture* 8.1 *be fulfilled.*

1. *If $p_1 = p_2$, then $T_0 = +\infty$, and we have the following upper estimate:*

$$\Phi(t) \le \Phi_0 e^{c_2 t}.$$

2. *If $p_1 > p_2$, then $T_0 = +\infty$, and we have the following upper estimate:*

$$\Phi(t) \le \left[\Phi_0^{1-\alpha_1} + (1-\alpha_1)c_2 t\right]^{1/(1-\alpha_1)}, \quad \alpha_1 = \frac{p_2}{p_1}.$$

3. *If $p_1 < p_2$ and the conditions*

$$\|\nabla u_0\|_{p_2}^{p_2} > \frac{p_2}{2}\|\nabla u_0\|_2^2,$$

$$\|\nabla u_0\|_{p_2}^{p_2} > \left[1 + \frac{\gamma}{2}\right]\|\nabla u_0\|_2^2 + \gamma\frac{p_1 - 1}{p_1}\|\nabla u_0\|_{p_1}^{p_1}$$

are fulfilled, then $T_0 \in (0, T_1]$, and the limit relation (8.197) *holds.*

Here

$$T_1 = \frac{\Phi_0^{1+\alpha}}{(\alpha-1)^2[(\Phi'(0))^2 - \frac{\beta}{\alpha-1}\Phi(0)^2]}, \quad \gamma = \frac{(p_2-2)\sqrt{2p_1}}{(p_2-p_1)},$$

$$\Phi'(0) = \|\nabla u_0\|_{p_2}^{p_2} - \|\nabla u_0\|_2^2,$$

$$\Phi_0 = \Phi(0) = \frac{1}{2}\|\nabla u_0\|_2^2 + \frac{p_1-1}{p_1}\|\nabla u_0\|_{p_1}^{p_1},$$

$$\alpha = \frac{p_2+p_1}{2p_1}, \quad \beta = \frac{(p_2-2)^2}{p_2-p_1}, \quad c_2 = c_1^{p_2}\left(\frac{p_1}{p_1-1}\right)^{p_2/p_1},$$

and c_1 *is the best constant of the embedding* $\mathbb{W}_0^{1,p_1}(\Omega) \subset \mathbb{W}_0^{1,p_2}(\Omega)$ *under the condition* $p_1 \geq p_2$.

Proof. In equation (8.196), as w, we choose a function

$$u(x,t) \in \mathbb{C}^{(1)}([0,T_0); \mathbb{W}_0^{1,p}(\Omega)).$$

Then, after integration by parts, we obtain the first energy equality

$$\frac{d\Phi}{dt} + \|\nabla u\|_2^2 = \|\nabla u\|_{p_2}^{p_2}. \tag{8.198}$$

Taking as w the function u' and integrating by parts, we obtain the second energy equality

$$\|\nabla u'\|_2^2 + (p_1-1)\int_\Omega dx\, |\nabla u|^{p_1-2}|\nabla u'|^2 = \frac{1}{p_2}\frac{d}{dt}\|\nabla u\|_{p_2}^{p_2} - \frac{1}{2}\frac{d}{dt}\|\nabla u\|_2^2. \tag{8.199}$$

We impose the condition

$$\frac{1}{2}\|\nabla u_0\|_2^2 < \frac{1}{p_2}\|\nabla u_0\|_{p_2}^{p_2}. \tag{8.200}$$

Integrating (8.199) by time, we get

$$\frac{1}{2}\|\nabla u\|_2^2 < \frac{1}{p_2}\|\nabla u\|_{p_2}^{p_2},$$

which implies the estimate

$$\|\nabla u\|_2^2 < \|\nabla u\|_{p_2}^{p_2}.$$

Thus, from the first energy equality (8.198) and (8.200) we obtain

$$\Phi'(t) > 0, \quad t \in [0,T_0). \tag{8.201}$$

Finally, since $u(x,t) \in \mathbb{C}^{(1)}([0,T_0); \mathbb{W}_0^{1,p}(\Omega))$, equation (8.198) implies $\Phi(t) \in \mathbb{C}^{(2)}[0,T_0)$.

Now we consider $(\Phi')^2$:

$$(\Phi')^2(t) \le \left(\int_\Omega dx\, (\nabla u', \nabla u) + (p_1 - 1) \int_\Omega dx\, |\nabla u|^{p_1-2} (\nabla u', \nabla u) \right)^2. \tag{8.202}$$

We estimate the terms in the right-hand side. For the first term, we have the inequality

$$\left| \int_\Omega (\nabla u, \nabla u')\, dx \right| \le \|\nabla u'\|_2 \|\nabla u\|_2.$$

The second term can be estimated in two ways: either

$$\left| \int_\Omega dx\, |\nabla u|^{p_1-2} (\nabla u', \nabla u) \right| \le \left(\int_\Omega dx\, |\nabla u|^{p_1-2} |\nabla u'|^2 \right)^{1/2} \left(\int_\Omega dx\, |\nabla u|^{p_1} \right)^{1/2},$$

or

$$\begin{aligned} \left| \int_\Omega dx\, |\nabla u|^{p_1-2} (\nabla u', \nabla u) \right| &\le \left(\int_\Omega dx\, |\nabla u|^{p_1-4} |(\nabla u', \nabla u)|^2 \right)^{1/2} \\ &\quad \times \left(\int_\Omega dx\, |\nabla u|^{p_1} \right)^{1/2}. \end{aligned}$$

Taking these estimates into account, from inequality (8.202) we obtain

$$\begin{aligned} (\Phi')^2 \le \Bigg(&\|\nabla u'\|_2 \|\nabla u\|_2 + (p_1 - 2)^{1/2} \left(\int_\Omega dx\, |\nabla u|^{p_1-4} |(\nabla u', \nabla u)|^2 \right)^{1/2} \\ &\times (p_1 - 2)^{1/2} \left(\int_\Omega dx\, |\nabla u|^{p_1} \right)^{1/2} \\ &+ \left(\int_\Omega dx\, |\nabla u|^{p_1-2} |\nabla u'|^2 \right)^{1/2} \left(\int_\Omega dx\, |\nabla u|^{p_1} \right)^{1/2} \Bigg)^2. \end{aligned}$$

This implies the inequalities

$$\begin{aligned} (\Phi')^2 \le{} & \|\nabla u'\|_2^2 \|\nabla u\|_2^2 + \int_\Omega dx\, |\nabla u|^{p_1-2} |\nabla u'|^2 \int_\Omega dx\, |\nabla u|^{p_1} \\ & + (p_1 - 2) \int_\Omega dx\, |\nabla u|^{p_1-4} |(\nabla u', \nabla u)|^2 (p_1 - 2) \int_\Omega dx\, |\nabla u|^{p_1} \\ & + 2\|\nabla u'\|_2 (p_1 - 1)^{1/2} \left(\int_\Omega dx\, |\nabla u|^{p_1} \right)^{1/2} \end{aligned}$$

$$\times \|\nabla u\|_2 (p_1 - 1)^{1/2} \left(\int_\Omega dx\, |\nabla u|^{p_1-2} |\nabla u'|^2 \right)^{1/2}$$
$$\leq \left(\|\nabla u'\|_2^2 + \int_\Omega dx\, |\nabla u|^{p_1-2} |\nabla u'|^2 + (p_1 - 2) \int_\Omega dx\, |\nabla u|^{p_1-4} |(\nabla u', \nabla u)|^2 \right)$$
$$\times \left(\|\nabla u\|_2^2 + (p_1 - 1) \int_\Omega dx\, |\nabla u|^{p_1} \right) \leq p_1 \Phi(t) \mathrm{J}(t), \tag{8.203}$$

where

$$\mathrm{J} \equiv \|\nabla u'\|_2^2 + \int_\Omega dx\, |\nabla u|^{p_1-2} |\nabla u'|^2 + (p_1 - 2) \int_\Omega dx\, |\nabla u|^{p_1-4} |(\nabla u', \nabla u)|^2. \tag{8.204}$$

Taking into account (8.199), we rewrite expression (8.204) in the form

$$\mathrm{J} \leq \frac{1}{p_2} \frac{d}{dt} [\|\nabla u\|_{p_2}^{p_2} - \|\nabla u\|_2^2] - \frac{p_2 - 2}{2p_2} \frac{d}{dt} \|\nabla u\|_2^2. \tag{8.205}$$

Now we apply the inequality

$$\frac{1}{2} \frac{d}{dt} \|\nabla u\|_2^2 \leq \frac{\varepsilon}{2} \|\nabla u'\|_2^2 + \frac{1}{2\varepsilon} \|\nabla u\|_2^2;$$

from (8.199), (8.204), and (8.205) and from the definition of the function $\Phi(t)$ we obtain

$$\left[1 - \varepsilon \frac{p_2 - 2}{2p_2} \right] \mathrm{J} \leq \frac{1}{p_2} \Phi''(t) + \frac{p_2 - 2}{p_2} \frac{1}{\varepsilon} \Phi(t). \tag{8.206}$$

From (8.203) and (8.206) we obtain the following second-order ordinary differential inequality (see [37]):

$$\Phi'' \Phi - \alpha (\Phi')^2 + \beta \Phi^2 \geq 0, \tag{8.207}$$

where

$$\alpha = \frac{p_2}{p_1} \left(1 - \varepsilon \frac{p_2 - 2}{2p_2} \right),$$
$$\beta = \frac{p_2 - 2}{\varepsilon}, \quad \varepsilon \in \left(0, \frac{2(p_2 - p_1)}{p_2 - 2} \right), \quad p_1 < p_2.$$

Note that we have $\alpha > 1$ and $\beta > 0$ for $p_2 > p_1$.

Applying the results presented in the Appendix, we conclude that, under the condition

$$A = [(\alpha - 1)^2 \Phi_0^{-2\alpha} (\Phi'(0))^2 - (\alpha - 1) \beta \Phi_0^{2-2\alpha}]^{1/2} > 0,$$

which is fulfilled if

$$\Phi'(0) > \left(\frac{\beta}{\alpha-1}\right)^{1/2}\Phi(0),$$

we obtain the third assertion of Theorem 8.10.

Now let $p_1 \geq p_2$. Consider the first energy equality (8.198). Using Sobolev's embedding theorems, we obtain the inequalities

$$\frac{d\Phi}{dt} \leq \|\nabla u\|_{p_2}^{p_2} \leq c_1^{p_2}(\|\nabla u\|_{p_1})^{p_2} \leq \left(\frac{p_1}{p_1-1}\right)^{p_2/p_1} c_1^{p_2}\Phi^{\alpha_1} = c_2\Phi^{\alpha_1}, \tag{8.208}$$

where $\alpha_1 = p_2/p_1$. From (8.208) we easily deduce the following estimates:

$$\Phi(t) \leq \Phi_0 \exp(c_2 t), \quad p_2 = p_1,$$

$$\Phi(t) \leq \left[\Phi_0^{\frac{p_1-p_2}{p_1}} + \frac{p_1-p_2}{p_1}c_2 t\right]^{\frac{p_1}{p_1-p_2}}, \quad p_1 > p_2.$$

These estimates complete the proof of the theorem since the first two assertions are their direct consequences. □

We have shown that the modified energy method is applicable to pseudo-parabolic problems. In the following section, we prove the blow-up for a whole class of initial-boundary-value problems.

8.3.2 Class of pseudo-parabolic equations

In this section, we consider the initial-boundary-value problem of the following general form:

$$\frac{\partial}{\partial t}(\Delta u - \varphi(x,u)) + \operatorname{div}(h(x,|\nabla u|)\nabla u) - g(x,u) + f(x,u) = 0, \tag{8.209}$$

$$\varphi(x,u) = u + \sum_{k=1}^{n} a_k(x)|u|^{p_k-2}u, \tag{8.210}$$

$$u|_{\partial\Omega} = 0, \quad u(x,0) = u_0(x), \tag{8.211}$$

where $\Omega \subset \mathbb{R}^N$ is a bounded domain with smooth boundary $\partial\Omega \in \mathbb{C}^{2,\delta}$, $\delta \in (0,1]$. On the functions $\varphi(x,s)$, $h(x,s)$, $g(x,s)$, and $f(x,s)$, we impose the following conditions.

Conditions for the function $\varphi(x,s)$

$(\mathrm{i})_1$ $a_k(x) \geq 0$ for almost all $x \in \Omega$, and $a_k(x) \in \mathbb{L}^{\infty}(\Omega)$;

(ii)$_1$ let

$$p_{k_0} = \max_{k \in \overline{1,n}} p_k, \quad p_{k_0} < p^*,$$

and $a_{k_0}(x) \geq a_0 > 0$;

(iii)$_1$ $p_k > 2$ for all $k = \overline{1, n}$.

Conditions for the function $h(x, s)$

(i)$_2$ $h(x, s) : \Omega \times \mathbb{R}^1_+ \to \mathbb{R}^1$ is a Carathéodori function;

(ii)$_2$ for almost all $x \in \Omega$,

$$c_2 s^{p-2} \leq h(x, s) \leq c_1 + c_2 s^{p-2}, \quad p \geq 2; \tag{8.212}$$

(iii)$_2$ for any $v(x) \in \mathbb{W}^{1,p}_0(\Omega)$,

$$0 \leq \int_\Omega h(x, |\nabla v(x)|)|\nabla v(x)|^2 \, dx \leq \theta_1 \int_\Omega \mathcal{H}(x, |\nabla v(x)|) \, dx, \tag{8.213}$$

where

$$\mathcal{H}(x, s) = \int_0^s h(x, \sigma)\sigma \, d\sigma, \quad \theta_1 > 0.$$

Note that, due to condition (8.212),

$$\operatorname{div}(h(x, |\nabla u|)\nabla u) : \mathbb{W}^{1,p}_0(\Omega) \to \mathbb{W}^{-1,p'}(\Omega),$$

where the operator $\operatorname{div}(\cdot)$ is meant in the weak sense. We have the following equality:

$$\langle \operatorname{div}(v), w \rangle_0 = -\int_\Omega dx \, (v, \nabla w), \quad v \in \mathbb{L}^{p'}(\Omega) \otimes \cdots \otimes \mathbb{L}^{p'}(\Omega), \quad p \geq 2,$$

for all $w \in \mathbb{W}^{1,p}_0(\Omega)$, where $\langle \cdot, \cdot \rangle_0$ is the duality bracket between the Banach spaces $\mathbb{W}^{1,p}_0(\Omega)$ and $\mathbb{W}^{-1,p'}(\Omega)$.

Conditions for the function $g(x, s)$

(i)$_3$ $g(x, s) : \Omega \times \mathbb{R}^1 \to \mathbb{R}^1$ is a Carathéodori function;

(ii)$_3$ for almost all $x \in \Omega$,

$$|g(x, s)| \leq c_4 + c_5 |s|^{q_1+1}, \quad q_1 \in [0, p^* - 2); \tag{8.214}$$

(iii)$_3$ for any $v(x) \in \mathbb{L}^{q_1+2}(\Omega)$,

$$0 \leq \int_\Omega v(x)g(x, v(x))\, dx \leq \theta_2 \int_\Omega \mathcal{G}(x, v(x))\, dx, \tag{8.215}$$

where

$$\mathcal{G}(x, s) = \int_0^s g(x, \sigma)\, d\sigma, \quad \theta_2 > 0.$$

Conditions for the function $f(x, s)$

(i)$_4$ $f(x, s) : \Omega \times \mathbb{R}^1 \to \mathbb{R}^1$ is a Carathéodori function;

(ii)$_4$ for almost all $x \in \Omega$,

$$|f(x, s)| \leq c_6 + c_7|s|^{q_2+1}, \quad q_2 \in (0, p^* - 2); \tag{8.216}$$

(iii)$_4$ for any $v(x) \in \mathbb{L}^{q_2+2}(\Omega)$,

$$\int_\Omega f(x, v(x))v(x) \geq \theta_3 \int_\Omega \mathcal{F}(x, v(x))\, dx, \quad \theta_3 > 0, \tag{8.217}$$

where

$$\mathcal{F}(x, s) = \int_0^s f(x, \sigma)\, d\sigma,$$

and

$$p^* = \begin{cases} \frac{Np}{N-p} & \text{for } N > p, \\ +\infty & \text{for } N \leq p; \end{cases}$$

obviously, $p^* > 2$ for $p \geq 2$.

We treat solutions of problem (8.209)–(8.211) in the weak generalized sense.

Definition 8.10. A *weak generalized solution* of problem (8.209)–(8.211) is a function $u(x)(t)$ of the class

$$u(x)(t) \in \mathbb{L}^\infty(0, T; \mathbb{W}_0^{1,p}(\Omega)), \quad u'(x)(t) \in \mathbb{L}^2(0, T; \mathbb{H}_0^1(\Omega)),$$
$$(|u|^{p_k/2})' \in \mathbb{L}^2(Q_T),$$

satisfying the equalities

$$\int_\Omega (\nabla u', \nabla v)\,dx + \int_\Omega \left[u' + \sum_{k=1}^{n}(p_k - 1)\frac{2}{p_k} a_k(x)|u|^{p_k/2-2}u(|u|^{p_k/2})'\right] v\,dx$$
$$+ \int_\Omega h(x, |\nabla u|)(\nabla u, \nabla v)\,dx + \int_\Omega [g(x,u) - f(x,u)]v\,dx = 0 \tag{8.218}$$

and

$$u(x)(0) = u_0(x) \in \mathbb{W}_0^{1,p}(\Omega)$$

for all $v \in \mathbb{L}^\infty(0, T; \mathbb{W}_0^{1,p}(\Omega))$.

Definition 8.10 implies that, after a possible redefinition on a subset of zero Lebesgue measure, a weak generalized solution belongs to the class

$$u(x)(t) \in \mathbb{C}([0,T]; \mathbb{H}_0^1(\Omega)).$$

We assume that a weak generalized solution exists and belongs to the class

$$u(x)(t) \in \mathbb{C}^{(1)}([0, T_0); \mathbb{W}_0^{1,p}(\Omega))$$

for some maximal time $T_0 > 0$ of existence of a nonextendable solution.

Denote

$$\Phi(t) \equiv \frac{1}{2}\int_0^t \|\nabla u\|_2^2\,ds + \frac{1}{2}\int_0^t \|u\|_2^2\,ds + \sum_{l=1}^{n}\frac{p_l - 1}{p_l}\int_0^t\int_\Omega a_l(x)|u|^{p_l}\,dx\,ds$$
$$+ \frac{1}{2p_{k_0}}\|\nabla u_0\|_2^2 + \frac{1}{2p_{k_0}}\|u_0\|_2^2 + \sum_{l=1}^{n}\frac{p_l - 1}{p_{k_0}p_l}\int_\Omega a_l(x)|u_0|^{p_l}\,dx, \tag{8.219}$$

$$\mathrm{J}(t) \equiv \int_0^t \|\nabla u'\|_2^2(s)\,ds + \int_0^t \|u'\|_2^2(s)\,ds$$
$$+ \sum_{l=1}^{n}(p_l - 1)\int_0^t\int_\Omega a_l(x)|u'|^2|u|^{p_l-2}\,dx\,ds$$
$$+ \frac{1}{2}\|\nabla u_0\|_2^2 + \frac{1}{2}\|u_0\|_2^2 + \sum_{l=1}^{n}\frac{p_l - 1}{p_l}\int_\Omega a_l(x)|u_0|^{p_l}\,dx. \tag{8.220}$$

For the functions introduced, we have the following important auxiliary result.

Lemma 8.3. *For all $t \in [0, T_0)$, we have the following differential inequality:*

$$(\Phi')^2 \le p_{k_0}\Phi\mathrm{J}, \quad p_{k_0} = \max_{k\in\overline{1,n}} p_k. \tag{8.221}$$

The proof of this lemma is similar to that of Lemma 8.2 (see p. 174).

Now we obtain the energy equalities. Setting $v = u(x)(t)$ in definition (8.218) of solutions, from (8.219) we obtain the following equality:

$$\frac{d^2\Phi(t)}{dt^2} + \int_\Omega h(x,|\nabla u|)|\nabla u|^2\,dx + \int_\Omega g(x,u)u\,dx = \int_\Omega f(x,u)u\,dx. \tag{8.222}$$

Further, setting $v = u'(x)(t)$ in (8.218), we obtain the second energy equality

$$\frac{dJ}{dt} + \frac{d}{dt}\int_\Omega \mathcal{H}(x,|\nabla u|)\,dx + \frac{d}{dt}\int_\Omega \mathcal{G}(x,u)\,dx = \frac{d}{dt}\int_\Omega \mathcal{F}(x,u)\,dx. \tag{8.223}$$

Integrating (8.223) by time, we obtain the equation

$$J + \int_\Omega \mathcal{H}(x,|\nabla u|)\,dx + \int_\Omega \mathcal{G}(x,u)\,dx - E(0) = \int_\Omega \mathcal{F}(x,u)\,dx, \tag{8.224}$$

where

$$\begin{aligned} E(0) \equiv{}& \frac{1}{2}\|\nabla u_0\|_2^2 + \frac{1}{2}\|u_0\|_2^2 + \sum_{l=1}^{n}\frac{p_l-1}{p_l}\int_\Omega a_l(x)|u_0|^{p_l}\,dx \\ &+ \int_\Omega \mathcal{H}(x,|\nabla u_0|)\,dx + \int_\Omega \mathcal{G}(x,u_0)\,dx - \int_\Omega \mathcal{F}(x,u_0)\,dx \end{aligned} \tag{8.225}$$

is the integration constant. Applying condition (8.217), from (8.224) we obtain the inequality

$$\begin{aligned} &\theta_3 J + \theta_3\int_\Omega \mathcal{H}(x,|\nabla u|)\,dx + \theta_3\int_\Omega \mathcal{G}(x,u)\,dx \\ &\quad - \theta_3 E(0) \le \int_\Omega f(x,u)u\,dx. \end{aligned} \tag{8.226}$$

Combining this with (8.222), we obtain

$$\begin{aligned} &\frac{d^2\Phi(t)}{dt^2} + \int_\Omega h(x,|\nabla u|)|\nabla u|^2\,dx \\ &\quad + \int_\Omega g(x,u)u\,dx \ge \theta_3 J + \theta_3\int_\Omega \mathcal{H}(x,|\nabla u|)\,dx + \theta_3\int_\Omega \mathcal{G}(x,u)\,dx - \theta_3 E(0). \end{aligned} \tag{8.227}$$

Using (8.213) and (8.215), we deduce the estimate

$$\frac{d^2\Phi(t)}{dt^2} + \theta_3 E(0) \ge (\theta_3 - \theta_1)\int_\Omega \mathcal{H}(x,|\nabla u|)\,dx$$

$$+ (\theta_3 - \theta_2) \int_\Omega \mathcal{G}(x,u)\,dx + \theta_3 J. \tag{8.228}$$

Impose the additional conditions

$$\theta_3 \geq \theta_1, \quad \theta_3 \geq \theta_2,$$

which allow us to eliminate two terms in the right-hand side of the inequality (8.228):

$$\frac{d^2\Phi(t)}{dt^2} + \theta_3 E(0) \geq \theta_3 J. \tag{8.229}$$

Using the differential inequality (8.221) from Lemma 8.3, we obtain from (8.229) the standard ordinary differential inequality

$$\Phi\Phi'' - \alpha(\Phi')^2 + \beta\Phi \geq 0, \tag{8.230}$$

where the coefficients are defined by the formulas

$$\alpha = \frac{\theta_3}{p_{k_0}}, \quad \beta = \theta_3 E(0). \tag{8.231}$$

We can apply Theorem A.1 from the Appendix if

$$\theta_3 > p_{k_0}, \quad p_{k_0} = \max_{l=\overline{1,n}} p_l. \tag{8.232}$$

We consider separately the cases where $\beta > 0$ and $\beta \leq 0$.

First, we consider the former case. Theorem A.1 implies that if the initial data satisfy the condition

$$\frac{1}{2}\|\nabla u_0\|_2^2 + \frac{1}{2}\|u_0\|_2^2 + \sum_{l=1}^{n} \frac{p_l - 1}{p_l} \int_\Omega a_l(x)|u_0|^{p_l}\,dx > \frac{2\theta_3}{2\theta_3 - p_{k_0}} E(0) > 0, \tag{8.233}$$

then we have the lower estimate

$$\Phi(t) \geq \frac{1}{[\Phi^{1-\alpha}(0) - At]^{1/(\alpha-1)}}, \tag{8.234}$$

where

$$A^2 = (\alpha - 1)^2 \Phi^{-2\alpha}(0) \left[(\Phi'(0))^2 - \frac{2\beta}{2\alpha - 1}\Phi(0) \right] > 0. \tag{8.235}$$

Condition (8.233) can be slightly modified:

$$\int_\Omega \mathcal{F}(x,u_0)\,dx > \int_\Omega \mathcal{H}(x,|\nabla u_0|)\,dx + \int_\Omega \mathcal{G}(x,u_0)\,dx$$

$$+\frac{p_{k_0}}{2\theta_3}\left(\frac{1}{2}\|\nabla u_0\|_2^2+\frac{1}{2}\|u_0\|_2^2\right.$$
$$\left.+\sum_{l=1}^{n}\frac{p_l-1}{p_l}\int_\Omega a_l(x)|u_0|^{p_l}\,dx\right),\quad p_{k_0}<\theta_3, \tag{8.236}$$

supplementing (8.236) by the condition $\beta > 0$:

$$\frac{1}{2}\|\nabla u_0\|_2^2+\frac{1}{2}\|u_0\|_2^2+\sum_{l=1}^{n}\frac{p_l-1}{p_l}\int_\Omega a_l(x)|u_0|^{p_l}\,dx$$
$$+\int_\Omega \mathcal{H}(x,|\nabla u_0|)\,dx+\int_\Omega \mathcal{G}(x,u_0)\,dx>\int_\Omega \mathcal{F}(x,u_0)\,dx. \tag{8.237}$$

The case where $\beta \le 0$ is easier:

$$\int_\Omega \mathcal{F}(x,u_0)\,dx \ge \frac{1}{2}\|\nabla u_0\|_2^2+\frac{1}{2}\|u_0\|_2^2+\sum_{l=1}^{n}\frac{p_l-1}{p_l}\int_\Omega a_l(x)|u_0|^{p_l}\,dx$$
$$+\int_\Omega \mathcal{H}(x,|\nabla u_0|)\,dx+\int_\Omega \mathcal{G}(x,u_0)\,dx. \tag{8.238}$$

Then the blow-up result immediately follows from inequality (8.234), in which we must set $\beta = 0$.

Thus, we have proved the following sufficient condition for blow-up.

Theorem 8.11. *Let all conditions for the functions $\varphi(x,s)$, $h(x,s)$, $g(x,s)$, and $f(x,s)$ be fulfilled. Moreover, let the first group of the conditions of Theorem 8.1 hold. Then, under the condition*

$$\int_\Omega \mathcal{F}(x,u_0)\,dx>\int_\Omega \mathcal{H}(x,|\nabla u_0|)\,dx+\int_\Omega \mathcal{G}(x,u_0)\,dx$$
$$+\frac{p_{k_0}}{2\theta_3}\left(\frac{1}{2}\|\nabla u_0\|_2^2+\frac{1}{2}\|u_0\|_2^2+\sum_{l=1}^{n}\frac{p_l-1}{p_l}\int_\Omega a_l(x)|u_0|^{p_l}\,dx\right), \tag{8.239}$$

where

$$p_{k_0}<\theta_3,\quad \theta_3\ge\theta_1,\quad \theta_3\ge\theta_2, \tag{8.240}$$

there are no global-in-time weak generalized solutions of problem (8.209)–(8.211), the limit relation

$$\limsup_{t\uparrow T_0}\|\nabla u\|_p=+\infty$$

holds, and we have the upper estimate

$$T_0\le T_\infty=\Phi^{1-\alpha}(0)A^{-1},$$

where the constant $A > 0$ *is defined by formula* (8.235),

$$\beta = \begin{cases} \theta_3 E(0) & \text{if } E(0) > 0, \\ 0 & \text{if } E(0) \leq 0, \end{cases}$$

and

$$E(0) \equiv \frac{1}{2}\|\nabla u_0\|_2^2 + \frac{1}{2}\|u_0\|_2^2 + \sum_{l=1}^{n} \frac{p_l - 1}{p_l} \int_\Omega a_l(x)|u_0|^{p_l}\, dx + \int_\Omega \mathcal{H}(x, |\nabla u_0|)\, dx + \int_\Omega \mathcal{G}(x, u_0)\, dx - \int_\Omega \mathcal{F}(x, u_0)\, dx.$$

This theorem basically answers the solvability question problem for initial-boundary-value problems for a wide class of nonlinear partial differential equations.

8.3.3 Nonlinear nonlocal equation of spin waves

Consider another class of problems that admits application of the modified energy method. Nonlocal parabolic equations have been discusses earlier; however, the nonlocality was related to a spatial integral, and now we examine the following equation containing an integral by time:

$$\frac{\partial}{\partial t}(-\Delta^2 u + \Delta u + \Delta_{p_1} u) + \int_0^t ds\, h(t-s)\Delta u(s) + \Delta u + \alpha_1 \frac{\partial}{\partial x_1}\left(\frac{\partial u}{\partial x_2}\frac{\partial u}{\partial x_3}\right) + \alpha_2 \frac{\partial}{\partial x_2}\left(\frac{\partial u}{\partial x_3}\frac{\partial u}{\partial x_1}\right) + \alpha_3 \frac{\partial}{\partial x_3}\left(\frac{\partial u}{\partial x_1}\frac{\partial u}{\partial x_2}\right) - \Delta_{p_2} u = 0, \tag{8.241}$$

$$u|_{\partial\Omega} = \left.\frac{\partial u}{\partial n_x}\right|_{\partial\Omega} = 0, \quad u(x,0) = u_0(x), \tag{8.242}$$

$$\Delta_p \psi \equiv \operatorname{div}(|\nabla\psi|^{p-2}\nabla\psi), \quad p \geq 2,\ p_1 \geq 2,\ p_2 > 2,$$

where $\alpha_1 + \alpha_2 + \alpha_3 = 0$ but $|\alpha_1| + |\alpha_2| + |\alpha_3| > 0$, $\Omega \subset \mathbb{R}^3$ is a bounded domain with smooth boundary $\partial\Omega \in \mathbb{C}^{4,\delta}$, $\delta \in (0,1]$, $p_1 \in [3,6]$, and $p_2 \in [2,6]$. Assume that the function $h(t)$ possesses property (A.53) (Appendix, p. 304) and hence $|h(t)|$ is bounded from above by the derivative of a nonpositive function.

We state the definition of a strong generalized solution of the problem.

Definition 8.11. A function $u(x)(t)$ of the class $\mathbb{C}^{(1)}([0,T]; \mathbb{H}_0^2(\Omega))$, where $T > 0$, is called a *strong generalized solution* of problem (8.241)–(8.242) if

$$\langle \mathbb{D}(u), w\rangle = 0 \quad \text{for all } w \in \mathbb{H}_0^2(\Omega),\ t \in [0,T], \tag{8.243}$$

where

$$\mathbb{D}(u) \equiv \frac{\partial}{\partial t}(-\Delta^2 u + \Delta u + \Delta_{p_1} u) + \int_0^t ds\, h(t-s)\Delta u(s) + \Delta u$$
$$+ \alpha_1 \frac{\partial}{\partial x_1}\left(\frac{\partial u}{\partial x_2}\frac{\partial u}{\partial x_3}\right) + \alpha_2 \frac{\partial}{\partial x_2}\left(\frac{\partial u}{\partial x_3}\frac{\partial u}{\partial x_1}\right) + \alpha_3 \frac{\partial}{\partial x_3}\left(\frac{\partial u}{\partial x_1}\frac{\partial u}{\partial x_2}\right) - \Delta_{p_2} u,$$

and $\langle\cdot,\cdot\rangle$ is the duality bracket between the Banach spaces $\mathbb{H}_0^2(\Omega)$ and $\mathbb{H}^{-2}(\Omega)$.

There exists another, equivalent definition of the strong generalized solution of the problem (8.241)–(8.242).

Definition 8.12. A function $u(x)(t) \in \mathbb{C}^{(1)}([0,T];\mathbb{H}_0^2(\Omega))$, where $T > 0$, is called a *strong generalized solution* of problem (8.241)–(8.242) if

$$\int_0^T dt \langle \mathbb{D}(u), v\rangle = 0 \quad \text{for all } v \in \mathbb{L}^2(0,T;\mathbb{H}_0^2(\Omega)), \tag{8.244}$$

where

$$\mathbb{D}(u) \equiv \frac{\partial}{\partial t}(-\Delta^2 u + \Delta u + \Delta_{p_1} u) + \int_0^t ds\, h(t-s)\Delta u(s) + \Delta u$$
$$+ \alpha_1 \frac{\partial}{\partial x_1}\left(\frac{\partial u}{\partial x_2}\frac{\partial u}{\partial x_3}\right) + \alpha_2 \frac{\partial}{\partial x_2}\left(\frac{\partial u}{\partial x_3}\frac{\partial u}{\partial x_1}\right) + \alpha_3 \frac{\partial}{\partial x_3}\left(\frac{\partial u}{\partial x_1}\frac{\partial u}{\partial x_2}\right) - \Delta_{p_2} u.$$

Let $u(x)(t) \in \mathbb{C}^{(1)}([0,T];\mathbb{H}_0^2(\Omega))$ be a nonextendable strong generalized solution of problem (8.241)–(8.242) for some $T_0 > 0$.

Now we obtain the energy equalities. In equation (8.243), we set $w = u$ and, after integration by parts, obtain the first energy equality

$$\frac{d\Phi}{dt} = -\|\nabla u\|_2^2 - \int_0^t ds \int_\Omega (\nabla u(s), \nabla u(t))\, dx + \|\nabla u\|_{p_2}^{p_2}. \tag{8.245}$$

Now, setting $w = u'$ in equation (8.243) and integrating by parts, we obtain the second energy equality

$$\mathbb{J}(t) + \frac{1}{2}\frac{d}{dt}\|\nabla u\|_2^2 = \frac{1}{p_2}\frac{d}{dt}\|\nabla u\|_{p_2}^{p_2} - \int_0^t ds\, h(t-s)\int_\Omega (\nabla u(s), \nabla u'(t))\, dx$$
$$+ \int_\Omega \left[\alpha_1 \frac{\partial}{\partial x_1}\left(\frac{\partial u}{\partial x_2}\frac{\partial u}{\partial x_3}\right) + \alpha_2 \frac{\partial}{\partial x_2}\left(\frac{\partial u}{\partial x_3}\frac{\partial u}{\partial x_1}\right) + \alpha_3 \frac{\partial}{\partial x_3}\left(\frac{\partial u}{\partial x_1}\frac{\partial u}{\partial x_2}\right)\right] u'\, dx, \tag{8.246}$$

where

$$\mathbb{J}(t) \equiv \|\Delta u'\|_2^2 + \|\nabla u'\|_2^2 + \int_{\Omega} |\nabla u'|^2 |\nabla u|^{p_1-2}\,dx + (p_1-2)\int_{\Omega} |\nabla u|^{p_1-4} |(\nabla u', \nabla u)|^2\,dx. \tag{8.247}$$

Note that it suffices to estimate the integral

$$\alpha_i \int_{\Omega} \frac{\partial}{\partial x_i}\left(\frac{\partial u}{\partial x_j}\frac{\partial u}{\partial x_k}\right) u'\,dx.$$

Indeed, all other estimates are similar to the estimates from the previous section. We have the following inequalities:

$$\left|\alpha_i \int_{\Omega} \frac{\partial}{\partial x_i}\left(\frac{\partial u}{\partial x_j}\frac{\partial u}{\partial x_k}\right) u'\right| \leq |\alpha_i| \int_{\Omega} |\nabla u|^2 |\nabla u'|\,dx \leq \frac{\varepsilon}{6}\|\nabla u'\|_2^2 + \frac{3}{2\varepsilon}|\alpha_i|^2 \|\nabla u\|_4^4. \tag{8.248}$$

Thus, we obtain the estimate

$$\begin{aligned}
&\left|\alpha_1 \int_{\Omega} \frac{\partial}{\partial x_1}\left(\frac{\partial u}{\partial x_2}\frac{\partial u}{\partial x_3}\right) u'\,dx + \alpha_2 \int_{\Omega} \frac{\partial}{\partial x_2}\left(\frac{\partial u}{\partial x_3}\frac{\partial u}{\partial x_1}\right) u'\,dx \right. \\
&\quad \left. + \alpha_3 \int_{\Omega} \frac{\partial}{\partial x_3}\left(\frac{\partial u}{\partial x_1}\frac{\partial u}{\partial x_2}\right) u'\,dx\right| \\
&\leq \frac{\varepsilon}{2}\|\nabla u'\|_2^2 + \frac{3}{2}\frac{1}{\varepsilon}\sum_{i=1}^{3} |\alpha_i^2| \|\nabla u\|_4^4 \\
&\leq \frac{\varepsilon}{2}\mathbb{J}(t) + \frac{3}{2\varepsilon}\sum_{i=1}^{3} |\alpha_i^2| c_1^4 \|\Delta u\|_2^4 \leq \frac{\varepsilon}{2}\mathbb{J}(t) + \frac{c_2}{\varepsilon}\Phi^2(t),
\end{aligned} \tag{8.249}$$

where

$$c_2 = 6\sum_{i=1}^{3} |\alpha_i|^2 c_1^4.$$

From these estimates we obtain the following upper estimate for $\mathbb{J}(t)$:

$$\begin{aligned}
\mathbb{J}(t) &\leq \frac{\varepsilon}{2}\mathbb{J}(t) + \frac{1}{\varepsilon}\left(\frac{p_2-2}{p_2}\right)^2 \Phi(t) + \frac{\varepsilon}{2}\mathbb{J}(t) \\
&\quad + \frac{a}{\varepsilon}\left(\frac{p_2-1}{p_2}\right)^2 \int_0^t ds\,|h(t-s)|\Phi(s) + \frac{1}{p_2}\Phi'' + \frac{2|h(0)|}{p_2}\Phi(t) + \varepsilon\Phi(t)
\end{aligned}$$

$$+\frac{b}{\varepsilon}\frac{1}{p_2^2}\int_0^t ds\,|h'(t-s)|\Phi(s)+\frac{\varepsilon}{2}\mathbb{J}(t)+\frac{c_2}{\varepsilon}\Phi^2(t)=\frac{3\varepsilon}{2}\mathbb{J}(t)+\frac{1}{p_2}\Phi''$$
$$+\left[\frac{1}{\varepsilon}\left(\frac{p_2-2}{p_2}\right)^2+\frac{2|h(0)|}{p_2}+\varepsilon\right]\Phi(t)$$
$$+\int_0^t ds\,h_\varepsilon(t-s)\Phi(s)+\frac{c_2}{\varepsilon}\Phi^2(t), \tag{8.250}$$

where

$$h_\varepsilon(t)\equiv\frac{a}{\varepsilon}\left(\frac{p_2-1}{p_2}\right)^2|h(t)|+\frac{b}{\varepsilon}\frac{1}{p_2^2}|h'(t)|.$$

The derivation of an inequality similar to (8.221) is not very difficult. Under the assumption $\varepsilon\in(0,2/3)$, we can obtain the inequality

$$\Phi\Phi''-\alpha(\Phi')^2+\gamma_1\Phi^2+\gamma_2\int_0^t ds\,h_\varepsilon(t-s)\Phi(s)\Phi(t)+\gamma_3\Phi^3\geq 0, \tag{8.251}$$

similar to an ordinary differential inequality used before. Here

$$\gamma_1=\frac{1}{\varepsilon}\frac{(p_2-2)^2}{p_2}+2|h(0)|+\varepsilon p_2,\quad \gamma_2=p_2,\quad \gamma_3=\frac{p_2c_2}{\varepsilon}, \tag{8.252}$$
$$\alpha=\left(1-\frac{3}{2}\varepsilon\right)\frac{p_2}{p_1},$$
$$h_\varepsilon(t)\equiv\frac{a}{\varepsilon}\left(\frac{p_2-1}{p_2}\right)^2|h(t)|+\frac{b}{\varepsilon}\frac{1}{p_2^2}|h'(t)|. \tag{8.253}$$

Despite the presence of an integral term, we can obtain the blow-up result from inequality (8.251). The corresponding theorem is proved in the Appendix (Theorem A.2, p. 304), and now we apply this theorem.

First, we verify the condition $\alpha>1$. Indeed, we have

$$\alpha=\left(1-\frac{3}{2}\varepsilon\right)\frac{p_2}{p_1}>1\quad\Rightarrow\quad 1-\frac{3}{2}\varepsilon>\frac{p_1}{p_2}\quad\Rightarrow$$
$$\Rightarrow\quad\frac{3}{2}\varepsilon<1-\frac{p_1}{p_2}\quad\Rightarrow\quad\varepsilon\in\left(0,\frac{2}{3}\frac{p_2-p_1}{p_1}\right),$$

and therefore $p_2>p_1$. Now we verify the condition $\lambda<2\alpha-1$. In our case, $\lambda=2$, and we have

$$2<2\alpha-1\quad\Rightarrow\quad\alpha>\frac{3}{2}\quad\Rightarrow\quad 1-\frac{3}{2}\varepsilon>\frac{3}{2}\frac{p_1}{p_2}\quad\Rightarrow\quad\varepsilon<\frac{2}{3}-\frac{p_1}{p_2},$$

and hence

$$p_2 > \frac{3}{2}p_1.$$

Thus, by Theorem A.2 we obtain the following blow-up result.

Lemma 8.4. *Assume that a strong generalized solution*

$$u(x)(t) \in \mathbb{C}^{(1)}([0, T_0); \mathbb{H}_0^2(\Omega))$$

satisfies the initial conditions

$$\Phi'(0) > \left(\frac{\gamma_4}{\alpha - 1}(\Phi(0))^2 + \frac{2\gamma_3}{(\alpha - 1)\delta}(\Phi(0))^3\right)^{1/2}, \quad \Phi(0) > 0, \tag{8.254}$$

and the condition

$$p_2 > \frac{3}{2}p_1. \tag{8.255}$$

Then

$$T_0 \le T_1, \quad T_1 = \frac{\Phi^{1-\alpha}(0)}{A},$$

and

$$\limsup_{t\uparrow T_0} \Phi(t) = +\infty, \tag{8.256}$$

where

$$\alpha > 1, \quad \gamma_4 = \gamma_1 + \gamma_2|\chi(0)|, \quad \delta = 1 + \frac{2-\alpha}{1-\alpha},$$

$$A^2 = (\alpha - 1)^2\Phi^{-2\alpha}(0)\left[(\Phi'(0))^2 - \frac{\gamma_4}{\alpha - 1}(\Phi(0))^2 - \frac{2\gamma_3}{(\alpha - 1)\delta}(\Phi(0))^3\right],$$

$$\Phi'(0) = \|\nabla u_0\|_{p_2}^{p_2} - \|\nabla u_0\|_2^2,$$

$$\Phi_0 = \Phi(0) = \frac{1}{2}\|\Delta u_0\|_2^2 + \frac{1}{2}\|\nabla u_0\|_2^2 + \frac{p_1 - 1}{p_1}\|\nabla u_0\|_{p_1}^{p_1}.$$

This theorem confirms the possibility of using the modified energy method for the study the blow-up effect in initial-boundary-value integro-differential problems.

8.3.4 Equations of a tunnel diode

In this section, we consider problems for pseudo-parabolic equations in which the nonlinear source is a *noncoercive* term of the form

$$\frac{\partial^2}{\partial t \partial x}\left(\frac{\partial u}{\partial x}\right)^2 \quad \text{or} \quad \frac{\partial u^2}{\partial t}.$$

Nevertheless, the modified energy method proposed by Korpusov and Sveshnikov is applicable. The first problem is

$$\frac{\partial^2}{\partial t^2}\left(\frac{\partial^2 u}{\partial x^2} - u\right) + \frac{\partial^2}{\partial t \partial x}\left(\frac{\partial u}{\partial x} - \left(\frac{\partial u}{\partial x}\right)^2\right) + \frac{\partial^2 u}{\partial x^2} = 0, \tag{8.257}$$

$$u|_{x=0} = u|_{x=1} = 0, \quad u(x,0) = u_0(x), \quad u'(x,0) = u_1(x); \tag{8.258}$$

and the second problem is

$$\frac{\partial^2}{\partial t^2}\left(\frac{\partial^2 u}{\partial x^2} - u\right) + \frac{\partial}{\partial t}(u - u^2) + \frac{\partial^2 u}{\partial x^2} = 0, \tag{8.259}$$

$$u|_{x=0} = u|_{x=1} = 0, \quad u(x,0) = u_0(x), \quad u'(x,0) = u_1(x), \tag{8.260}$$

where $x \in [0,1]$. Note the following continuous and dense embeddings:

$$\mathbb{W}_0^{1,3}(0,1) \overset{ds}{\subset} \mathbb{H}_0^1(0,1) \overset{ds}{\subset} \mathbb{H}^{-1}(0,1) \overset{ds}{\subset} \mathbb{W}^{-1,3/2}(0,1),$$

$$\mathbb{H}_0^1(0,1) \overset{ds}{\subset} \mathbb{L}^3(0,1) \overset{ds}{\subset} \mathbb{L}^2(0,1) \overset{ds}{\subset} \mathbb{L}^{3/2}(0,1) \overset{ds}{\subset} \mathbb{H}^{-1}(0,1).$$

Therefore, we have the following equalities for the duality brackets:

$$\langle f, u\rangle_1 = \langle f, u\rangle \qquad \text{for all } f \in \mathbb{H}^{-1}(0,1),\ u \in \mathbb{W}_0^{1,3}(0,1),$$

$$\langle f, u\rangle = \int_0^1 dx\, f(x)u(x) \quad \text{for all } f(x) \in \mathbb{L}^{3/2}(0,1),\ u(x) \in \mathbb{H}_0^1(0,1),$$

$$\langle f, u\rangle = \int_0^1 dx\, f(x)u(x) \quad \text{for all } f(x) \in \mathbb{L}^{2}(0,1),\ u(x) \in \mathbb{H}_0^1(0,1),$$

where $\langle \cdot,\cdot\rangle$ is the duality bracket between $\mathbb{H}_0^1(0,1)$ and $\mathbb{H}^{-1}(0,1)$, and $\langle \cdot,\cdot\rangle_1$ is the duality bracket between $\mathbb{W}_0^{1,3}(0,1)$ and $\mathbb{W}^{-1,3/2}(0,1)$.

We give the definition of strong generalized solutions.

Definition 8.13. A *strong generalized solution* of problem (8.257)–(8.258) is a function $u(x)(t) \in \mathbb{C}^{(2)}([0,T]; \mathbb{W}_0^{1,3}(0,1))$ satisfying the equality

$$\langle \mathbb{D}(u), w\rangle_1 = 0 \quad \text{for all } w \in \mathbb{W}_0^{1,3}(0,1),\ t \in [0,T], \tag{8.261}$$

where

$$\mathbb{D}(u) \equiv \frac{\partial^2}{\partial t^2}\left(\frac{\partial^2 u}{\partial x^2} - u\right) + \frac{\partial^2}{\partial t \partial x}\left(\frac{\partial u}{\partial x} - \left(\frac{\partial u}{\partial x}\right)^2\right) + \frac{\partial^2 u}{\partial x^2},$$

$$u(x)(0) = u_0(x), \quad u'(x)(0) = u_1(x), \quad u_0(x), u_1(x) \in \mathbb{W}_0^{1,3}(0,1).$$

Definition 8.14. A *strong generalized solution* of the problem (8.259)–(8.260) is a function $u(x)(t) \in \mathbb{C}^{(2)}([0,T]; \mathbb{W}_0^{1,3}(0,1))$ satisfying the equality

$$\langle \mathbb{D}(u), w \rangle = 0 \quad \text{for all } w \in \mathbb{H}_0^1(0,1),\ t \in [0,T], \tag{8.262}$$

where

$$\mathbb{D}(u) \equiv \frac{\partial^2}{\partial t^2}\left(\frac{\partial^2 u}{\partial x^2} - u\right) + \frac{\partial}{\partial t}(u - u^2) + \frac{\partial^2 u}{\partial x^2},$$
$$u(x)(0) = u_0(x), \quad u'(x)(0) = u_1(x), \quad u_0(x), u_1(x) \in \mathbb{H}_0^1(0,1).$$

We cannot say anything about the solvability of problem (8.257), (8.258) in the strong generalized sense; therefore, further, in the proof of the blow-up of a strong generalized solution, we assume the local solvability in the sense mentioned. On the contrary, to problem (8.259)–(8.260), we can apply the following result of [1].

Theorem 8.12. *For any initial functions $u_0(x), u_1(x) \in \mathbb{H}_0^1(0,1)$, there exists a unique strong generalized solution of problem* (8.259)–(8.260) *that belongs to the class $u(x)(t) \in \mathbb{C}^{(2)}([0,T_0); \mathbb{H}_0^1(0,1))$, where either $T_0 = +\infty$ or $0 < T_0 < +\infty$. In the latter case,*

$$\limsup_{t \uparrow T_0} \Phi(t) = +\infty, \quad \textit{where } \Phi(t) \equiv \frac{1}{2}\|u\|_2^2 + \frac{1}{2}\|u_x\|_2^2. \tag{8.263}$$

First, we analyze the blow-up of a strong generalized solution of problem (8.257)–(8.258). Note that we must present a sufficiently wide class of initial functions $u_0(x)$, $u_1(x) \in \mathbb{H}_0^1(0,1)$ for which solutions blow up in finite time. We assume that the initial functions are related by the formula

$$u_{1xx} - u_1 = -u_{0xx} + ((u_{0x})^2)_x, \quad u_0(x), u_1(x) \in \mathbb{W}_0^{1,3}(0,1). \tag{8.264}$$

Clearly, equation (8.264) has a solution in the class specified for sufficiently smooth $u_0(x)$. Thus we further consider initial functions $u_0(x)$ and $u_1(x)$ of the class (8.264).

Integrating equation (8.261) by time, we obtain the following problem in the class of initial functions (8.264):

$$\langle \mathbb{R}(u), w \rangle_1 = 0 \quad \text{for all } w \in \mathbb{W}_0^{1,3}(0,1),\ t \in [0,T], \tag{8.265}$$

where

$$\mathbb{R}(u) \equiv \frac{\partial}{\partial t}(u_{xx} - u) + u_{xx} + \int_0^t ds\, u_{xx} = ((u_x)^2)_x.$$

As usual, we derive the first and second energy equalities. Denote

$$\Phi(t) \equiv \frac{1}{2}\|u\|_2^2 + \frac{1}{2}\|u_x\|_2^2, \quad \mathbb{J}(t) \equiv \|u'\|_2^2 + \|u_x'\|_2^2. \tag{8.266}$$

In equation (8.265), we set $w = u(x)(t)$; by the assumption, this function is a strong generalized solution of problem (8.257)–(8.258) for some $T > 0$. Integrating by parts, we obtain the first equality

$$\frac{d\Phi}{dt} + \|u_x\|_2^2 + \int_0^t ds \int_0^1 u_x(s)u_x(t) = \int_0^1 dx\,(u_x)^3. \tag{8.267}$$

To obtain the second energy equality, we set $w = u'(x)(t)$ in (8.265). Integrating by parts, we obtain

$$\mathbb{J}(t) + \frac{1}{2}\frac{d}{dt}\|u_x\|_2^2 + \int_0^t ds \int_0^1 u_x(s)u_x'(t)\,dx = \frac{1}{3}\frac{d}{dt}\int_0^1 dx\,(u_x)^3. \tag{8.268}$$

We need the following auxiliary lemma (cf. [1]).

Lemma 8.5. *Let $u(x)(t) \in \mathbb{C}^{(1)}([0,T]; \mathbb{H}_0^1(0,1))$. Then*

$$(\Phi'(t))^2 \le 2\Phi(t)\mathbb{J}(t), \quad t \in [0,T]. \tag{8.269}$$

We substitute the expression of the integral

$$\int_0^1 dx\,(u_x)^3$$

from the first energy equality (8.267) into the second one (8.268). We can rewrite the expression for $\mathbb{J}(t)$ as follows:

$$\mathbb{J}(t) = \frac{1}{3}\Phi'' - \frac{1}{6}\frac{d}{dt}\|u_x\|_2^2 + \frac{1}{3}\|u_x\|_2^2 - \frac{2}{3}\int_0^t ds \int_0^1 u_x(s)u_x'(t)\,dx. \tag{8.270}$$

Now we estimate the last three terms in the right-hand side. Obviously, we have the following inequalities:

$$\frac{1}{3}\|u_x\|_2^2 \le \frac{2}{3}\Phi(t), \tag{8.271}$$

$$\begin{aligned}\left|\frac{1}{6}\frac{d}{dt}\|u_x\|_2^2\right| &\le \frac{1}{3}\|u_x'\|_2\|u_x\|_2 \\ &\le \frac{\varepsilon}{2}\|u_x'\|_2^2 + \frac{1}{2\varepsilon}\frac{1}{9}\|u_x\|_2^2 \le \frac{\varepsilon}{2}\mathbb{J}(t) + \frac{1}{9\varepsilon}\Phi(t),\end{aligned} \tag{8.272}$$

$$\left|\frac{2}{3}\int_0^t ds \int_0^1 u_x(s)u_x'(t)\,dx\right| \le \frac{2}{3}\int_0^t ds\,\|u_x'(t)\|_2\|u_x(s)\|_2$$

$$\leq \frac{\varepsilon}{2}\|u_x'\|_2^2 + \frac{1}{2\varepsilon}\frac{4}{9}T\int_0^t ds\,\|u_x\|_2^2(s)$$

$$\leq \frac{\varepsilon}{2}\mathbb{J}(t) + \frac{4T}{9\varepsilon}\int_0^t ds\,\Phi(s), \quad t \in [0,T]. \tag{8.273}$$

Substituting estimates (8.272)–(8.273) into equation (8.270), we obtain the inequality

$$\mathbb{J}(t) \leq \frac{1}{3}\Phi'' + \frac{\varepsilon}{2}\mathbb{J}(t) + \frac{1}{9\varepsilon}\Phi(t) + \frac{2}{3}\Phi(t) + \frac{\varepsilon}{2}\mathbb{J}(t) + \frac{4T}{9\varepsilon}\int_0^t ds\,\Phi(s), \tag{8.274}$$

which implies the estimate $\mathbb{J}(t)$ through Φ:

$$(1-\varepsilon)\mathbb{J}(t) \leq \frac{1}{3}\Phi'' + \left[\frac{1}{9\varepsilon} + \frac{2}{3}\right]\Phi(t) + \frac{4T}{9\varepsilon}\int_0^t ds\,\Phi(s). \tag{8.275}$$

Imposing the condition $\varepsilon \in (0,1)$, from inequality (8.275) and Lemma 8.5 (see inequality (8.269)) we obtain the required integro-differential inequality

$$\Phi\Phi'' - \alpha(\Phi')^2 + \beta\Phi^2 + \gamma T\int_0^t ds\,\Phi(s)\Phi(t) \geq 0, \quad t \in [0,T], \tag{8.276}$$

whose coefficients are defined by the formulas

$$\alpha = (1-\varepsilon)\frac{3}{2}, \quad \beta = \frac{1}{3\varepsilon} + 2, \quad \gamma = \frac{4}{3\varepsilon}.$$

Assuming that $\alpha > 1$ or, equivalently,

$$\varepsilon \in \left(0, \frac{1}{3}\right), \tag{8.277}$$

we apply Theorem A.2 (Appendix, p. 304) to the integro-differential inequality (8.276). The result can be formulated as follows.

Theorem 8.13. *Let initial conditions $u_0(x), u_1(x) \in \mathbb{W}_0^{1,3}(0,1)$ satisfy condition* (8.264) *and the inequalities*

$$\Phi'(0) \geq \left(\frac{\beta + \gamma T_1^2}{\alpha - 1} + \frac{1}{T_1^2(\alpha-1)^2}\right)^{1/2}\Phi(0), \quad \Phi(0) > 0, \tag{8.278}$$

where

$$\Phi'(0) = \int_0^1 dx\,(u_{0x})^3 - \|u_{0x}\|_2^2, \quad \Phi(0) = \frac{1}{2}\|u_0\|_2^2 + \frac{1}{2}\|u_{0x}\|_2^2.$$

Then there are no global-in-time strong generalized solutions of problem (8.257)–(8.258). *Moreover, there exists a moment* $T_0 \leq T_1$ *such that*

$$\limsup_{t \uparrow T_0} \Phi(t) = +\infty. \tag{8.279}$$

Remark. Note that condition (8.278) makes sense since the asymptotic growth order of $\Phi'(0)$ is 3 and that of $\Phi(0)$ is 2.

For problem (8.259)–(8.260), the reader can similarly prove the following blow-up result.

Theorem 8.14. *For any initial function* $u_0(x), u_1(x) \in \mathbb{H}_0^1(0,1)$, *there exists a unique strong generalized solution of problem* (8.259)–(8.260) *of the class* $\mathbb{C}^{(2)}([0,T_0); \mathbb{H}_0^1(0,1))$, *where* $T_0 > 0$. *Moreover, under the sufficient condition*

$$\Phi'(0) \geq \left(\frac{\beta + \gamma T_1^2}{\alpha - 1} + \frac{1}{T_1^2(\alpha-1)^2}\right)^{1/2} \Phi(0), \quad \Phi(0) > 0, \tag{8.280}$$

$$u_{1xx} - u_1 = -u_0 + (u_0)^2, \quad u_0(x), u_1(x) \in \mathbb{H}_0^1(0,1), \tag{8.281}$$

there are no global-in-time strong generalized solutions. Here

$$\Phi'(0) = \int_0^1 dx\,(u_0)^3 - \|u_0\|_2^2, \quad \Phi(0) = \frac{1}{2}\|u_0\|_2^2 + \frac{1}{2}\|u_{0x}\|_2^2.$$

Further examples of application of the energy method to pseudo-parabolic problems can be found in [47] (see also the references therein).

8.4 Pseudo-hyperbolic equations

In this section, we consider model pseudo-hyperbolic equations, in particular, the equation of internal waves in stratified fluids and the equation of ion-acoustic waves in plasma. Using the modified energy method, we obtain sufficient conditions of blow-up in finite time of solutions of these equations.

8.4.1 Equation of internal waves

We obtain sufficient conditions of blow-up of solutions of the following initial-boundary-value problem:

$$\frac{\partial^2}{\partial t^2}(\Delta_3 u - \gamma^2(x_3)u) + \omega^2(x_3)\Delta_2 u + f(x,u) = 0, \tag{8.282}$$

$$u|_{\partial\Omega} = 0, \quad u(x,0) = u_0(x), \quad u'(x,0) = u_1(x), \tag{8.283}$$

where $\Omega \subset \mathbb{R}^3$ is a bounded domain with smooth boundary $\partial\Omega \in \mathbb{C}^{2,\delta}$, $\delta \in (0,1]$,

$$\Delta_3 \equiv \frac{\partial^2}{\partial x_1^2} + \frac{\partial^2}{\partial x_2^2} + \frac{\partial^2}{\partial x_3^2}, \quad \Delta_2 \equiv \frac{\partial^2}{\partial x_1^2} + \frac{\partial^2}{\partial x_2^2}, \quad x = (x_1, x_2, x_3).$$

Assume that the function $f(x,u) : \Omega \times \mathbb{R}^1 \to \mathbb{R}^1$ satisfies the following conditions.

Conditions for the function $f(x,u)$

(i) $f(x,u)$ is a Carathéodori function;

(ii) the function $f(x,u)$ satisfies the following growth condition:

$$|f(x,u)| \le a_1(x) + c_1|u|^{q+1}, \quad a_1(x) \in \mathbb{L}_+^{(q+2)/(q+1)}(\Omega), \quad q \in (0,4]; \tag{8.284}$$

(iii) for any function $v(x) \in \mathbb{L}^{q+2}(\Omega)$ and some $\theta > 2$, we the inequality

$$\int_\Omega v(x) f(x, v(x))\, dx \ge \theta \int_\Omega dx \int_0^{v(x)} f(x,s)\, ds. \tag{8.285}$$

Note that the set of functions $f(x,u)$ satisfying these conditions is nonempty. Indeed, for example, the functions $f(x,u) = \chi(x)|u|^q u$ and $f(x,u) = \chi(x)|u|^{q+1}$, where $q \in (0,4]$, satisfy conditions (i)–(iii). Here $\chi(x)$ is the characteristic function of an arbitrary measurable subdomain $\Omega_0 \subseteq \Omega$:

$$\chi(x) = \begin{cases} 1 & \text{if } x \in \Omega_0, \\ 0 & \text{if } x \in \Omega\backslash\Omega_0. \end{cases}$$

Thus, considering problem (8.282)–(8.283) with functions $f(x,u)$ satisfying conditions (i)–(iii), we take into account the realistic physical situation where sources described by the function $f(x,u)$ are spatially localized.

Moreover, we assume that the following properties are fulfilled:

$$\gamma(x_3), \omega(x_3) \in \mathbb{C}(\mathbb{R}^1), \tag{8.286}$$

$$\gamma(x_3) \in [0, \gamma_0], \quad \omega(x_3) \in [0, \omega_0], \quad \gamma_0, \omega_0 \in (0, +\infty). \tag{8.287}$$

We treat solutions of problem (8.282)–(8.283) in the following sense.

Definition 8.15. A *strong generalized solution* of problem (8.282)–(8.283) is a function $u(x,t) \in \mathbb{C}^{(2)}([0,T]; \mathbb{H}_0^1(\Omega))$, where $T > 0$, satisfying the equality

$$\begin{gathered} \langle \mathrm{D}(u), w\rangle = 0 \quad \text{for all } t \in [0,T],\ w \in \mathbb{H}_0^1(\Omega), \\ u(x,0) = u_0(x) \in \mathbb{H}_0^1(\Omega), \quad u'(0,x) = u_1(x) \in \mathbb{H}_0^1(\Omega), \\ \mathrm{D}(u) \equiv \frac{\partial^2}{\partial t^2}(\Delta_3 u - \gamma^2(x_3)u) + \omega^2(x_3)\Delta_2 u + f(x,u), \end{gathered} \tag{8.288}$$

where $\langle \cdot, \cdot \rangle$ is the duality bracket between the Hilbert spaces $\mathbb{H}_0^1(\Omega)$ and $\mathbb{H}^{-1}(\Omega)$, and the prime $'$ means the derivative by time.

We assume that a nonextendable strong generalized solution of problem (8.282)–(8.283) exists for some $T > 0$, that is, $u(x,t) \in \mathbb{C}^{(2)}([0,T]; \mathbb{H}_0^1(\Omega))$, and prove its blow-up. Denote

$$\Phi(t) \equiv \Phi[u](t) \equiv \frac{1}{2} \int\limits_\Omega |\nabla u|^2 \, dx + \frac{1}{2} \int\limits_\Omega \gamma^2(x_3)|u|^2 \, dx, \tag{8.289}$$

$$\mathrm{J}(t) \equiv \mathrm{J}[u](t) \equiv \int\limits_\Omega |\nabla u'|^2 \, dx + \int\limits_\Omega \gamma^2(x_3)|u'|^2 \, dx. \tag{8.290}$$

We need the following auxiliary result.

Lemma 8.6. *Let $u(x,t) \in \mathbb{C}^{(1)}([0,T]; \mathbb{H}_0^1(\Omega))$ for some $T > 0$. Then we have the inequality*

$$(\Phi')^2(t) \le 2\Phi(t)\mathrm{J}(t) \quad \textit{for all } t \in [0,T]. \tag{8.291}$$

Proof. Since $u(x,t) \in \mathbb{C}^{(1)}([0,T]; \mathbb{H}_0^1(\Omega))$, we have the equality

$$\Phi'(t) = \int\limits_\Omega (\nabla u', \nabla u) \, dx + \int\limits_\Omega \gamma^2(x_3) u' u \, dx.$$

The terms in the right-hand side possess the following estimates:

$$\left| \int\limits_\Omega (\nabla u', \nabla u) \, dx \right| \le \|\nabla u'\|_2 \|\nabla u\|_2,$$

$$\left| \int\limits_\Omega \gamma^2(x_3) u' u \, dx \right| \le \left(\int\limits_\Omega \gamma^2(x_3)|u'|^2 \, dx \right)^{1/2} \left(\int\limits_\Omega \gamma^2(x_3)|u|^2 \, dx \right)^{1/2}.$$

Therefore we have

$$\begin{aligned}
(\Phi')^2 &\le \left(\|\nabla u'\|_2 \|\nabla u\|_2 + \left(\int\limits_\Omega \gamma^2(x_3)|u'|^2 \, dx \right)^{1/2} \left(\int\limits_\Omega \gamma^2(x_3)|u|^2 \, dx \right)^{1/2} \right)^2 \\
&\le \|\nabla u'\|_2^2 \|\nabla u\|_2^2 + \int\limits_\Omega \gamma^2(x_3)|u'|^2 \, dx \int\limits_\Omega \gamma^2(x_3)|u|^2 \, dx \\
&\quad + 2\|\nabla u'\|_2 \|\nabla u\|_2 \left(\int\limits_\Omega \gamma^2(x_3)|u'|^2 \, dx \right)^{1/2} \left(\int\limits_\Omega \gamma^2(x_3)|u|^2 \, dx \right)^{1/2} \\
&\le \|\nabla u'\|_2^2 \|\nabla u\|_2^2 + \int\limits_\Omega \gamma^2(x_3)|u'|^2 \, dx \int\limits_\Omega \gamma^2(x_3)|u|^2 \, dx
\end{aligned}$$

$$+ \|\nabla u'\|_2^2 \int_\Omega \gamma^2(x_3)|u|^2\,dx + \|\nabla u\|_2^2 \int_\Omega \gamma^2(x_3)|u'|^2\,dx$$

$$= \Big(\|\nabla u'\|_2^2 + \int_\Omega \gamma^2(x_3)|u'|^2\,dx\Big)\Big(\|\nabla u\|_2^2 + \int_\Omega \gamma^2(x_3)|u|^2\,dx\Big) = 2J(t)\Phi(t)$$

where we have used the inequality $2ab \le a^2 + b^2$ with

$$a = \|\nabla u'\|_2\Big(\int_\Omega \gamma^2(x_3)|u|^2\,dx\Big)^{1/2}, \quad b = \|\nabla u\|_2\Big(\int_\Omega \gamma^2(x_3)|u'|^2\,dx\Big)^{1/2}.$$

The lemma is proved. □

Now we deduce the basic differential inequality. We set $w = u$ in definition (8.288), where $u(x,t) \in \mathbb{C}^{(2)}([0,T_0);\mathbb{H}_0^1(\Omega))$ is a strong generalized solution for some $T_0 > 0$ (T_0 is the maximal time of existence of a nonextendable solution). Integrating by parts, we obtain

$$\int_\Omega (\nabla u'', \nabla u)\,dx + \int_\Omega \gamma^2(x_3)u''u\,dx + \int_\Omega \omega^2(x_3)(\nabla_2 u, \nabla_2 u)\,dx = \int_\Omega u(x)f(x,u(x))\,dx, \tag{8.292}$$

where $\nabla_2 \equiv (\partial_{x_1}, \partial_{x_2})$. We perform the following transformation:

$$\int_\Omega (\nabla u'', \nabla u)\,dx = \frac{d}{dt}\int_\Omega (\nabla u', \nabla u)\,dx - \|\nabla u'\|_2^2 = \frac{1}{2}\frac{d^2}{dt^2}\|\nabla u\|_2^2 - \|\nabla u'\|_2^2, \tag{8.293}$$

$$\int_\Omega \gamma^2(x_3)u''u\,dx = \frac{d}{dt}\int_\Omega \gamma^2(x_3)u'u\,dx - \int_\Omega \gamma^2(x_3)|u'|^2\,dx$$
$$= \frac{1}{2}\frac{d^2}{dt^2}\int_\Omega \gamma^2(x_3)|u|^2\,dx - \int_\Omega \gamma^2(x_3)|u'|^2\,dx. \tag{8.294}$$

From (8.293), (8.294), and (8.292), taking into account (8.289) and (8.290), we arrive at the first energy equality

$$\Phi'' - J + \int_\Omega \omega^2(x_3)|\nabla_2 u|^2 = \int_\Omega u(x)f(x,u(x))\,dx. \tag{8.295}$$

We further need the following auxiliary reasoning. Introduce the functional

$$\psi(v) \equiv \int_\Omega \mathcal{F}(x,v(x))\,dx, \quad \mathcal{F}(x,v) = \int_0^v f(x,s)\,ds. \tag{8.296}$$

For almost all $x \in \Omega$, we have the equality

$$\frac{d}{dt}\psi(v) = \int_{\Omega} v' f(x,v)\,dx \quad \text{for all } v(x,t) \in \mathbb{C}^{(1)}([0,T]; \mathbb{H}_0^1(\Omega)). \tag{8.297}$$

To obtain the second energy equality, we set $w = u'$ in definition (8.288), and, taking into account (8.297) and integrating by parts, we get

$$\int_{\Omega} (\nabla u'', \nabla u')\,dx + \int_{\Omega} \gamma^2(x_3) u' u''\,dx + \int_{\Omega} \omega^2(x_3)(\nabla_2 u, \nabla_2 u') = \frac{d}{dt}\int_{\Omega} \mathcal{F}(x,u)\,dx,$$

which by (8.290) can be rewritten in the form

$$\frac{1}{2}\frac{d}{dt}\mathrm{J} + \frac{1}{2}\frac{d}{dt}\int_{\Omega} \omega^2(x_3)|\nabla_2 u|^2\,dx = \frac{d}{dt}\int_{\Omega} \mathcal{F}(x,u)\,dx.$$

Integrating by $t \in [0,T]$, we obtain the second energy equality

$$E(t) \equiv \mathrm{J}(t) + \int_{\Omega} \omega^2(x_3)|\nabla_2 u|^2\,dx - 2\int_{\Omega} \mathcal{F}(x,u)\,dx = E(0). \tag{8.298}$$

Assuming that the initial data satisfy the inequality

$$E(0) \le 0, \tag{8.299}$$

from (8.298) we obtain the inequality $E(t) \le 0$ for all $t \in [0,T]$. Therefore,

$$\mathrm{J}(t) + \int_{\Omega} \omega^2(x_3)|\nabla_2 u|^2\,dx \le 2\int_{\Omega} \mathcal{F}(x,u)\,dx \quad \text{for all } t \in [0,T]. \tag{8.300}$$

Remark. In fact, condition (8.299), which due to the initial data can be rewritten in the form

$$\int_{\Omega} |\nabla u_1|^2\,dx + \int_{\Omega} \gamma^2(x_3)|u_1|^2\,dx + \int_{\Omega} \omega^2(x_3)|\nabla_2 u_0|^2\,dx \le 2\int_{\Omega} \mathcal{F}(x,u_0)\,dx \tag{8.301}$$

means that the initial data must be sufficiently large. Indeed, taking, for example, $f(x,u) = \chi(x)u^2$, we obtain the condition

$$\int_{\Omega} |\nabla u_1|^2\,dx + \int_{\Omega} \gamma^2(x_3)|u_1|^2\,dx + \int_{\Omega} \omega^2(x_3)|\nabla_2 u_0|^2\,dx \le \frac{2}{3}\int_{\Omega_0} u_0^3\,dx,$$

which always holds for almost all $x \in \Omega$ for fixed $u_1(x) \in \mathbb{H}_0^1(\Omega)$ and sufficiently large and nonnegative $u_0(x) \in \mathbb{H}_0^1(\Omega)$.

Now from (8.300) by condition (iii) we obtain the inequality

$$\mathrm{J}(t) + \int_\Omega \omega^2(x_3)|\nabla_2 u|^2\,dx \le \frac{2}{\theta}\int_\Omega u(x)f(x,u(x))\,dx \quad \text{for all } t \in [0,T].$$

Therefore

$$\int_\Omega u(x)f(x,u(x))\,dx - \frac{\theta}{2}\mathrm{J}(t) - \frac{\theta}{2}\int_\Omega \omega^2(x_3)|\nabla_2 u|^2\,dx \ge 0; \tag{8.302}$$

then equation (8.295) implies the estimate

$$\begin{aligned}\Phi'' &- \left(1+\frac{\theta}{2}\right)\mathrm{J} + \left(1-\frac{\theta}{2}\right)\int_\Omega \omega^2(x_3)|\nabla_2 u|^2\,dx \\ &= \int_\Omega u(x)f(x,u(x))\,dx - \frac{\theta}{2}\mathrm{J}(t) - \frac{\theta}{2}\int_\Omega \omega^2(x_3)|\nabla_2 u|^2\,dx \ge 0.\end{aligned} \tag{8.303}$$

By condition (iii) we have $\theta > 2$; therefore from (8.303) we obtain the inequality

$$\Phi'' \ge \left(1+\frac{\theta}{2}\right)\mathrm{J} \quad \text{for all } t \in [0,T].$$

This and the auxiliary lemma (see (8.291)) imply the required differential inequality

$$\Phi\Phi'' - \alpha(\Phi')^2 \ge 0, \quad \alpha = \frac{1}{2}\left(1+\frac{\theta}{2}\right), \tag{8.304}$$

where by condition (iii) we have $\alpha > 1$.

To solve this differential inequality, we impose the additional condition

$$\Phi'(0) > 0 \quad \Leftrightarrow \quad \int_\Omega (\nabla u_1, \nabla u_0)\,dx + \int_\Omega \gamma^2(x_3)u_1u_0\,dx > 0. \tag{8.305}$$

Then there exists a time moment $T_0 \in [0, T_1]$ such that

$$\limsup_{t\uparrow T_0} \Phi(t) = +\infty,$$

where

$$T_1 = \frac{1}{\alpha-1}\frac{\Phi(0)}{\Phi'(0)} = \frac{4}{\theta-2}\frac{\Phi(0)}{\Phi'(0)}, \quad \theta > 2. \tag{8.306}$$

The theorem is proved.

8.4.2 Equation if ion-acoustic waves in plasma

In this section, we consider the initial-boundary-value problem for a wide class of pseudo-hyperbolic equations and obtain sufficient conditions of blow-up of their solutions. The problem is of the form

$$\frac{\partial^2}{\partial t^2}(\Delta u - u) + \frac{\partial}{\partial t}(\Delta u - g(x,u)) + \Delta u + f(x,u) = 0, \tag{8.307}$$

$$u|_{\partial\Omega} = 0, \quad u(x,0) = u_0(x), \quad u'(x,0) = u_1(x). \tag{8.308}$$

First, introduce certain conditions for the functions $g(x,u)$ and $f(x,u)$. Assume that the function $g(x,u) : \Omega \times \mathbb{R}^1 \to \mathbb{R}^1$ satisfies the following conditions.

Conditions for the function $g(x,u)$

(i)$_1$ $g(x,u)$ is a Carathéodori function;

(ii)$_1$ the function $g(x,u)$ satisfies the following growth conditions:

$$|g(x,u)| \le a_1(x) + c_1|u|^{q_1+1}, \quad a_1(x) \in \mathbb{L}_+^{(q_1+2)/(q_1+1)}(\Omega) \tag{8.309}$$

for almost all $x \in \Omega$ and $q_1 \in (0,4]$;

(iii)$_1$ the function $g(x,s)$ belongs to the class $\mathbb{C}^{(1)}(\mathbb{R}^1)$ with respect to the variable $s \in \mathbb{R}^1$, and

$$|g'_u(x,u)| \le a_3 + c_3|u|^{q_1}, \quad a_3, c_3 > 0, \tag{8.310}$$

$$g'_u(x,u) \ge -\lambda_1 \quad \text{for all } u \in \mathbb{R}^1 \text{ and almost all } x \in \Omega; \tag{8.311}$$

here $\lambda_1 > 0$ is the first eigenvalue of the Laplace operator in the bounded domain Ω with homogeneous Dirichlet condition on the boundary, and, moreover, $g'_u(x,u)$ is a Carathéodori function.

Assume that the function $f(x,u) : \Omega \times \mathbb{R}^1 \to \mathbb{R}^1$ satisfies the following conditions.

Conditions for the function $f(x,u)$

(i)$_2$ $f(x,u)$ is a Carathéodori function;

(ii)$_2$ the function $f(x,u)$ satisfies the growth condition

$$|f(x,u)| \le a_2(x) + c_2|u|^{q_2+1}, \quad a_2(x) \in \mathbb{L}_+^{(q_2+2)/(q_2+1)}(\Omega), \quad q_2 \in (0,4]; \tag{8.312}$$

$(\mathrm{iii})_2$ for any function $v(x) \in \mathbb{L}^{q_2+2}(\Omega)$ and some $\theta > 2$,

$$\int_\Omega v(x)f(x,v(x))\,dx \geq \theta \int_\Omega dx \int_0^{v(x)} f(x,s)\,ds. \tag{8.313}$$

The set of functions $f(x,u)$ and $g(x,u)$ satisfying these conditions is nonempty. Indeed, for example, the functions $f(x,u) = \chi_1(x)|u|^{q_1}u$ and $f(x,u) = \chi_1(x)|u|^{q_1+1}$, where $q_1 \in (0,4]$, satisfy these conditions. Here $\chi_1(x)$ is the characteristic function of an arbitrary measurable subdomain $\Omega_1 \subseteq \Omega$:

$$\chi_1(x) = \begin{cases} 1 & \text{if } x \in \Omega_1, \\ 0 & \text{if } x \in \Omega\backslash\Omega_1. \end{cases}$$

As $g(x,u)$, we can take the function $g(x,u) = \chi_2(x)|u|^{q_2}u$, where $q_2 \in (0,4]$, and $\chi_2(x)$ is the characteristic function of the measurable subdomain $\Omega_2 \subset \Omega$.

Thus, considering problem (8.307)–(8.308) with functions $f(x,u)$ and $g(x,u)$ satisfying conditions $(\mathrm{i})_2$–$(\mathrm{iii})_2$ and $(\mathrm{i})_1$–$(\mathrm{iii})_1$, we take into account the realistic physical situation where sources described by the function $f(x,u)$ and sinks described by the function $g(x,u)$ are spatially localized.

We formulate the definition of a strong generalized solution.

Definition 8.16. A *strong generalized solution* of problem (8.307)–(8.308) is a function $u(x,t) \in \mathbb{C}^{(2)}([0,T];\mathbb{H}_0^1(\Omega))$, where $T > 0$, satisfying the equality

$$\begin{gathered} \langle \mathrm{D}(u), w\rangle = 0 \quad \text{for all } t \in [0,T],\ w \in \mathbb{H}_0^1(\Omega), \\ u(x,0) = u_0(x) \in \mathbb{H}_0^1(\Omega), \quad u'(0,x) = u_1(x) \in \mathbb{H}_0^1(\Omega), \\ \mathrm{D}(u) \equiv \frac{\partial^2}{\partial t^2}(\Delta u - u) + \frac{\partial}{\partial t}(\Delta u - g(x,u)) + \Delta u + f(x,u), \end{gathered} \tag{8.314}$$

where $\langle\cdot,\cdot\rangle$ is the duality bracket between the Hilbert spaces $\mathbb{H}_0^1(\Omega)$ and $\mathbb{H}^{-1}(\Omega)$, and the prime $'$ means the derivative by time.

Definition 8.17. The *weak derivative* of a function $v \in \mathbb{L}^2(\Omega)$ with respect to the duality bracket $\langle\cdot,\cdot\rangle$ between the Hilbert spaces $\mathbb{H}_0^1(\Omega)$ and $\mathbb{H}^{-1}(\Omega)$ is defined as follows:

$$\left\langle \frac{\partial v}{\partial x_i}, w \right\rangle \equiv \left\langle v, -\frac{\partial w}{\partial x_i} \right\rangle = -\int_\Omega v(x)\frac{\partial w}{\partial x_i}\,dx \quad \forall w \in \mathbb{H}_0^1(\Omega),\ i = 1,2,3.$$

Now we obtain the energy equalities. Assume that there exists a strong generalized solution of problem (8.307)–(8.308) for a certain maximal lifetime $T_0 > 0$ of a nonextendable solution:

$$u(x,t) \in \mathbb{C}^{(2)}([0,T_0);\mathbb{H}_0^1(\Omega)).$$

Denote

$$\Phi(t) \equiv \Phi[u](t) \equiv \frac{1}{2}\int_{\Omega} |\nabla u|^2\,dx + \frac{1}{2}\int_{\Omega} |u|^2\,dx, \tag{8.315}$$

$$\mathrm{J}(t) \equiv \mathrm{J}[u](t) \equiv \int_{\Omega} |\nabla u'|^2\,dx + \int_{\Omega} |u'|^2\,dx. \tag{8.316}$$

We further need the following auxiliary result.

Lemma 8.7. *Let $u(x,t) \in \mathbb{C}^{(1)}([0,T_0);\mathbb{H}_0^1(\Omega))$. Then, for all $t \in [0,T_0)$, we have the inequality*

$$(\Phi')^2(t) \le 2\Phi(t)\mathrm{J}(t). \tag{8.317}$$

The proof is similar to that of Lemma 8.6 (p. 227), where we must set $y(x_3) \equiv 1$. Setting $w = u$ in definition (8.314) and integrating by parts, we obtain the equation

$$\int_{\Omega} (\nabla u'', \nabla u)\,dx + \int_{\Omega} u'' u\,dx + \int_{\Omega} (\nabla u', \nabla u)\,dx \\ + \int_{\Omega} g_u'(x,u)u'u\,dx + \int_{\Omega} |\nabla u|^2\,dx = \int_{\Omega} f(x,u)u\,dx. \tag{8.318}$$

Note that

$$\int_{\Omega} (\nabla u'', \nabla u)\,dx = \frac{d}{dt}\int_{\Omega} (\nabla u', \nabla u)\,dx \\ - \|\nabla u'\|_2^2 = \frac{1}{2}\frac{d^2}{dt^2}\|\nabla u\|_2^2 - \|\nabla u'\|_2^2, \tag{8.319}$$

$$\int_{\Omega} u'' u\,dx = \frac{d}{dt}\int_{\Omega} u' u\,dx - \int_{\Omega} |u'|^2\,dx = \frac{1}{2}\frac{d^2}{dt^2}\int_{\Omega} |u|^2\,dx - \int_{\Omega} |u'|^2\,dx. \tag{8.320}$$

Therefore, taking into account (8.319), (8.320), (8.315), and (8.316), we can represent (8.318) in the form

$$\Phi'' - \mathrm{J} + \int_{\Omega} (\nabla u', \nabla u)\,dx \\ + \int_{\Omega} g_u'(x,u)u'u\,dx + \int_{\Omega} |\nabla u|^2\,dx = \int_{\Omega} f(x,u)u\,dx. \tag{8.321}$$

We estimate the terms in the right-hand side of (8.321):

$$\left|\int_{\Omega} (\nabla u', \nabla u)\,dx\right| \le \|\nabla u'\|_2 \|\nabla u\|_2$$

$$\leq \frac{\varepsilon}{2}\|\nabla u'\|_2^2 + \frac{1}{2\varepsilon}\|\nabla u\|_2^2 \leq \frac{\varepsilon}{2}\mathrm{J} + \frac{1}{\varepsilon}\Phi. \tag{8.322}$$

By the growth condition (8.310) we obtain the estimate

$$\begin{aligned}\left|\int_\Omega g'_u(x,u)u'u\,dx\right| &\leq a_3\int_\Omega |uu'|\,dx + c_3\int_\Omega |u|^{q_1+1}|u'|\,dx \\ &\leq a_3\|u'\|_2\|u\|_2 + c_3\|u'\|_{q_1+2}\left(\int_\Omega |u|^{q_1+2}\,dx\right)^{(q_1+1)/(q_1+2)}.\end{aligned} \tag{8.323}$$

Consider the expression

$$\left(\int_\Omega |u|^{q_1+2}\,dx\right)^{1/2}$$

separately. Let $q \in [0,4]$; then by property (8.309) and the fact that the space is three-dimensional we obtain the continuous embedding

$$\mathbb{H}_0^1(\Omega) \subset \mathbb{L}^q(\Omega).$$

Therefore, there exists a constant $c_4 > 0$ such that

$$\|w\|_q \leq c_4\|\nabla w\|_2 \quad \text{for all } w \in \mathbb{H}_0^1(\Omega).$$

Therefore, for $q_1 \in [0,4]$, we have the inequality

$$\left(\int_\Omega |u|^{q_1+2}\,dx\right)^{(q_1+1)/(q_1+2)} \leq c_4^{q_1+1}\|\nabla u\|_2^{q_1+1}. \tag{8.324}$$

Then from (8.323) we deduce the following inequalities:

$$\begin{aligned}\left|\int_\Omega g'_u(x,u)u'u\,dx\right| &\leq \frac{\varepsilon}{2}\|u'\|_2^2 + \frac{a_3^2}{2\varepsilon}\|u\|_2^2 + \frac{\varepsilon}{2}\|\nabla u'\|_2^2 + \frac{c_3^2}{2\varepsilon}c_4^{2(q_1+1)}\|\nabla u\|_2^{2(q_1+1)} \\ &\leq \varepsilon\mathrm{J} + \frac{c_5}{\varepsilon}\Phi + \frac{c_6}{\varepsilon}\Phi^{1+q_1},\end{aligned} \tag{8.325}$$

where

$$c_5 = a_3^2, \quad c_6 = 2^{q_1}c_3^2c_4^{2(q_1+1)}.$$

Combining estimates (8.322) and (8.325) and substituting them into (8.321), we obtain

$$\Phi'' - \mathrm{J} + \frac{3\varepsilon}{2}\mathrm{J} + \frac{c_7}{\varepsilon}\Phi + \frac{c_6}{\varepsilon}\Phi^{1+q_1} + \int_\Omega |\nabla u|^2\,dx \geq \int_\Omega uf(x,u)\,dx, \tag{8.326}$$

where $c_7 = 1 + c_5$ and $c_6 = 2^{q_1}c_3^2c_4^{2(q_1+1)}$.

Now we obtain the second energy inequality. Recall the definition of the Nemytsky operator and one auxiliary result. Introduce the functional

$$\psi(v) \equiv \int_{\Omega} \mathcal{F}(x, v(x))\, dx, \quad \mathcal{F}(x, v) = \int_{0}^{v} f(x, s)\, ds. \tag{8.327}$$

For almost all $x \in \Omega$, we have the equality

$$\frac{d}{dt}\psi(v) = \int_{\Omega} v' f(x, v)\, dx \quad \text{for all } v(x,t) \in \mathbb{C}^{(1)}([0,T]; \mathbb{H}_0^1(\Omega)). \tag{8.328}$$

Note that by condition (8.311) property $(\mathrm{ii})_2$ implies the inequality

$$\int_{\Omega} [|\nabla u'|^2 + g_u'(x,u)|u'|^2]\, dx \geq 0, \tag{8.329}$$

since by the assumption $u(x,t) \in \mathbb{C}^{(2)}([0,T_0); \mathbb{H}_0^1(\Omega))$ and $u' \in \mathbb{C}^{(1)}([0,T_0); \mathbb{H}_0^1(\Omega))$ and hence by the Friedrichs inequality we have the estimate

$$\int_{\Omega} [|\nabla u'|^2 + g_u'(x,u)|u'|^2]\, dx \geq \int_{\Omega} [g_u'(x,u) + \lambda_1]|u'|^2\, dx \geq 0.$$

Setting $w = u'$ in definition (8.314), integrating by parts, and applying (8.329), we obtain the inequality

$$\frac{1}{2}\frac{d\mathrm{J}}{dt} + \frac{1}{2}\frac{d}{dt}\int_{\Omega} |\nabla u|^2\, dx \leq \frac{d}{dt}\int_{\Omega} \mathcal{F}(x,u)\, dx.$$

Integrating by time, we obtain the second energy inequality

$$E(t) \geq E(0), \quad E(t) \equiv \int_{\Omega} \mathcal{F}(x,u)\, dx - \frac{1}{2}\mathrm{J} - \frac{1}{2}\int_{\Omega} |\nabla u|^2\, dx. \tag{8.330}$$

Assume that the initial data satisfy the condition

$$\begin{aligned} E(0) \equiv \int_{\Omega} \mathcal{F}(x,u_0)\, dx &- \frac{1}{2}\int_{\Omega} |\nabla u_1|^2\, dx \\ &- \frac{1}{2}\int_{\Omega} |u_1|^2\, dx - \frac{1}{2}\int_{\Omega} |\nabla u_0|^2\, dx \geq 0. \end{aligned} \tag{8.331}$$

Then inequality (8.330) implies the estimate

$$E(t) \equiv \int_{\Omega} \mathcal{F}(x,u)\, dx - \frac{1}{2}\mathrm{J} - \frac{1}{2}\int_{\Omega} |\nabla u|^2\, dx \geq 0. \tag{8.332}$$

Applying condition (iii)$_2$, from (8.332) we deduce the inequality

$$\int_{\Omega} u(x)f(x,u)\,dx - \frac{\theta}{2}J - \frac{\theta}{2}\int_{\Omega} |\nabla u|^2\,dx \geq 0, \quad \theta > 2. \tag{8.333}$$

Then from (8.326) and (8.333) we obtain

$$\Phi'' - \left(1+\frac{\theta}{2}\right)J + \frac{3\varepsilon}{2}J + \frac{c_7}{\varepsilon}\Phi + \frac{c_6}{\varepsilon}\Phi^{1+q_1} + \left(1-\frac{\theta}{2}\right)\int_{\Omega} |\nabla u|^2\,dx$$
$$\geq \int_{\Omega} u(x)f(x,u)\,dx - \frac{\theta}{2}J - \frac{\theta}{2}\int_{\Omega} |\nabla u|^2\,dx \geq 0.$$

Since $\theta > 2$, the last inequality can be rewritten in the form

$$\left(1+\frac{\theta}{2}-\frac{3\varepsilon}{2}\right)J \leq \Phi'' + \frac{c_7}{\varepsilon}\Phi + \frac{c_6}{\varepsilon}\Phi^{1+q_1}, \quad \theta > 2. \tag{8.334}$$

Finally, assuming that $\varepsilon > 0$ satisfies the inequality

$$\frac{1}{2}\left(1+\frac{\theta}{2}-\frac{3\varepsilon}{2}\right) > 1 \quad \Rightarrow \quad \varepsilon \in \left(0, \frac{\theta-2}{3}\right), \quad \theta > 2,$$

from (8.317) and (8.334) we obtain the following second-order ordinary differential inequality:

$$\Phi\Phi'' - \alpha(\Phi')^2 + \beta\Phi^2 + \gamma\Phi^{2+q_1} \geq 0, \tag{8.335}$$

where the coefficients are defined by the formulas

$$\alpha = \frac{1}{2}\left(1+\frac{\theta}{2}-\frac{3\varepsilon}{2}\right), \quad \beta = \frac{c_7}{\varepsilon}, \quad \gamma = \frac{c_6}{\varepsilon}, \quad \varepsilon \in \left(0, \frac{\theta-2}{3}\right), \quad \theta > 2.$$

To analyze (8.335), introduce the new function

$$\Psi = \Phi^{1-\alpha}. \tag{8.336}$$

Dividing both sides of inequality (8.335) by $\Phi^{1+\alpha} > 0$, we have

$$\frac{\Phi''}{\Phi^{\alpha}} - \frac{\alpha(\Phi')^2}{\Phi^{1+\alpha}} + \beta\Phi^{1-\alpha} + \gamma\Phi^{1+q_1-\alpha} \geq 0 \quad \Rightarrow$$
$$\Rightarrow \quad \frac{d}{dt}\left[\frac{\Phi'}{\Phi^{\alpha}}\right] + \beta\Phi^{1-\alpha} + \gamma\Phi^{1+q_1-\alpha} \geq 0 \quad \Rightarrow$$
$$\Rightarrow \quad \frac{1}{1-\alpha}\frac{d^2}{dt^2}\Phi^{1-\alpha} + \beta\Phi^{1-\alpha} + \gamma\Phi^{1+q_1-\alpha} \geq 0.$$

Taking into account the notation (8.336), we obtain the inequality

$$\frac{1}{1-\alpha}\Psi'' + \beta\Psi + \gamma\Psi^{(1+q_1-\alpha)/(1-\alpha)} \geq 0. \tag{8.337}$$

We make one more assumption concerning the initial data:

$$\Phi'(0) = \int_\Omega (\nabla u_1, \nabla u_0)\,dx + \int_\Omega u_1 u_0\,dx > 0. \tag{8.338}$$

Since $u(x,t) \in \mathbb{C}^{(2)}([0,T]; \mathbb{H}_0^1(\Omega))$, we conclude that $\Phi(t) \in \mathbb{C}^{(1)}([0,T])$, and hence there exists a time moment $t_1 > 0$ such that

$$\Phi'(t) \geq 0 \quad \text{for all } t \in [0, t_1]. \tag{8.339}$$

By definition (8.336) of the function $\Psi(t)$ we have

$$\Psi' = (1-\alpha)\Phi^{-\alpha}\Phi', \tag{8.340}$$

and hence by (8.339) and the inequality $\alpha > 1$ we obtain

$$\Psi'(t) \leq 0 \quad \text{for all } t \in [0, t_1]. \tag{8.341}$$

Multiplying (8.337) by Ψ', we arrive at the inequality

$$\frac{1}{1-\alpha}\Psi'\Psi'' + \beta\Psi'\Psi + \gamma\Psi'\Psi^{(1+q_1-\alpha)/(1-\alpha)} \leq 0 \quad \text{for all } t \in [0, t_1]$$

and further to the inequality

$$\Psi'\Psi'' \geq \beta(\alpha-1)\Psi'\Psi + \gamma(\alpha-1)\Psi'\Psi^{(1+q_1-\alpha)/(1-\alpha)} \quad \text{for all } t \in [0, t_1].$$

Therefore we have

$$\frac{1}{2}\frac{d}{dt}(\Psi')^2 \geq \frac{\beta(\alpha-1)}{2}\frac{d}{dt}\Psi^2 + \frac{\gamma(\alpha-1)}{\delta}\frac{d}{dt}\Psi^\delta, \tag{8.342}$$

where

$$\delta = 1 + \frac{1+q_1-\alpha}{1-\alpha}.$$

Let $\delta > 0$. Then we have

$$\begin{aligned} \delta > 0 \quad &\Rightarrow \quad \frac{1-\alpha+1+q_1-\alpha}{1-\alpha} > 0 \quad \Rightarrow \quad \frac{2\alpha-2-q_1}{\alpha-1} > 0 \quad \Rightarrow \quad 2\alpha > q_1 + 2 \\ &\Rightarrow \quad 1 + \frac{\theta - 3\varepsilon}{2} > q_1 + 2 \quad \Rightarrow \quad \theta - 3\varepsilon > 2q_1 + 2 \quad \Rightarrow \quad \varepsilon \in \left(0, \frac{\theta - 2q_1 - 2}{3}\right). \end{aligned}$$

Thus, in addition to condition $(\text{iii})_2$, we obtain the following condition for $\theta > 2$:

$$\theta > 2 + 2q_1. \tag{8.343}$$

Remark. Note that there exist functions $f(x,u)$ for which condition (8.343) is fulfilled. Indeed, let

$$f(x,u) = \chi(x)|u|^{q_2}u, \quad q_2 > 2q_1.$$

For this function, $\theta = q_2 + 2$, and hence (8.343) holds under the condition $q_2 > 2q_1$.

Now we return to the proof of blow-up. Let condition (8.343) be fulfilled. Integrating inequality (8.342), we obtain

$$(\Psi')^2 \geq A^2 + \beta(\alpha-1)\Psi^2 + \frac{2\gamma(\alpha-1)}{\delta}\Psi^\delta \geq A^2, \tag{8.344}$$

where

$$A^2 \equiv (\Psi'(0))^2 - \beta(\alpha-1)\Psi^2(0) - \frac{2\gamma(\alpha-1)}{\delta}\Psi^\delta(0). \tag{8.345}$$

Assume that $A > 0$. In view of definition (8.336) of the function $\Psi(t)$, this is equivalent to the inequality

$$(\alpha-1)^2\Phi_0^{-2\alpha}(\Phi'(0))^2 - \beta(\alpha-1)\Phi_0^{-2\alpha}\Phi_0^2 - \frac{2\gamma(\alpha-1)}{\delta}\Phi_0^{-\delta\alpha}\Phi_0^\delta > 0,$$

which we rewrite in the form

$$(\Phi'(0))^2 > \frac{\beta}{\alpha-1}\Phi_0^2 + \frac{2\gamma}{(\alpha-1)\delta}\Phi_0^{\alpha(2-\delta)+\delta}. \tag{8.346}$$

We have

$$\alpha(2-\delta) + \delta = 2\alpha + (1-\alpha)\delta = 2\alpha + 1 - \alpha + 1 + q_1 - \alpha = 2 + q_1,$$
$$\delta(\alpha-1) = \alpha - 1 - 1 - q_1 + \alpha = 2\alpha - q_1 - 2.$$

Thus from (8.346) we obtain the inequality

$$(\Phi'(0))^2 > \frac{\beta}{\alpha-1}\Phi_0^2 + \frac{2\gamma}{2\alpha - q_1 - 2}\Phi_0^{2+q_1}; \tag{8.347}$$

moreover, $2\alpha - q_1 - 2 > 0$ under the condition $\theta > 2 + 2q_1$ and

$$\varepsilon \in \left(0, \frac{\theta - 2q_1 - 2}{3}\right).$$

Let us show that the set of functions $f(x,u)$ and $g(x,u)$ and the set of initial conditions $u_0(x)$ and $u_1(x)$ satisfying conditions (8.331), (8.338), and (8.347) is nonempty.

Indeed, let $2q_1 < q_2$ and $q_3 \in (2q_1, q_2)$. For a fixed function $u_0(x) \in \mathbb{C}_0^\infty(\Omega) \subset \mathbb{H}_0^1(\Omega)$ such that $\|\nabla u_0\|_2 > 0$, we set

$$u_1(x) = |u_0|^{q_3/2}u_0 \in \mathbb{H}_0^1(\Omega).$$

Then condition (8.338) takes the form

$$\left(1+\frac{q_3}{2}\right)\int_\Omega dx|\nabla u_0|^2|u_0|^{q_3/2}\,dx+\int_\Omega |u_0|^{2+q_3/2}\,dx>0,$$

which obviously holds. Now we verify condition (8.347) substituting into it the initial functions chosen:

$$\begin{aligned}&\left(\left(1+\frac{q_3}{2}\right)\int_\Omega dx|\nabla u_0|^2|u_0|^{q_3/2}\,dx+\int_\Omega |u_0|^{2+q_3/2}\,dx\right)^2\\ &\quad>\frac{\beta}{\alpha-1}\Phi_0^2+\frac{2\gamma}{2\alpha-q_1-2}\Phi_0^{2+q_1}.\end{aligned}$$

Now, instead of $u_0(x)$, we take the function $ru_0(x)$ with $r>0$. Then condition (8.338) remains unchanged, but the last inequality takes the form

$$\begin{aligned}&r^{4+q_3}\left(\left(1+\frac{q_3}{2}\right)\int_\Omega dx|\nabla u_0|^2|u_0|^{q_3/2}\,dx+\int_\Omega |u_0|^{2+q_3/2}\,dx\right)^2\\ &\quad>r^4\frac{\beta}{\alpha-1}\Phi_0^2+r^{4+2q_1}\frac{2\gamma}{2\alpha-q_1-2}\Phi_0^{2+q_1}.\end{aligned}$$

Since $q_3>2q_1$ by definition, we see that inequality (8.347) is fulfilled for sufficiently large $r>0$. Finally, consider condition (8.331), which we rewrite in the form

$$\int_\Omega \mathcal{F}(x,u_0)\,dx\geq\frac{1}{2}\int_\Omega|\nabla u_1|^2\,dx+\frac{1}{2}\int_\Omega|u_1|^2\,dx+\frac{1}{2}\int_\Omega|\nabla u_0|^2\,dx.$$

Taking $f(x,u)=|u|^{q_2}u$ and integrating, we obtain

$$\int_\Omega \mathcal{F}(x,u_0)\,dx=\frac{1}{q_2+2}\int_\Omega|u_0|^{q_2+2}\,dx.$$

Then (8.331) becomes

$$\frac{1}{2}\int_\Omega|\nabla u_1|^2\,dx+\frac{1}{2}\int_\Omega|u_1|^2\,dx+\frac{1}{2}\int_\Omega|\nabla u_0|^2\,dx\leq\frac{1}{q_2+2}\int_\Omega|u_0|^{q_2+2}\,dx.$$

Substituting $u_1(x)=|u_0|^{q_3/2}u_0$ and performing the change $u_0\to ru_0$, we obtain the inequality

$$\begin{aligned}&\frac{1}{2}\left(1+\frac{q_3}{2}\right)^2r^{2+q_3}\int_\Omega|u_0|^{q_3}|\nabla u_0|^2\,dx+\frac{1}{2}r^{2+q_3}\int_\Omega|u_0|^{2+q_3}\,dx\\ &\quad+\frac{1}{2}r^2\int_\Omega|\nabla u_0|^2\,dx\leq r^{q_2+2}\frac{1}{q_2+2}\int_\Omega|u_0|^{q_2+2}\,dx.\end{aligned}$$

Obviously, the last inequality holds for sufficiently large $r > 0$ since $q_3 < q_2$. Thus, conditions (8.331), (8.338), and (8.347) are compatible.

From inequality (8.344) we have

$$|\Psi'| \geq A > 0 \quad \text{for all } t \in [0, t_1]. \tag{8.348}$$

Therefore

$$-\Psi' \geq A > 0 \quad \Rightarrow \quad \Psi' \leq -A < 0 \quad \text{for all } t \in [0, t_1].$$

Recall that

$$\Psi'(t) = (1 - \alpha)\Phi^{-\alpha}\Phi'(t).$$

Hence

$$(1 - \alpha)\Phi^{-\alpha}\Phi'(t) \leq -A \quad \Rightarrow \quad \Phi' \geq \frac{A}{\alpha - 1}\Phi^{\alpha} > 0.$$

This implies that, for the initial condition (8.338), $\Phi'(t)$ becomes positive on the whole interval of the existence of solution of the problem. Thus, from (8.348) we obtain

$$\Psi'(t) \leq -A < 0 \quad \text{for all } t \in [0, T_0),$$

where T_0 is the lifetime of the solution. Integrating the last inequality, we obtain

$$\Psi(t) \leq \Psi_0 - At;$$

therefore, for a finite time $T_0 \in [0, T_1]$, the function $\Psi(t)$ vanishes:

$$T_1 = \frac{\Phi_0^{1-\alpha}}{A} > 0.$$

Hence

$$\limsup_{t \uparrow T_0} \Phi(t) = +\infty.$$

Now we must choose an optimal value of the parameter

$$\varepsilon \in \left(0, \frac{\theta - 2q_1 - 2}{3}\right).$$

We take $\varepsilon_0 \in (0, (\theta - 2q_1 - 2)/3)$ so that the function

$$Q_0 = Q_0(\varepsilon) \equiv \frac{\beta}{\alpha - 1} + \frac{2\gamma}{2\alpha - q_1 - 2}\Phi_0^{q_1}$$

attains the minimum. Clearly, this number ε_0 is a function of Φ_0.

Thus, under conditions (8.331), (8.338), (8.343), and (8.347), we have proved the blow-up of a strong generalized solution $u(x, t) \in \mathbb{C}^{(2)}([0, T_0); \mathbb{H}_0^1(\Omega))$ of problem (8.307)–(8.308) in the sense of Definition 8.16.

8.5 Bibliographical notes

The material for this chapter was taken from [1, 44–48]. Note that the method presented can be applied to a sufficiently wide class of equations with nonpotential terms and strong nonlinear dissipation

$$\frac{\partial^2 u}{\partial t^2} - \frac{\partial}{\partial t}\left(\Delta u + \Delta_{p_1} u - u + c_1 \frac{\partial u}{\partial x}\right) - \Delta_{p_2} u + c_2 \frac{\partial |u|^r}{\partial x} = f(x, u),$$

where $c_1, c_2 \neq 0$, $p_1, p_2 > 2$, $r > 0$, and the Carathéodori function $f(x, v)$ satisfies the standard growth condition and the energy inequality of the form (8.313). This direction is being intensively developed by the authors of this monograph together with A. A. Panin.

9 Nonlinear Schrödinger equation

In this chapter, we examine the global insolvability of the Cauchy problem for the following second-order nonlinear equation, which plays an important role in the theory of nonlinear waves:

$$iu_t = -\Delta u - |u|^{p-1}u.$$

This equation, called the nonlinear Schrödinger equation (briefly NSE), has applications in diverse areas of physics. For example, it describes envelopes of wave packets in dispersive nonlinear media during propagation of electromagnetic waves in plasma or waves in crystals with dispersion and central symmetry.

The study of the nonlinear Schrödinger equation is closely related to the theory of the well-known Korteweg–de Vries equation; both equations possess soliton solutions. We have already mentioned that in the general case the question on the global solvability of the classical Cauchy problem for the Korteweg–de Vries equation still remains open (see Chapter 3). It should be noted that, for the Schrödinger equation, this question has been solved. Moreover, a successful research is facilitated by the fact that from the physical point of view, the blow-up phenomenon is similar in some sense to that of self-focusing of laser beams. For certain particular cases, this question is carefully studied, and there exists a large number of theoretical and numerical results on the dynamics of solutions. We refer the reader to [30, 72, 83, 9] and the references therein.

9.1 Virial law. Result of Glassey

9.1.1 Conservation laws

First, we consider the general Cauchy problem for the nonlinear Schrödinger equation:

$$iu_t = -\Delta u - |u|^{p-1}u, \quad (t,x) \in [0,T) \times \mathbb{R}^N, \tag{9.1}$$

$$u(x,0) = u_0(x), \quad u_0 : \mathbb{R}^N \to \mathbb{C}. \tag{9.2}$$

For the parameter $p > 1$ and sufficiently smooth, decreasing at infinity initial data $u_0(x)$, for example,

$$u_0(x) \in \mathbb{H}^1(\mathbb{R}^N) = \{u, \ \nabla u \in \mathbb{L}^2(\mathbb{R}^N)\},$$

we examine the existence of classical solutions that blow up in finite time. For brevity, instead of $(x, \nabla v)$ and $\operatorname{div}(v\nabla w)$, we will write $x\nabla v$ and $\nabla(v\nabla w)$, respectively, where $(\cdot,\cdot)$ is the scalar product in the Euclidean space $\mathbb{R}^N$.

https://doi.org/10.1515/9783110602074-009

Assume that $u(x,t)$ is a solution of problem (9.1)–(9.2). Multiplying equation (9.1) by $2u^*$ and taking the imaginary part, we arrive at the differential equation

$$\frac{\partial |u|^2}{\partial t} = -\nabla(2\,\mathrm{Im}\, u^* \nabla u).$$

Integrating over the domain, we obtain the conservation law for the $\mathbb{L}^2$-norm:

$$\int_{\mathbb{R}^N} |u(x,t)|^2\,dx = \int_{\mathbb{R}^N} |u_0(x)|^2\,dx \equiv \|u\|_{\mathbb{L}^2} = \mathrm{const}. \tag{9.3}$$

Similarly, multiplying by $2u_t^*$, we obtain the energy conservation law:

$$\int_{\mathbb{R}^n} \left(|\nabla u|^2 - \frac{2}{p+1}|u|^{p+1} \right) dx = E_0 = \mathrm{const}. \tag{9.4}$$

Assume that in equation (9.1) the following upper estimate of the degree of nonlinearity is fulfilled:

$$p + 1 < \frac{2N}{N-2} \quad \text{for } N \geq 3,$$

or

$$p \in \mathbb{R}_+^1 \quad \text{for } N = 1, 2.$$

Then we have the following Sobolev interpolation inequality:

$$\|u\|_{\mathbb{L}^{p+1}} \leq c\|u(t)\|_{\mathbb{L}^2}^{1-q} \|\nabla u(t)\|_{\mathbb{L}^2}^{q},$$

where

$$\frac{1}{p+1} = q\left(\frac{1}{2} - \frac{1}{N}\right) + \frac{1-q}{2}.$$

In other words, by the boundedness of the norm $\|u\|_{\mathbb{L}^2}$ we can obtain the estimate

$$\|u(t)\|_{\mathbb{L}^{p+1}}^{p+1} \leq c\|\nabla u(t)\|_{\mathbb{L}^2}^{N(p-1)/2},$$

and setting $|E_0| < \infty$, the estimate of the norm $\|\nabla u\|_{\mathbb{L}^2}$:

$$\|\nabla u(t)\|_{\mathbb{L}^2}^2 \leq |E_0| + c\|\nabla u(t)\|_{\mathbb{L}^2}^{N(p-1)/2}.$$

The last inequality implies that for

$$\frac{N(p-1)}{2} < 2, \quad \text{that is, } p < 1 + \frac{4}{N},$$

the norm $\|\nabla u(t)\|_{\mathbb{L}^2}$ is bounded from above and cannot blow up in finite time. Moreover, in 1977, Ginibre and Velo [30] proved both local and global solvability of problem (9.1)–(9.2) under the condition that the norm $\|\nabla u\|_{\mathbb{L}^2}$ is bounded. This implies that, for small p, the problem itself is globally solvable.

Similar reasoning for large p is invalid, and the boundedness may occur only for small initial conditions. This fact aroused suspicion on the global insolvability. Following Glassey [32], we show that, for $p > 1 + 4/N$ and $E_0 \leq 0$, a finite time moment T can exist for which

$$\|\nabla u(t)\|_{\mathbb{L}^2} \to +\infty \quad \text{as } t \to T - 0.$$

9.1.2 Theorem on blow-up

First, we prove an auxiliary lemma for a problem more general than (9.1)–(9.2):

$$iu_t = \Delta u + F(|u|^2)u, \quad x \in \mathbb{R}^N,\ t > 0, \tag{9.5}$$

$$u(x,0) = u_0(x), \quad u_0 : \mathbb{R}^N \to \mathbb{C}, \tag{9.6}$$

where F is a sufficiently smooth real function. Such a lemma, which states the existence of a unique classical local-in-time solution belonging to the Sobolev space of a sufficiently high order, can be obtained by using the existence theorems proved in [97].

We denote

$$G(u) \equiv \int_0^u F(s)\,ds$$

and state the following properties of solutions of problem (9.5)–(9.6).

Lemma 9.1. *Any solution $u(x,t)$ of the Cauchy problem* (9.5)–(9.6) *on the interval* $0 \leq t < t_1$ *possesses the following properties*:

$$\|u(t)\|_{\mathbb{L}^2} = \|u_0\|_{\mathbb{L}^2} = \text{const}, \tag{9.7}$$

$$\int_{\mathbb{R}^N} (|\nabla u|^2 - G(|u|^2))\,dx \equiv E_0 = \text{const}, \tag{9.8}$$

$$\frac{d}{dt}\int_{\mathbb{R}^N} x^2|u|^2\,dx = 4\,\mathrm{Im}\int_{\mathbb{R}^N} xu^*\nabla u\,dx, \tag{9.9}$$

$$\frac{d}{dt}\,\mathrm{Im}\int_{\mathbb{R}^N} xu^*\nabla u\,dx = 2\int_{\mathbb{R}^N} |\nabla u|^2\,dx - N\int_{\mathbb{R}^N} (|\nabla u|^2 + G(|u|^2))\,dx. \tag{9.10}$$

Proof. Multiplying (9.5) by $(2u^*)$, extracting the imaginary part

$$\frac{\partial}{\partial t}|u|^2 = -\nabla(2\,\mathrm{Im}\,u^*\nabla u),$$

multiplying by $|x|^2$, and integrating by parts the right-hand side, we obtain expression (9.9).

To obtain (9.10), we multiply (9.5) by $(2x\nabla u^*)$,

$$2ix\nabla u^* u_t = -2x\nabla u^*\Delta u - 2F(|u|^2)xu\nabla u^*,$$

and integrate the real part of this equality. We represent the result obtained as the sum $I + II + III$ where

$$I = -\,\mathrm{Re}\left[i\int\limits_{\mathbb{R}^N}\sum_{k=1}^{N} x_k(u^*_{x_k}u_t - u_{x_k}u^*_t)\,dx\right],$$

$$II = 2\,\mathrm{Re}\int\limits_{\mathbb{R}^N} x\nabla u^*\Delta u\,dx, \quad III = 2\,\mathrm{Re}\int\limits_{\mathbb{R}^N} F(|u|^2)xu\nabla u^*\,dx.$$

Integrating II and III by parts, we obtain

$$\begin{aligned}
II &= -2\,\mathrm{Re}\int\limits_{\mathbb{R}^N}\sum_{i,j=1}^{N,N}(x_i u^*_{x_i})_{x_j}u_{x_j}\,dx \\
&= -2\,\mathrm{Re}\int\limits_{\mathbb{R}^N}\sum_{i,j=1}^{N,N} x_i u^*_{x_i x_j}u_{x_j}\,dx - 2\,\mathrm{Re}\int\limits_{\mathbb{R}^N}|\nabla u|^2\,dx \\
&= -\,\mathrm{Re}\int\limits_{\mathbb{R}^N}\sum_{i,j=1}^{N,N} x_i u^*_{x_i x_j}u_{x_j}\,dx + N\,\mathrm{Re}\int\limits_{\mathbb{R}^N}|\nabla u|^2\,dx \\
&\quad + \mathrm{Re}\int\limits_{\mathbb{R}^N}\sum_{i,j=1}^{N,N} x_i u^*_{x_j}u_{x_j x_i}\,dx - 2\,\mathrm{Re}\int\limits_{\mathbb{R}^N}|\nabla u|^2\,dx \\
&= 2\,\mathrm{Re}\int\limits_{\mathbb{R}^N}\sum_{i,j=1}^{N,N} x_i(\mathrm{Re}\,u_{x_j}\,\mathrm{Re}\,u_{x_i x_j} + \mathrm{Im}\,u_{x_j}\,\mathrm{Im}\,u_{x_i x_j} \\
&\quad - \mathrm{Re}\,u_{x_i x_j}\,\mathrm{Re}\,u_{x_j} - \mathrm{Im}\,u_{x_i x_j}\,\mathrm{Im}\,u_{x_j})\,dx \\
&\quad + (N-2)\int\limits_{\mathbb{R}^N}|\nabla u|^2 dx = (N-2)\int\limits_{\mathbb{R}^N}|\nabla u|^2\,dx
\end{aligned}$$

and

$$III = \mathrm{Re}\int\limits_{\mathbb{R}^N} xF(|u|^2)\nabla|u|^2\,dx = -N\int\limits_{\mathbb{R}^N} G(|u|^2)\,dx.$$

Now we rewrite the expression I in the form

$$I = \mathrm{Re}\Big[i \int\limits_{\mathbb{R}^N} \sum_{k=1}^{N} x_k\Big(-\frac{\partial}{\partial t}(u^*_{x_k} u) + \frac{\partial}{\partial x_k}(uu^*_t)\Big)\,dx\Big]$$

$$= -\frac{d}{dt}\mathrm{Re}\Big[i \int\limits_{\mathbb{R}^N} xu\nabla u^*\,dx\Big] - N\,\mathrm{Re}\Big[i \int\limits_{\mathbb{R}^N} uu^*_t\,dx\Big].$$

Substituting u^*_t from (9.5), we arrive at the equality

$$I = \frac{d}{dt}\mathrm{Im} \int\limits_{\mathbb{R}^N} xu^*\nabla u\,dx + N\int |\nabla u|^2\,dx - N\int |u|^2F(|u|^2)\,dx.$$

Finally, equating I to the sum $II + III$, we obtain property (9.10). The lemma is proved. □

Using Lemma 9.1, we recall and prove Glassey's blow-up result.

Theorem 9.1. *Assume that, for $p > 1 + 4/N$, there exists a classical solution of problem (9.1)–(9.2) satisfying the following three conditions:*
(i) *the initial energy is nonpositive, $E_0 \le 0$;*
(ii) *the following inequality holds:*

$$\mathrm{Im} \int\limits_{\mathbb{R}^N} xu_0\nabla u_0^*dx < 0;$$

(iii) *there exists a constant $c_N > 1 + 2/N$ such that $sF(s) \ge c_N G(s)$ for all $s \ge 0$.*

Then the lifetime T of the solution is finite, and

$$\lim_{t\to T^-} \|\nabla u(t)\|_{\mathbb{L}^2} = +\infty.$$

Proof. On the domain of existence of a solution $u(x,t)$, we denote the notation

$$y(t) = -\,\mathrm{Im} \int\limits_{\mathbb{R}^N} xu\nabla u^*\,dx.$$

Then, by assumption (ii) we have $y(0) > 0$, and the last assertion of Lemma (9.10) can be rewritten in the form

$$\frac{dy}{dt} = -2\int\limits_{\mathbb{R}^N} |\nabla u|^2\,dx + N\int\limits_{\mathbb{R}^N} (|u|^2F(|u|^2) - G(|u|^2))\,dx. \tag{9.11}$$

Using estimate (iii), we rewrite (9.11) in the following form:

$$\frac{dy}{dt} \ge -2\int\limits_{\mathbb{R}^N} |\nabla u|^2\,dx + N(c_N - 1)\int\limits_{\mathbb{R}^N} G(|u|^2)\,dx. \tag{9.12}$$

Substituting (9.8) into (9.12), under condition (i), we obtain

$$\frac{dy}{dt} \geq -2 \int_{\mathbb{R}^N} |\nabla u|^2\, dx + N(c_N - 1)\left[\int_{\mathbb{R}^N} |\nabla u|^2\, dx - E_0 \right]$$
$$= [N(c_N - 1) - 2] \int_{\mathbb{R}^N} |\nabla u|^2\, dx - N(c_N - 1)E_0 \geq k_N \|\nabla u(t)\|^2_{\mathbb{L}^2}. \tag{9.13}$$

Here $k_N = [N(c_N - 1) - 2] > 0$ due to (iii). Thus, the function $y(t)$ increases at the points where the solution $u(x, t)$ does. Then from expression (9.9) we obtain

$$\frac{d}{dt} \int_{\mathbb{R}^N} x^2 |u|^2\, dx = 4\,\mathrm{Im} \int_{\mathbb{R}^N} x u \nabla u^*\, dx = -4y(t) < 0,$$

which yields the following upper estimate for the integral:

$$\int_{\mathbb{R}^N} x^2 |u|^2\, dx < \int_{\mathbb{R}^N} x^2 |u_0|^2\, dx \equiv d_0^2 < \infty.$$

Applying the Young inequality and the estimate

$$|y(t)| \leq \left| \int_{\mathbb{R}^N} x^2 |u|^2\, dx \right|^{1/2} \left| \int_{\mathbb{R}^N} |\nabla u|^2\, dx \right|^{1/2} < d_0 \|\nabla u\|_{\mathbb{L}^2},$$

we rewrite (9.13) in the form of ordinary differential inequality,

$$\frac{dy}{dt} > \frac{k_N}{d_0^2} y^2(t),$$

which can be easily integrated under the condition $y(0) > 0$. Then, on the interval

$$0 \leq t < \frac{d_0^2}{k_N y(0)},$$

we get the following lower estimate of the norm of the solution:

$$\|\nabla u(t)\|^2_{\mathbb{L}^2} \geq \frac{1}{d_0} y(t) \geq \frac{y(0) d_0}{d_0^2 - k_N y(0) t}. \tag{9.14}$$

Therefore, for $T \leq T_0 < \infty$, the blow-up occurs:

$$\lim_{t \to T^-} \|\nabla u(t)\|_{\mathbb{L}^2} = +\infty, \quad T = \frac{d_0^2}{k_N y(0)}.$$

Now we can specify the expression of the function $F(s)$ according to equation (9.1):

$$F(s) = s^{(p-1)/2}, \quad G(|u|^2) = \frac{2}{p+1} |u|^{p+1}.$$

In this particular case, condition (iii) takes the form

$$s \cdot s^{(p-1)/2} \geq c_N\left(\frac{p+1}{2} s^{(p+1)/2}\right),$$

which provides an appropriate choice of the constant $c_N = (p+1)/2$. Then (iii) is replaced by the condition $p > 1 + 4/N$, and the blow-up is a direct consequence of inequality (9.14). The theorem is proved. □

The particular case where $F(s) = s^{(p-1)/2}$ is also important since this form of the nonlinearity allows us to prove the blow-up in other $\mathbb{L}^q$-norms. For example, we have the following:

Corollary 9.1. *Assume that the initial data of problem* (9.1)–(9.2) *satisfy the following conditions*:

(i) *the initial energy is nonpositive*:

$$E_0 \equiv \int_{\mathbb{R}^N} \left(|\nabla u_0|^2 - \frac{2}{p+1}|u_0|^{p+1}\right) dx \leq 0;$$

(ii) *the following inequality holds*:

$$\operatorname{Im} \int_{\mathbb{R}^N} x u_0 \nabla u_0^* dx < 0;$$

(iii) $p > 1 + 4/N$.

Then the solution of the problem blows up in finite time in the $\mathbb{L}^q$*-norm, that is,*

$$\|u(t)\|_{\mathbb{L}^q} \to \infty,$$

under the following conditions:

(a) $q \geq p+1$ *and* $N \geq 1$;

(b) $N(p-1)/2 < q < p+1$ *for* $N = 1, 2$, *and* $p < \infty$;

(c) $N(p-1)/2 < q < p+1$ *for* $N \geq 3$, *and* $p < (N+1)/(N-2)$.

Proof. Indeed, to prove (a), it suffices to note that the condition $E_0 \leq 0$ and the Young inequality imply the following inequalities:

$$\|\nabla u\|^2_{\mathbb{L}^2} \leq \frac{2}{p+1}\|u\|^{p+1}_{\mathbb{L}^{p+1}} \leq \frac{2}{p+1}\left(\|u\|^r_{\mathbb{L}^2}\|u\|^{1-r}_{\mathbb{L}^q}\right)^{p+1},$$

$$\frac{1}{p+1} = \frac{r}{2} + \frac{1-r}{q}.$$

Thus the blow-up in the $\mathbb{L}^q$-norm immediately follows from the conservation law

$$\|u(t)\|_{\mathbb{L}^2} = \|u_0\|_{\mathbb{L}^2}.$$

Cases (b) and (c) can be proved, for example, by using the Sobolev inequality:

$$\|u\|_{\mathbb{L}^{p+1}} \le \|\nabla u\|_{\mathbb{L}^2}^{r} \|u\|_{\mathbb{L}^q}^{1-r}, \qquad \frac{1}{p+1} = r\left(\frac{1}{2} - \frac{1}{N}\right) + \frac{1-r}{q}.$$

The condition $E_0 \le 0$ implies

$$\|u\|_{\mathbb{L}^{p+1}} \le c\|u\|_{\mathbb{L}^{p+1}}^{(p+1)r/2} \|u\|_{\mathbb{L}^q}^{1-r};$$

therefore, the solution blows up in the $\mathbb{L}^q$-norm under the condition

$$\frac{(p+1)r}{2} < 1 \quad \Rightarrow \quad q > \frac{N(p-1)}{2}.$$

The corollary is proved. □

Now we discuss the behavior of collapsing solutions of the nonlinear Schrödinger equation by using the concept of basic states.

9.2 Critical case. Concept of basic states

In this section, we examine in detail a critical nonlinearity that leads to the disappearance of the global solvability and demonstrate a close connection between the classical Gagliardo–Nirenberg inequality and the blow-up in the following problem for the nonlinear Schrödinger equation:

$$iu_t = -\Delta u - |u|^{p-1}u, \quad (t, x) \in [0, T) \times \mathbb{R}^N, \tag{9.15}$$

$$u(x, 0) = u_0(x) \in \mathbb{H}^1(\mathbb{R}^N), \quad u_0 : \mathbb{R}^N \to \mathbb{C}. \tag{9.16}$$

We use results obtained by Weinstein [107], who found necessary conditions of blow-up in the critical case. First, we clarify the definitions.

We have proved before that, for $p > 1 + 4/N$, solution of (9.15)–(9.16) blows up in finite time, whereas, for $p < 1 + 4/N$, it is bounded everywhere and globally defined in time. The following question appears: Does a classical global solution in the critical case (i. e., for $p = 1 + 4/N$) exist?

The first answer was given in 1979 by Ginibre and Velo [30]. They showed that the problem is globally solvable for sufficiently small $\|u_0\|_{\mathbb{L}^2}$. The boundary state leading to the blow-up of the solution was found by Weinstein [107] in 1983.

Theorem 9.2. *Let $u_0 \in \mathbb{H}^1(\mathbb{R}^N)$ and $p = 1+4/N$. Then a sufficient condition of the global-in-time solvability of the Cauchy problem (9.15)–(9.16) is the following upper estimate for the initial function:*

$$\|u_0\|_{\mathbb{L}^2} < \|\psi\|_{\mathbb{L}^2}, \tag{9.17}$$

where the function $v = \psi$ is a positive solution of the equation

$$\Delta v - v + v^p = 0 \tag{9.18}$$

with the minimal $\mathbb{L}^2$-norm. Moreover, for any $\delta > 0$, there exist a function

$$\xi(x) : \|\xi - \psi\|_{\mathbb{L}^2} < \delta$$

such that the solution of problem (9.15)–(9.16) *with $u_0(x) = \xi(x)$ blows up in finite time:*

$$\lim_{t \to T-0} \|\nabla u(t)\|_{\mathbb{L}^2} = \infty, \quad 0 < T < \infty.$$

Proof. Note that if $\psi(x)$ is a solution of (9.18), then the function $\psi(x)\exp(it)$ is a solution of the original equation (9.15). In this connection, in the literature, ψ is usually called the basic state (or the ground state) of the nonlinear Schrödinger equation.

Following Weinstein's ideas, we first obtain the exact constant in the Gagliardo–Nirenberg inequality ($N \geq 2$):

$$\|v\|^{2(\sigma+1)}_{\mathbb{L}^{2(\sigma+1)}} \leq c^{2(\sigma+1)}_{\sigma,N} \|\nabla v\|^{\sigma N}_{\mathbb{L}^2} \|v\|^{2+\sigma(2-N)}_{\mathbb{L}^2}, \quad 0 < \sigma < \frac{2}{N-2}. \tag{9.19}$$

We can show that the constant

$$c_{\sigma,N} = \left(\frac{\sigma + 1}{\|\psi\|^{2\sigma}_{\mathbb{L}^2}} \right)^{\frac{1}{2\sigma+2}},$$

where $\psi(x)$ is a solution of the equation

$$\frac{\sigma N}{2} \Delta\psi - \left(1 + \frac{\sigma}{2}(2 - N)\right)\psi + \psi^{2\sigma+1} = 0. \tag{9.20}$$

We pose the following natural question: *Find the value of the constant $c^{2(\sigma+1)}_{\sigma,N}$ as the minimum of the functional*

$$J^{\sigma,N}(v) = \frac{\|\nabla v\|^{\sigma N}_{\mathbb{L}^2} \|v\|^{2+\sigma(2-N)}_{\mathbb{L}^2}}{\|v\|^{2(\sigma+1)}_{\mathbb{L}^{2(\sigma+1)}}};$$

the function that proves the minimum must satisfy the Euler–Lagrange equation, which implies equation (9.20).

Remark 9.1. Note that there exists many works devoted to estimates of the constant in the Gagliardo–Nirenberg inequality. For example, the case $N = 1$ was discussed in [75]; Payne [81] obtained the estimate $c_{1,2} \leq 2^{-1/4}$ for the case where $N = 2$ and $\sigma = 1$; in [59], H. A. Levine clarified the estimate $c_{1,2} \leq \pi^{-1/4}$ based on [19, 20]. A finite exact estimate was obtained by Weinstein: $c_{1,2} \leq (\pi \cdot 1{,}86\ldots)^{-1/4}$.

For the study of equation (9.20), we must discuss the question on its solvability. For example, in [69] the existence of a unique solution with minimal $\mathbb{L}^2$-norm was proved for the equation

$$\Delta v - v + v^{2\sigma+1} = 0 \tag{9.21}$$

under the following conditions:

$$\begin{aligned} 0 < \sigma < \infty \quad & \text{for } 1 \le N \le 2, \\ 0 < \sigma < \frac{2}{N-2} \quad & \text{for } 2 \le N \le 4, \\ 0 < \sigma < \frac{8-N}{2N} \quad & \text{for } 4 < N < 8. \end{aligned}$$

However, these results do not cover all cases of the existence of basic solutions (see [77, 92]).

We need an auxiliary lemma.

Lemma 9.2. *The minimum of the functional $J^{\sigma,N}(v)$ is attained at the function $v = \psi \in \mathbb{H}^1(\mathbb{R}^N)$ that possesses the following three properties:*
(i) *$\psi > 0$ depends only on $|x|$;*
(ii) *$\psi \in \mathbb{H}^1(\mathbb{R}^N) \cap \mathbb{C}^\infty(\mathbb{R}^N)$;*
(iii) *ψ is a solution of (9.20) with the minimal $\mathbb{L}^2$-norm.*

To prove the lemma, we use the idea of Strauss [99] on compact properties of the space

$$\mathbb{H}^1_{\text{radial}}(\mathbb{R}^N) \subset \mathbb{H}^1(\mathbb{R}^N)$$

of radially symmetric functions.

Proposition 9.1 (on compactness). *If the parameter σ belongs to the interval*

$$0 < \sigma < \frac{2}{N-2},$$

then the embedding of the space

$$\mathbb{H}^1_{\text{radial}}(\mathbb{R}^N) \hookrightarrow \mathbb{L}^{2\sigma+2}(\mathbb{R}^N)$$

is compact.

Proof. The result follows from the following interpolation inequality in a bounded domain Ω:

$$\|v\|^{2\sigma+2}_{\mathbb{L}^{2\sigma+2}} \le c\|v\|^{\sigma N}_{\mathbb{H}^1}\|v\|^{2+\sigma(2-N)}_{\mathbb{L}^2}, \quad 0 < \sigma < \frac{2}{N-2},$$

for all $v \in \mathbb{H}^1(\Omega)$ [99]. To extend the inequality to an unbounded domain, it suffices to prove that a bounded sequence from $\mathbb{H}^1_{\text{radial}}(\Omega)$ uniformly decreases at infinity, which was done by Strauss:

$$|v(x)| \le \frac{c}{|x|^{(N-1)/2}} \|v\|_{\mathbb{H}^1}.$$

To minimize the functional $J^{\sigma,N}$, for certain positive parameters λ and μ, we introduce the function

$$v^{\lambda,\mu}(x) = \mu v(\lambda x)$$

satisfying the following relations:

$$J^{\sigma,N}(v^{\lambda,\mu}) = J^{\sigma,N}(v), \tag{9.22}$$

$$\|v^{\lambda,\mu}\|^2_{\mathbb{L}^2} = \lambda^{-N}\mu^2\|v\|^2_{\mathbb{L}^2}, \tag{9.23}$$

$$\|\nabla_x v^{\lambda,\mu}\|^2_{\mathbb{L}^2} = \lambda^{2-N}\mu^2\|\nabla v\|^2_{\mathbb{L}^2}. \tag{9.24}$$

The nonnegativity of the functional $J^{\sigma,N}(v) \ge 0$ implies that there exists a minimizing sequence

$$\{v_\nu\} \in \mathbb{H}^1(\mathbb{R}^N) \cap \mathbb{L}^{2\sigma+2}(\mathbb{R}^N)$$

such that

$$\alpha = \inf J^{\sigma,N}(v) = \lim_{\nu\to\infty} J(v_\nu).$$

We can assume that $v_\nu > 0$ and, after symmetrization (see, e. g., [99]), take $v_\nu = v_\nu(|x|)$.

Using the function $v^{\lambda,\mu}$, we define the sequence

$$\psi_\nu(x) = v_\nu^{\lambda_\nu,\mu_\nu}(x) \quad \text{for } \lambda_\nu = \frac{\|v_\nu\|_{\mathbb{L}^2}}{\|\nabla v_\nu\|_{\mathbb{L}^2}}, \; \mu_\nu = \frac{\|v_\nu\|^{N/2-1}_{\mathbb{L}^2}}{\|\nabla v_\nu\|^{N/2}_{\mathbb{L}^2}}.$$

We can easily verify the this sequence satisfies the following conditions:

$$\psi_\nu \ge 0, \quad \psi_\nu = \psi_\nu(|x|), \tag{9.25}$$

$$\psi_\nu \in \mathbb{H}^1(\mathbb{R}^N), \tag{9.26}$$

$$\|\psi_\nu\|_{\mathbb{L}^2} = 1, \quad \|\nabla\psi_\nu\|_{\mathbb{L}^2} = 1, \tag{9.27}$$

$$J^{\sigma,N}(\psi_\nu) \to \alpha \quad \text{as } \nu \to \infty. \tag{9.28}$$

Since the sequence $\{\psi_\nu\}$ is bounded in the space $\mathbb{H}^1(\mathbb{R}^N)$, we can extract a subsequence such that

$$\psi_\nu \rightharpoonup \psi^* \quad \text{weakly in } \mathbb{H}^1(\mathbb{R}^N) \text{ as } \nu \to +\infty.$$

Since ψ_ν is a radial function uniformly bounded in $\mathbb{H}^1(\mathbb{R}^N)$, due to Proposition 9.1 on compactness, for all $0 < \sigma < 2/(N-2)$, we can choose a subsequence $\{\psi_\nu\}$ such that

$$\psi_\nu \to \psi^* \quad \text{strongly in } \mathbb{L}^{2\sigma+2}(\mathbb{R}^N) \text{ as } \nu \to +\infty.$$

Due to the weak convergence of the sequence $\{\psi_\nu\}$ in $\mathbb{H}^1(\mathbb{R}^N)$ and equation (9.27), we have

$$\|\psi_\nu^*\|_{\mathbb{H}^1} \leq \liminf_{\nu\to+\infty} \|\psi_\nu\|_{\mathbb{H}^1} < +\infty.$$

Therefore

$$\begin{aligned} \alpha \equiv \inf_{v\in\mathbb{H}^1(\mathbb{R}^N)} J^{\sigma,N}(v) &\leq J^{\sigma,N}(\psi^*) \\ &\leq \left| \int_{\mathbb{R}^N} |\psi^*|^{2\sigma+2} dx \right|^{-1} = \lim_{\nu\to\infty} J(\psi_\nu) = \alpha. \end{aligned} \tag{9.29}$$

This implies that

$$\|\nabla\psi^*\|_{\mathbb{L}^2}^{\sigma N} \|\psi^*\|_{\mathbb{L}^2}^{2+\sigma(2-N)} = 1, \quad \|\psi^*\|_{\mathbb{L}^2} = \|\nabla\psi^*\|_{\mathbb{L}^2} = 1. \tag{9.30}$$

Thus

$$\psi_\nu \to \psi^* \quad \text{strongly in } \mathbb{H}^1(\mathbb{R}^N) \text{ as } \nu \to +\infty,$$

which implies properties (i) and (ii) of the auxiliary lemma.

Property (iii) follows from the fact that the function ψ^* is a minimizing function for the functional $J^{\sigma,N}(v)$ in the Hilbert space $\mathbb{H}^1(\mathbb{R}^N)$ and, therefore, is a solution of the Euler–Lagrange equation (see [14]):

$$\frac{d}{d\varepsilon} J^{\sigma,N}(\psi^* + \varepsilon\eta)\Big|_{\varepsilon=0} = 0 \quad \text{for all } \eta \in \mathbb{C}_0^\infty(\mathbb{R}^N).$$

Taking into account (9.30), we differentiate with respect to ε all norms that are involved in the expression of the functional $J^{\sigma,N}$. For the norm

$$\|\nabla v\|_{\mathbb{L}^2}^{\sigma N} \|v\|_{\mathbb{L}^2}^{2+\sigma(2-N)},$$

we obtain the following expression:

$$\begin{aligned} \frac{d}{d\varepsilon}\|\nabla(\psi^* + \varepsilon\eta)\|_{\mathbb{L}^2}^{\sigma N} &= \frac{\sigma N}{2}\|\nabla(\psi^* + \varepsilon\eta)\|_{\mathbb{L}^2}^{\sigma N-2} 2\int_{\mathbb{R}^N} \nabla(\psi^* + \varepsilon\eta)\nabla\eta \, dx \\ &= -\frac{\sigma N}{2}\|\nabla(\psi^* + \varepsilon\eta)\|_{\mathbb{L}^2}^{\sigma N-2} 2\int_{\mathbb{R}^N} \Delta(\psi^* + \varepsilon\eta)\eta \, dx. \end{aligned}$$

Similarly, we obtain expressions for the other two norms. Finally, we obtain the required Euler–Lagrange equation. Thus, the function ψ^* satisfies the equation

$$\frac{\sigma N}{2}\Delta\psi^* - \left(1 + \frac{\sigma}{2}(2-N)\right)\psi^* + \alpha(1+\sigma)(\psi^*)^{2\sigma+1} = 0, \tag{9.31}$$

where

$$\alpha = J^{\sigma,N}(\psi^*).$$

Using results of [4], we can prove that the function possesses the required smoothness. The change of variables

$$\psi = \psi^*(\alpha(1+\sigma))^{1/2\sigma}$$

reduces equation (9.31) to equation (9.20) for the function ψ:

$$\alpha = \frac{\|\psi\|_{\mathbb{L}^2}^{2\sigma}}{\sigma+1},$$

and we complete the proof of the lemma. □

Lemma 9.2 immediately implies the global-in-time existence of a solution.

Theorem 9.3 (on the global solvability). *Let the initial function u_0 belong to the class $\mathbb{H}^1(\mathbb{R}^N)$. Then:*

(a) *if $0 < p < 1 + 4/N$, then there exists a unique solution of the Cauchy problem (9.15)–(9.16) of the class*

$$u(x,t) \in \mathbb{C}([0,\infty); \mathbb{H}^1(\mathbb{R}^N));$$

(b) *if $p = 1 + 4/N$ and the norm $\|u_0\|_{\mathbb{L}^2}$ is sufficiently small, then the previous assertion is also valid;*

(c) *for all $u \in \mathbb{H}^1(\mathbb{R}^N)$, the functionals*

$$N(u) \equiv \int_{\mathbb{R}^N} |u(x,t)|^2\,dx, \quad E(u) \equiv \int_{\mathbb{R}^N} \left(|\nabla u|^2 - \frac{2}{p+1}|u|^{p+1}\right)dx$$

are independent of time.

Proof. We present here a brief sketch of the proof (a detailed proof can be found in [107]). First, the theorem on the local solvability (see [30]) implies that the interval $[t_0, t_0 + T]$ of the existence of a solution depends only on the norm $\|u(t_0)\|_{\mathbb{H}^1}$ at the initial moment of time. Thus, if $u(x,t)$ is a solution defined on the largest interval $[t_0, t_{max}]$, then either $t_{max} = \infty$ or

$$\lim_{t\to t_{max}} \|u(t)\|_{\mathbb{H}^1} = +\infty.$$

The proof of the local solvability proposed in [30] was based on the conservation laws (9.7) and (9.8) and the a priori $\mathbb{H}^1$-estimate obtained for $0 < \sigma < 2/N$:

$$\|u(t)\|_{\mathbb{H}^1(\mathbb{R}^N)} \le \text{const}. \tag{9.32}$$

Indeed, using inequality (9.19), we obtain from the conservation laws and the Gagliardo–Nirenberg inequality with the best constant $c_{\sigma,N}^{2\sigma+2}$ the following inequality:

$$\|\nabla u\|^2_{\mathbb{L}^2} \le E + \frac{c_{\sigma,N}^{2\sigma+2}}{\sigma+1}\|u_0\|^{2+\sigma(2-N)}_{\mathbb{L}^2}\|\nabla u\|^{\sigma N}_{\mathbb{L}^2}, \quad p = 2\sigma+1. \tag{9.33}$$

In the case $0 < \sigma < 2/N$, estimate (9.32) immediately follows from (9.33). In the case where $\sigma = 2/N$ ($p = 1 + 4/N$), we can rewrite estimate (9.33) in the form

$$\left(1 - \frac{c_{\sigma,N}^{4/N+2}}{2/N+1}\|u_0\|^{4/N}_{\mathbb{L}^2}\right)\|\nabla u(t)\|^2_{\mathbb{L}^2} \le E.$$

Using the fact that $c_{\sigma,N}^{2\sigma+2} = 1/\alpha$, we reduce the inequality to the form

$$\left(1 - \left(\frac{\|u_0\|_{\mathbb{L}^2}}{\|\psi\|_{\mathbb{L}^2}}\right)^{4/N}\right)\|\nabla u(t)\|^2_{\mathbb{L}^2} \le E. \tag{9.34}$$

Clearly, taking $\|u_0\|_{\mathbb{L}^2} < \|\psi\|_{\mathbb{L}^2}$, we obtain an upper estimate for $\|u(t)\|_{\mathbb{L}^2}$ necessary for the proof of the global-in-time solvability. Since the mapping $f(x) \to \lambda^{1/\sigma} f(\lambda x)$ preserves the $\mathbb{L}^2$-norm, we can choose a function ψ satisfying (9.18), which immediately implies the theorem on the global solvability. □

Now we examine sufficient conditions of the blow-up. We formulate and prove an assertion that strengthens Glassey's result presented in Section 9.1.

Theorem 9.4 (on the global insolvability). *Let*

$$1 + \frac{4}{N} \le p < 1 + \frac{4}{N-2},$$

and let one of the following three conditions be fulfilled:

(i) $E(u_0) < 0$;

(ii) $E(u_0) = 0$ *and*

$$\operatorname{Im}\int_{\mathbb{R}^N} x u_0^* \nabla u_0 \, dx < 0;$$

(iii) $E(u_0) > 0$ *and*

$$\operatorname{Im}\int_{\mathbb{R}^N} x u_0^* \nabla u_0 \, dx \le -2\sqrt{E(u_0)}\|x u_0\|_{\mathbb{L}^2},$$

where

$$E(u_0) \equiv \int_{\mathbb{R}^n}\left(|\nabla u_0|^2 - \frac{2}{p+1}|u_0|^{p+1}\right) dx.$$

Then there exists a time moment T such that

$$\lim_{t\to T-0}\|\nabla u(t)\|_{\mathbb{L}^2}=\infty,\qquad 0<T<\infty.$$

Proof. Recall that equation (9.15) implies the formula

$$\frac{d}{dt}\int\limits_{\mathbb{R}^N}\left[|xu+2it\nabla u|^2-\frac{4t^2}{\sigma+1}|u|^{2\sigma+2}\right]dx=4t\left(\frac{\sigma N-2}{\sigma+1}\right)\int\limits_{\mathbb{R}^N}|u|^{2\sigma+2}dx, \tag{9.35}$$

which plays an important role in the study of the nonlinear Schrödinger equation; it is called the *pseudoconformal conservation law.*

This formula can be proved as follows:

$$\begin{aligned}
&\frac{d}{dt}\int\limits_{\mathbb{R}^N}\left[|xu+2it\nabla u|^2-\frac{4t^2}{\sigma+1}|u|^{2\sigma+2}\right]dx\\
&=\frac{d}{dt}\int\limits_{\mathbb{R}^N}\left[x^2|u|^2+2itx(u^*\nabla u-u\nabla u^*)+4t^2|\nabla u|^2-\frac{4t^2}{\sigma+1}|u|^{2\sigma+2}\right]dx\\
&=\int\limits_{\mathbb{R}^N}\left[\frac{d}{dt}x^2|u|^2+2ix(u^*\nabla u-u\nabla u^*)\right]dx\\
&\quad+t\int\limits_{\mathbb{R}^N}\left[2ix\frac{d}{dt}(u^*\nabla u-u\nabla u^*)+8|\nabla u|^2-\frac{8}{\sigma+1}|u|^{2\sigma+2}\right]dx.
\end{aligned} \tag{9.36}$$

Consider the first term in the right-hand side:

$$\begin{aligned}
&\int\limits_{\mathbb{R}^N}\left[\frac{d}{dt}x^2|u|^2+2ix(u^*\nabla u-u\nabla u^*)\right]dx\\
&\quad=\int\limits_{\mathbb{R}^N}[ix^2(u^*\Delta u-u\Delta u^*)+2ix(u^*\nabla u-u\nabla u^*)]\,dx=0.
\end{aligned} \tag{9.37}$$

We have

$$\begin{aligned}
&\int\limits_{\mathbb{R}^N}2ix\frac{d}{dt}(u^*\nabla u-u\nabla u^*)\,dx\\
&\quad=\int\limits_{\mathbb{R}^N}2ix(\bar{u}\nabla(i\Delta u+iu|u|^{2\sigma})-\nabla u(i\Delta u^*+iu^*|u|^{2\sigma})\\
&\qquad-\nabla u^*(i\Delta u+iu|u|^{2\sigma})+u\nabla(i\Delta u^*+iu^*|u|^{2\sigma}))\,dx\\
&\quad=\int\limits_{\mathbb{R}^N}\Big(2x\nabla u^*\Delta u+2x\nabla u\Delta u^*-2xu\nabla\Delta u^*\\
&\qquad-2xu^*\nabla\Delta u+\frac{4\sigma N}{\sigma+1}|u|^{2\sigma+2}\Big)\,dx
\end{aligned}$$

$$= \int_{\mathbb{R}^N} \left(-8|\nabla u|^2 + \frac{4\sigma N}{\sigma+1} |u|^{2\sigma+2} \right) dx; \tag{9.38}$$

this implies that the second term of (9.36) can be represented in the form

$$t \int_{\mathbb{R}^N} \left[2ix \frac{d}{dt}(u^* \nabla u - u \nabla u^*) + 8|\nabla u|^2 - \frac{8}{\sigma+1} |u|^{2\sigma+2} \right] dx$$
$$= 4t \frac{\sigma N - 2}{\sigma+1} \int_{\mathbb{R}^N} |u|^{2\sigma+2} dx,$$

which implies the pseudoconformal conservation law. Its direct consequence (see (9.37) and (9.38)) is the equation

$$\begin{aligned} \frac{d^2}{dt^2} \int_{\mathbb{R}^N} |x|^2 |u|^2 \, dx &= - \int_{\mathbb{R}^N} 2ix \frac{d}{dt}(u^* \nabla u - u \nabla u^*) \, dx \\ &= \int_{\mathbb{R}^N} \left(8|\nabla u|^2 - \frac{4\sigma N}{\sigma+1} |u|^{2\sigma+2} \right) dx \\ &= 8E(u_0) + \frac{4N}{\sigma+1} \left(\frac{2}{N} - \sigma \right) \int_{\mathbb{R}^N} |u|^{2\sigma+2} dx, \end{aligned} \tag{9.39}$$

obtained by Vlasov [105]. Conditions (i), (ii), or (iii) of the theorem on the global insolvability imply that, for $u(x,t) \in \mathbb{H}^1(\mathbb{R}^N)$, there exists a time moment T^* such that

$$\lim_{t \to T^*} \int_{\mathbb{R}^N} |x|^2 |u|^2 \, dx = 0.$$

Applying equation (9.9), we obtain

$$\frac{d}{dt} \int |x|^2 |u|^2 dx \bigg|_{t=0} = 4 \operatorname{Im} \int x u_0^* \nabla u_0 \, dx. \tag{9.40}$$

In [104], the following estimate was proved:

$$\|f\|_{L^2}^2 \le \frac{2}{N} \|\nabla f\|_{L^2} \|xf\|_{L^2} \tag{9.41}$$

for all functions $f(x)$ from the class

$$\{f \in \mathbb{L}^2(\mathbb{R}^N),\ |x|f \in \mathbb{L}^2(\mathbb{R}^N),\ \nabla f \in \mathbb{L}^2(\mathbb{R}^N)\}.$$

This estimate can be deduced from the Young inequality if we take into account that

$$-N \int_{\mathbb{R}^N} |f|^2 \, dx = 2 \operatorname{Re} \int_{\mathbb{R}^N} x \nabla f f^* \, dx.$$

Due to estimate (9.41) and the conservation law for the norm $\|u\|^2_{\mathbb{L}^2}$, we obtain the inequality

$$\|u_0\|^2_{\mathbb{L}^2} = \|u\|^2_{\mathbb{L}^2} \leq \frac{2}{N}\|\nabla u(t)\|_{\mathbb{L}^2}\|xu(t)\|_{\mathbb{L}^2}. \tag{9.42}$$

Thus, for some T^*, under the conditions of the theorem, we have proved the blow-up of a solution:

$$\lim_{t\to T^*} \|\nabla u(t)\|_{\mathbb{L}^2} = +\infty.$$

The theorem on the global insolvability is proved. □

To complete the proof of Theorem 9.2, we separately consider the critical case $p = 1+4/N$ ($\sigma = 2/N$). In the right-hand side of expression (9.39), only the term $2E(u_0)$ does not vanish:

$$\frac{d^2}{dt^2}\int_{\mathbb{R}^N} |x|^2|u|^2\,dx = 2E(u_0). \tag{9.43}$$

If we take the basic state as a particular solution of problem (9.15)–(9.16), $v = \exp(it)\psi(x)$, where the function $\psi(x)$ satisfies equation (9.18), then the right-hand side of (9.43) vanishes, $E(\psi) = 0$. This leads us to the second part of the basic theorem, that is, to the assertion on the unstable boundary state.

In our case, to demonstrate the instability, it suffices to prove that there exists a solution of the Cauchy problem for the Schrödinger equation with an initial function arbitrarily close to ψ:

$$u_0(x) = \xi(x) \quad \text{for } \|\xi - \psi\|_{L^2} < \delta,$$

for any $\delta > 0$, and this solution becomes infinite in finite time. We take, for example, $\xi^\varepsilon(x) = (1+\varepsilon)\psi(x)$, where $\varepsilon > 0$; then the condition of global solvability is violated:

$$\|\xi^\varepsilon\|^2_{L^2} = (1+\varepsilon)^2\|\psi\|^2_{L^2} > \|\psi\|^2_{L^2}.$$

Thus

$$E(\xi^\varepsilon) = -2\varepsilon\|\psi\|^2_{L^2} + O(\varepsilon^2) < 0.$$

The blow-up result immediately follows from the theorem on global insolvability. Theorem 9.2 is proved. □

9.3 Bibliographical notes

The paper [32] is devoted to the global insolvability of the Cauchy problem for the nonlinear Schrödinger equation (see also the references therein). The question on local solvability is examined in [30]. A detailed discussion of the critical case can be found in [107]. In [83], asymptotics of solutions in the critical case are considered.

10 Variational method of L. E. Payne and D. H. Sattinger

The authors of the method of saddle points, L. E. Payne and D. H. Sattinger, examined the following problem for the nonlinear hyperbolic equation:

$$u_{tt} = \Delta u + f(u), \quad u(x,0) = u_0(x), \quad u_t(x,0) = v_0(x), \quad t > 0, \tag{10.1}$$

with the Dirichlet boundary condition $u = 0$ on the smooth boundary $\partial\Omega$ of a bounded domain $\Omega \subset \mathbb{R}^N$. They proposed a method that allows us to obtain conditions of blow-up in finite time for strong and weak solutions. The idea is based on the detailed analysis of saddle points.

10.1 Introduction

Consider the following one-dimensional mechanical analog of equation (10.1):

$$\ddot{x} = -x + f(x), \quad x = x(t), \tag{10.2}$$

for real numbers. Equation (10.2) describes a mechanical system with one degree of freedom, whereas the equation of problem (10.1) can be treated as a system with infinite number of degrees of freedom. Introduce the potential energy

$$V(x) = \frac{x^2}{2} - F(x), \quad F(x) = \int_0^x f(s)\,ds. \tag{10.3}$$

Assume that the potential energy $V(x)$ has a local minimum at the point $x = 0$ and a maximum at a point $x = x_1 > 0$, which is equal to d. Denote by

$$W = \{x : V(x) < d, \ x < x_1\} \neq \varnothing$$

the potential well. The total energy (potential and kinetic) is of the form

$$E(x) = \frac{(\dot{x})^2}{2} + V(x)$$

and is a conserved quantity.

Remark 10.1. To prove this, it suffices, under the assumption of required smoothness of a solution $x = x(t) \in \mathbb{C}^{(2)}(\mathbb{R}^1_+)$, to multiply both parts of equation (10.2) by $\dot{x}$ and to integrate by time t.

https://doi.org/10.1515/9783110602074-010

We have the following simple assertions.

Proposition 10.1. *If at the initial moment of time $t = 0$ we have $E(0) < d$ and $x(0) \in W$, then $x(t) \in W$ for all $t > 0$.*

Proof. Indeed, since the term $(\dot{x})^2/2 \geq 0$ in the total energy E is always positive, we have the inequalities

$$V(x(t)) \leq E(x(t)) = E(0) < d \quad \text{and} \quad V(x(0)) < d.$$

Therefore, the solution remains in the potential well W. □

Proposition 10.2. *If $E(0) < d$ and $x(0) > x_1$, then the solution never lies in the potential well, that is, $x(t) \notin W$ for all $t \geq 0$.*

Proof. Indeed, assume the contrary. For some $t = t_1 > 0$, we have $x(t_1) \in W$. This means that there exists a time moment $t_2 \in (0, t_1)$ for which $x(t_2) = x_1$; therefore, at this moment, we have $W(x(t_2)) = d$. Hence, for $t = t_2$, the total energy satisfies the inequality

$$d \leq E(t_2) = E(0) < d,$$

that is, the solution never lies in the potential well W. □

Now we complement equation (10.2) with the initial conditions

$$x(0) = a_0, \quad \dot{x}(0) = a_1 \tag{10.4}$$

and introduce the following functionals due to the modified method of Korpusov and Sveshnikov:

$$\Phi(t) = \frac{1}{2}x^2(t), \quad J(t) = (\dot{x})^2(t), \quad x = x(t). \tag{10.5}$$

Obviously, these functionals are related by the formula

$$(\dot{\Phi})^2 = 2\Phi(t)J(t). \tag{10.6}$$

Multiplying both sides of equation (10.2) by $x(t)$, we obtain the first energy equality

$$\frac{1}{2}\frac{d^2x^2}{dt^2} - (\dot{x})^2 + x^2 = xf(x). \tag{10.7}$$

Multiplying both sides of equation (10.2) by $\dot{x}$, we obtain the relation

$$\frac{d}{dt}\left(\frac{(\dot{x})^2}{2} + \frac{x^2}{2} - \int_0^x f(s)\,ds\right) = 0. \tag{10.8}$$

Integrating equation (10.8) by time, we obtain the second energy equality

$$\frac{1}{2}J(t) + \Phi(t) - E_0 = \int_0^x f(s)\,ds, \quad E_0 = \frac{a_1^2}{2} + \frac{a_0^2}{2} - \int_0^{a_0} f(s)\,ds, \tag{10.9}$$

where we have applied notation (10.5). Taking this notation into account, we can also rewrite the first energy equality (10.7) as follows:

$$\frac{d^2\Phi}{dt^2} - J(t) + 2\Phi(t) = xf(x). \tag{10.10}$$

Assume that

$$xf(x) \geq \theta \int_0^x f(s)\,ds \quad \text{for } \theta > 2 \text{ for all } x \in \mathbb{R}^1. \tag{10.11}$$

Using this condition, from the energy equalities (10.9) and (10.10) we obtain the inequality

$$\frac{d^2\Phi}{dt^2} - J(t) + 2\Phi(t) \geq \frac{\theta}{2}J(t) + \theta\Phi(t) - \theta E_0. \tag{10.12}$$

Applying equality (10.6), we obtain the required differential inequality

$$\Phi\ddot{\Phi} - \alpha(\dot{\Phi})^2 - \gamma_1\Phi^2 + \theta E_0\Phi \geq 0 \quad \text{for } \theta > 2, \tag{10.13}$$

where

$$\gamma_1 = \theta - 2, \quad \alpha = \frac{1}{2}\left(1 + \frac{\theta}{2}\right), \quad E_0 = \frac{a_1^2}{2} + \frac{a_0^2}{2} - \int_0^{a_0} f(s)\,ds.$$

Note that the differential inequality (10.13) is interesting in itself. However, we essentially simplify it under the assumption that $E_0 \leq 0$. This condition holds for any fixed a_1, sufficiently "large" $|a_0|$, and, for example, for

$$f(x) = |x|^{p-1}x, \quad p > 1, \quad \theta = p + 1.$$

We may also impose the conditions from Proposition 10.2:

$$|a_0| > \left(\frac{p+1}{2}\right)^{1/(p+1)} (a_1^2 + a_0^2)^{1/(p+1)}, \quad p > 1.$$

In this case, $\theta = p + 1 > 2$. Then the differential inequality (10.13) takes the form

$$\Phi\ddot{\Phi} - \alpha(\dot{\Phi})^2 \geq 0, \tag{10.14}$$

which implies the inequality

$$\Phi(t) \geq \frac{\Phi(0)}{[1-at]^{1/(\alpha-1)}}, \quad a = \frac{(p-1)a_1}{2a_0}, \quad \alpha = \frac{p+3}{4}, \tag{10.15}$$

under the additional condition for the initial data $a_0 a_1 > 0$. Clearly, solutions of problem (10.2), (10.4) do not exist globally in time. On the other hand, for "small" initial data, we can prove the global solvability.

This example shows the influence of the initial data on the boundedness or unboundedness of solutions of finite intervals of time.

To extend these results to the infinite-dimensional case (10.1), we introduce the functional of potential energy

$$J(u) = \int_\Omega \left(\frac{1}{2}|\nabla u|^2 - F(u)\right) dx, \quad F(u) = \int_0^u f(s)\, ds. \tag{10.16}$$

Under certain conditions on $f(u)$, the authors of the method proved the existence of a local minimum and a potential well W of positive depth d with a saddle point $u = w$ for (10.1). As in the finite-dimensional case, under initial conditions such that $E(0) < d$ and u_0 lie outside W, the solution of problem (10.1) blows up in finite time in the $\mathbb{L}^2$-norm.

10.2 Potential well in the functional space

First, we state conditions for the nonlinear function $f(u)$ under which all nontrivial critical points of the functional J are unstable equilibria points. We find these points as extremums of the variational problem and prove the uniqueness and sign-definiteness. Then we prove the existence of a potential well W of depth $d > 0$.

We further we assume that Ω is a smooth bounded domain of the space $\mathbb{R}^N$. The embedding theorems imply the existence of a constant c_p such that the following upper estimate holds for all $p \leq 2^*$:

$$|u|_p \leq c_p \|u\|, \quad |u|_p = \left|\int_\Omega |u|^p\, dx\right|^{1/p}, \quad \|u\| = \left|\int_\Omega |\nabla u|^2\, dx\right|^{1/2}$$

for all $u \in \mathbb{H}_0^1(\Omega)$. Moreover, the embedding

$$\mathbb{H}_0^1(\Omega) \hookrightarrow \mathbb{L}^p(\Omega)$$

is compact if $p < 2^*$, where by 2^* we denote the quantity

$$2^* = \begin{cases} 2N/(N-2) & \text{if } N \geq 3, \\ +\infty & \text{if } N = 1, 2. \end{cases}$$

In the class $\mathbb{H}_0^1(\Omega)$, we consider the functional

$$J(u) = \frac{1}{2}\|u\|^2 - \int_\Omega F(u)\,dx, \tag{10.17}$$

which can be considered as the potential energy for the infinite-dimensional dynamical system (10.1). For $p \le 2^*$, this functional possesses the property

$$|F(u)| = O(|u|^p) \quad \text{as } |u| \to \infty.$$

Thus critical points $w \in \mathbb{H}_0^1(\Omega)$ of the functional $J(u)$ for a regular function f must satisfy the Euler equation

$$\Delta w + f(w) = 0 \quad \text{for } x \in \Omega, \qquad w = 0 \quad \text{for } x \in \partial\Omega. \tag{10.18}$$

To arrive at expression (10.18), we must calculate the Fréchet derivative of the functional (10.17) and equate it to zero (the necessary condition of an extremum of the functional).

We assume that the function f satisfies the following conditions:

(i) the functional $F(u)$ defined by formula (10.16) is Fréchet-differentiable;

(ii) $f(x) \in \mathbb{C}^{(1)}(\mathbb{R}^1)$, $f(0) = f'(0) = 0$, but the function f does not vanish identically in a neighborhood of the point $x = 0$;

(iii)$_a$ $f(u)$ is increasing, convex for $u > 0$, and concave for $u < 0$;

or

(iii)$_b$ $f(u)$ is convex for all u;

(iv) we have

$$(p+1)F(u) \le uf(u);$$

moreover, there exists γ such that

$$2 < p + 1 \le \gamma \le 2^*$$

and the following estimate holds:

$$|uf(u)| \le \gamma|F(u)|. \tag{10.19}$$

Lemma 10.1. *Under conditions* (i)–(iv), *we have*

$$|F(u)| = O(|u|^\gamma). \tag{10.20}$$

Moreover, in case (iii), *we have the inequality*

$$u(uf' - f) \ge 0, \tag{10.21}$$

where the equality sign can be realized only for $u = 0$.

Proof. First, we prove the estimate of growth. The growth condition $|F(u)| = O(|u|^\gamma)$ is obtained by integration of the inequality. We must examine two cases, $u > 0$ and $u < 0$. We describe the case $u > 0$ in detail; the second case is proved similarly.

If $u > 0$, then for a positive function $f(u) > 0$ ($F(u) > 0$), we can remove modules and obtain the differential inequality

$$uf(u) \leq \gamma F(u), \quad f(u) = \frac{dF(u)}{du}.$$

It can be easily integrated, which leads to the growth estimate

$$F(u) \leq c|u|^\gamma, \quad c > 0.$$

For a negative function $f(u) < 0$ ($F(u) < 0$), the inequality is of the form

$$-uf(u) \leq -\gamma F(u).$$

Dividing both sides by $-uF(u) > 0$, we obtain the inequality

$$\frac{-F'(u)}{-F(u)} \leq \gamma u.$$

Integrating it, we obtain the upper estimate of the growth rate of $F(u)$:

$$-c|u|^\gamma \leq F(u), \quad c > 0.$$

Combining both estimates, we arrive at the inequality

$$|F(u)| \leq c|u|^\gamma.$$

Now we prove inequality (10.21). Consider the graph of the function $y = f(x)$. This graph is convex for $x > 0$; therefore the tangent to the graph passing through $x = u$ (its equation is $y(x) = f'(u)(x - u) + f(u)$) intersects the axis $x = 0$ for nonnegative y ($x = 0$):

$$y(x = 0) = f(u) - f'(u)u \geq 0.$$

Conversely, for $u < 0$, the graph is concave. Therefore

$$y(x = 0) = f(u) - f'(u)u \leq 0.$$

Combining these inequalities in the form $u(f(u) - f'(u)u) \geq 0$, we obtain the required assertion. □

Proposition 10.3. *Under conditions* (i)–(iv), *all nontrivial critical points of the hyperbolic problem* (10.1) *are a priori points of unstable equilibria of the functional $J(u)$ defined by formula* (10.17).

Proof. Calculate the second variation of the functional J at a critical point:

$$\delta^2 J[w](v) \equiv \langle J''_{ff}(w)v, v\rangle = \frac{1}{2}\Big(\|v\|^2 - \int_\Omega f'(w)v^2\,dx\Big), \tag{10.22}$$

where

$$J''_{ff}(\cdot) : \mathbb{H}^1_0(\Omega) \to \mathcal{L}(\mathbb{H}^1_0(\Omega); \mathbb{H}^{-1}(\Omega)),$$

and $\langle\cdot,\cdot\rangle$ is the duality bracket between the Hilbert spaces $\mathbb{H}^1_0(\Omega)$ and $\mathbb{H}^{-1}(\Omega)$. Indeed, the first and second Fréchet derivatives of the functional J are

$$J'_f(v) = -\Delta v - f(v), \quad J''_{ff}(w)v = -\Delta v - f'(w)v.$$

A necessary condition of minimum of J at a point w is the positive definiteness of $\delta^2 J$. At the point $w = 0$, the quadratic functional is

$$\delta^2 J[v] = \frac{\|v\|^2}{2} \geq 0.$$

However, at a nontrivial critical point $w \neq 0$, the functional is not necessarily positive definite. Since w satisfies the Dirichlet boundary conditions, we have

$$\begin{aligned}\delta^2 J[w](w) &= \frac{1}{2}\|w\|^2 - \frac{1}{2}\int_\Omega f'(w)w^2\,dx \\ &= -\frac{1}{2}\int_\Omega w(\Delta w + f'(w)w)\,dx \\ &= -\frac{1}{2}\int_\Omega w(f'(w)w - f(w))\,dx. \end{aligned}\tag{10.23}$$

If w satisfies condition (iii), then (10.21) implies that

$$\delta^2 J[w] < 0. \tag{10.24}$$

If f is convex, then, by the maximum principle, $w > 0$ and

$$\Delta w = -f(w) < 0.$$

Therefore, inequality (10.24) also holds.

Example 10.1. The reader can easily verify that, for the function $f(u) = u^p$, by (10.23) we the following relations:

$$wf'(w) = pf(w), \quad \delta^2 J[w] = -\frac{p-1}{2}\|w\|^2 < 0.$$

Thus, under conditions (i)–(iv), all nontrivial critical points of the hyperbolic problem (10.1) are a priori points of unstable equilibria.

Proposition 10.3 is proved. □

We show that, under condition (iv), the point $w = 0$ is a point of local minimum of the functional of potential energy J and the depth of the potential well is positive.

For all $u \in \mathbb{H}_0^1(\Omega)$, we introduce a function $j = j(\lambda)$ possessing the following properties:

$$j(\lambda) = J(\lambda u) = \frac{\lambda^2}{2}\|u\|^2 - \int_\Omega \int_0^{\lambda u} f(s)\,ds\,dx, \quad j(0) = 0,$$

$$\frac{dj(\lambda)}{d\lambda} = \lambda\|u\|^2 - \int_\Omega uf(\lambda u)\,dx, \quad \left.\frac{dj(\lambda)}{d\lambda}\right|_{\lambda=0} = 0,$$

$$\frac{d^2 j(\lambda)}{d\lambda^2} = \|u\|^2 - \int_\Omega u^2 f'(\lambda u)\,dx, \quad \left.\frac{d^2 j(\lambda)}{d\lambda^2}\right|_{\lambda=0} = \|u\|^2.$$

Therefore, for small λ and all $u \in \mathbb{H}_0^1(\Omega)$, the function $j(\lambda)$ is convex. We prove that the function $j = j(\lambda)$ under the assumptions with respect to f has a unique positive critical point $\lambda^* = \lambda^*(u)$.

Lemma 10.2. *Under conditions* (i)–(iv), *for all* $u \in \mathbb{H}_0^1(\Omega)$, $u \neq 0$, *we have*:
(a) $\lim_{\lambda\to\infty} j(\lambda) = -\infty$;
(b) *there exists a unique* $\lambda^* = \lambda^*(u) > 0$ *such that*

$$\left.\frac{dj}{d\lambda}\right|_{\lambda=\lambda^*} = 0;$$

(c) *at the point* λ^*,

$$\left.\frac{d^2 j}{d\lambda^2}\right|_{\lambda=\lambda^*} < 0.$$

Proof. Similarly to the proof of estimate (10.20) in Lemma 10.1, we can obtain from property (iv) the following growth condition:

$$F(u) \geq b|u|^{p+1} \quad \text{for } |u| \geq 1, \quad b = \min\{F(1), F(-1)\}. \tag{10.25}$$

For $p > 1$,

$$j(\lambda) \to -\infty \quad \text{as } \lambda \to \infty.$$

Indeed,

$$j(\lambda) = \frac{\lambda^2}{2}\|u\|^2 - \int_\Omega F(\lambda u)\,dx \leq \frac{\lambda^2}{2}\|u\|^2 - b|\lambda|^{p+1} \int_{\Omega\cap\{\lambda u \geq 1\}} |u|^{p+1}\,dx. \tag{10.26}$$

To prove assertion (b), we assume that there exist two roots $\lambda_1 < \lambda_2$ of the equation $j'(\lambda) = 0$. The existence of λ^* is provided by the fact that

$$j(\lambda) \to -\infty \quad \text{as } \lambda \to \infty$$

and the convexity of $j(\lambda)$ for small λ. For λ_1 and λ_2, we have

$$\lambda_{1,2}\|u\|^2 - \int_\Omega uf(\lambda_{1,2}u)\,dx = 0. \tag{10.27}$$

Excluding $\|u\|$ from these expressions, we obtain

$$\int_\Omega u\left(\frac{f(\lambda_2 u)}{\lambda_2} - \frac{f(\lambda_1 u)}{\lambda_1}\right)dx = 0. \tag{10.28}$$

Setting $w = \lambda_1 u$ and $\lambda = \lambda_2/\lambda_1$, we rewrite (10.28) in the form

$$\int_\Omega w(f(\lambda w) - \lambda f(w))\,dx = 0 \quad \text{for } \lambda > 1. \tag{10.29}$$

If f satisfies (iii), then the integrand does not change its sign. Therefore (10.29) cannot hold. The contradiction proved assertion (b).

To prove (c), we note that

$$0 = \left.\frac{dj(\lambda)}{d\lambda}\right|_{\lambda=\lambda^*} = \lambda^*\|u\|^2 - \int_\Omega uf(\lambda^* u)\,dx. \tag{10.30}$$

Then

$$\begin{aligned} j''(\lambda^*) &= \|u\|^2 - \int_\Omega u^2 f'(\lambda^* u)\,dx \\ &= \frac{1}{(\lambda^*)^2}\int_\Omega (\lambda^* uf(\lambda^* u) - (\lambda^* u)^2 f'(\lambda^* u))\,dx < 0. \end{aligned} \tag{10.31}$$

The lemma is proved. □

Having proved the existence of a unique $\lambda^* = \lambda^*(u)$ such that

$$\left.\frac{dJ(\lambda u)}{d\lambda}\right|_{\lambda=\lambda^*} = 0,$$

we define the functional

$$H(u) = J(\lambda^* u). \tag{10.32}$$

Denote

$$d = \inf_{u\neq 0} H(u).$$

Multiplying both sides of equation (10.30) by λ^*, we obtain

$$\|v\|^2 - \int_\Omega vf(v)\,dx = 0, \quad v = \lambda^* u.$$

Denote

$$K(v) \equiv \|v\|^2 - \int_\Omega v f(v)\,dx. \tag{10.33}$$

Then our variational problem can be represented as follows:

$$d = \{\inf J(v),\ \|v\| \neq 0,\ K(v) = 0\}. \tag{10.34}$$

Clearly, all nontrivial critical points of J are extremals of (10.34). The existence of such extremals will be proved further.

Now we prove the following:

Theorem 10.1. *Assume that the function f satisfies conditions* (i)–(iv). *Then all extremals* (10.34) *are critical points of J. In case* (iii), *the extremal does not change its sign; moreover, there exists no more than one extremal w_1 under the additional condition $u \geq 0$ and no more than one w_2 under the condition $u \leq 0$. If an extremal exists and the function f is even, then $w_1 = -w_2$.*

Proof. First, we prove the following lemma.

Lemma 10.3. *If f satisfies property* $(\text{iii})_\text{a}$ *or* $(\text{iii})_\text{b}$, *then the extremal* (10.34) *is sign-constant.*

Proof. Assume that u is an extremal; then

$$d = J(u) \quad \text{and} \quad K(u) = 0.$$

We need the following auxiliary assertion.

Proposition 10.4. *If condition* $(\text{iii})_\text{a}$ *is fulfilled, then*

$$v_1 f(v_1) \leq v_2 f(v_2) \quad \textit{for all } v_1 \leq v_2. \tag{10.35}$$

Proof. Introduce the function $L(s) = sf(s)$; its derivative is

$$L'(s) = f(s) + sf'(s).$$

Recall that $f(s)$ is increasing. Let $s \geq 0$; then $f(s) \geq f(0) = 0$ and $f'(s) \geq 0$. Therefore $L'(s) \geq 0$ for $s \geq 0$. If $s \leq 0$, then $f(s) \leq f(0) = 0$, and hence $L'(s) \leq 0$. By integration we conclude that

$$L(s_1) \leq L(0) = 0 \leq L(s_2) \quad \text{for } s_1 \leq 0 \leq s_2.$$

The proposition is proved. □

By Proposition 10.4 we have

$$|u|f(|u|) \geq uf(u) \quad \text{if } |u| \geq u.$$

Therefore, if u changes sign, then

$$\int_\Omega |u|f(|u|)\,dx > \int_\Omega uf(u)\,dx \quad \Rightarrow \quad K(|u|) < K(u) = 0$$

since $\||u|\| = \|u\|$.

Lemma 10.2(b) implies that there exists a unique

$$\bar{\lambda} = \bar{\lambda}(|u|) \geq 0 \quad \Rightarrow \quad K(\bar{\lambda}|u|) = 0,$$

since, by formula (10.30),

$$\bar{\lambda}\||u|\|^2 = \int_\Omega |u|f(\bar{\lambda}|u|)\,dx \geq 0 \quad \Rightarrow \quad \bar{\lambda} \geq 0.$$

Proposition 10.5. *We have* $\bar{\lambda} \in [0, 1)$.

Proof. The inequality $\bar{\lambda} \geq 0$ has been proved earlier. We verify that $\bar{\lambda} \leq 1$. Indeed, assume the contrary, that is, let $\bar{\lambda} > 1$. Then

$$\bar{\lambda}|u|f(\bar{\lambda}|u|) \geq |u|f(|u|) \quad \Rightarrow \quad \int_\Omega \bar{\lambda}|u|f(\bar{\lambda}|u|)\,dx > \int_\Omega |u|f(|u|)\,dx$$

$$\Rightarrow \quad 0 = K(\bar{\lambda}|u|) < K(|u|) = 0.$$

This contradiction implies $\bar{\lambda} \leq 1$. However, since $K(|u|) < K(u) = 0$, we conclude that $\bar{\lambda} < 1$.

For convex functions, we argue as follows. If a function u changes its sign, then $K(|u|) < K(u)$. To prove this, we rewrite it in the explicit form

$$\|u\|^2 - \int_\Omega |u|f(|u|)\,dx < \|u\|^2 - \int_\Omega uf(u)\,dx$$

and simplify:

$$\int_{u<0} (-u)f(-u)\,dx > \int_{u<0} uf(u)\,dx.$$

Therefore we obtain

$$\int_{u<0} u(f(u) + f(-u))\,dx > 0,$$

which is obvious since $(f(u) + f(-u)) > 0$ for convex functions and $u < 0$. The proposition is proved. □

This proposition implies the following.

Proposition 10.6. *For all* $\lambda \in (0,1)$, *we have*:

$$\frac{d}{d\lambda}J(\lambda u) > 0. \tag{10.36}$$

Proof. Let us prove that $K(\lambda u) > 0$ for $\lambda \in (0,1)$. To this end, we rewrite it in the form

$$\lambda^2\|u\|^2 - \int_\Omega \lambda u f(\lambda u)\,dx > 0$$

and subtract the obvious equality

$$\lambda^2 K(u) = \lambda^2\|u\|^2 - \lambda^2 \int_\Omega u f(u)\,dx = 0.$$

Then we have

$$\lambda \int_\Omega u(\lambda f(u) - f(\lambda u))\,dx = 0.$$

Using the definition of a convex function, for $\lambda \in (0,1)$, we have

$$f(\lambda x + (1-\lambda)y) \le \lambda f(x) + (1-\lambda)f(y),$$

and setting $x = u$ and $y = 0$, we see that $\lambda f(u) - f(\lambda u) \ge 0$ for $u > 0$. Conversely, for $u < 0$, the function $f(u)$ is concave, and therefore $\lambda f(u) - f(\lambda u) \le 0$. The two inequalities obtained can be combined,

$$u(\lambda f(u) - f(\lambda u)) \ge 0,$$

which implies that

$$K(\lambda u) > 0 \quad \text{for } \lambda < 1.$$

Finally,

$$\frac{d}{d\lambda}J(\lambda u) = \frac{K(\lambda u)}{\lambda} > 0 \quad \text{for } \lambda \in (0,1).$$

The proposition is proved. □

Integrating (10.36), we arrive at the inequality

$$d = J(u) > J(\bar{\lambda} u). \tag{10.37}$$

On the other hand, we can easily verify that

$$J(\bar{\lambda} u) > J(\bar{\lambda}|u|). \tag{10.38}$$

Thus, the function $\bar{\lambda}|u|$ satisfies the condition $K(\bar{\lambda}|u|) = 0$, whereas $J(\bar{\lambda}|u|) < d$, which contradicts the definition of d since d is the infimum of the functional $J(\cdot)$. Therefore, the function u does not change its sign. The lemma is proved. □

Remark 10.2. By the result obtained we can reformulate the variational problem (10.34) as follows:

$$\{d = \inf J(u),\ u \geq 0,\ K(u) = 0\}. \tag{10.39}$$

Let us prove that extremals of problem (10.34) correspond to extremals of the functional J. The Euler–Lagrange equation

$$J_f'(u) + \Lambda K_f'(u) = 0$$

corresponding to (10.34) is

$$(\Delta w + f(w)) + \Lambda(2\Delta w + wf'(w) + f(w)) = 0$$

or, after simple transformations,

$$(1 + 2\Lambda)(\Delta w + f(w)) - \Lambda(f(w) - wf'(w)) = 0, \tag{10.40}$$

where Λ is the Lagrange parameter. Multiplying (10.40) by w and integrating, we obtain the equality

$$\Lambda \int_\Omega w(f(w) - wf'(w))\, dx = 0 \tag{10.41}$$

since

$$\|w\|^2 - \int_\Omega wf(w)\, dx = 0.$$

Under conditions (iii), the integral in (10.41) does not vanish due to (10.21). We can argue similarly in the case where the function f is convex since $w \geq 0$

Therefore, in both cases, $\Lambda = 0$, and w satisfies the Euler equation

$$\Delta w + f(w) = 0.$$

If f is convex, then $\Delta w \leq 0$ and $w > 0$ by the strict maximum principle. If f is monotonic, then $w \geq 0$ and $\Delta w \leq 0$, and, by the strict maximum principle, $w > 0$. Similarly, for $w \leq 0$, the function w satisfies the strict minimum principle.

To prove the uniqueness in case (iii_b), we use the following well-known result (the so-called “Serrin's sweeping statement”).

Theorem 10.2. *Assume that $v(x, \lambda) = v_\lambda$ is an increasing family of upper solutions of* (10.18) *on $a \leq \lambda \leq b$, namely,*

$$\Delta v(x, \lambda) + f(v(x, \lambda)) \leq 0, \quad x \in \Omega. \tag{10.42}$$

If a function u is a solution of (10.18) *such that*

$$u(x) \le v_b \quad \text{and} \quad u(x) \le v_a \quad \text{on the boundary } \partial\Omega,$$

then either $u \equiv v_a$ or $u < v_a$ in the whole domain Ω. Similarly, if v_λ is an increasing family of lower solutions and u is a solution of (10.18) *such that*

$$u(x) \ge v_a \quad \text{and} \quad u(x) \ge v_b \quad \text{on the boundary } \partial\Omega,$$

then either $u \equiv v_b$ or $u > v_b$ in Ω.

To apply this theorem, we consider the family $\{\lambda w\}_{\lambda \ge 1}$. Since f is convex for $w > 0$ and $\lambda \ge 1$, we obtain the following equality:

$$\Delta(\lambda w) + f(\lambda w) \le \lambda(\Delta w + f(w)) = 0. \tag{10.43}$$

Let u be an arbitrary solution of (10.18). If

$$u(x) > w(x) \quad \text{in the domain } \Omega,$$

then we can choose λ so large that $u(x) \le \lambda w(x)$. Then the first part of Serrin's theorem implies

$$u(x) < w(x) \quad \text{since } v = \lambda w = 0 \quad \text{on the boundary } \partial\Omega.$$

Thus $u(x) \le w(x)$ everywhere in Ω. Applying the second assertion of Serrin's theorem for lower solutions $\{\lambda w\}_{\lambda \le 1}$, we obtain $u(x) \ge w(x)$, and therefore $u \equiv w$ everywhere in Ω.

For case (iii$_a$), the same reasoning can be used for the proof of the uniqueness of positive and negative extremals. The theorem is proved. □

The following assertion can be applied independently in both cases (iii$_a$) and (iii$_b$).

Lemma 10.4. *Assume that f satisfies conditions* (i) *and* (ii) *and also the growth condition $|f(u)| = O(|u|^p)$ for $p + 1 \le 2^*$. If the function w is an extremal of* (10.34) *such that*

$$\left.\frac{d^2 J(\lambda w)}{d\lambda^2}\right|_{\lambda=1} < 0, \tag{10.44}$$

then w is a critical point of J.

Proof. For the extremal w, we have:

$$\|w\|^2 - \int_\Omega w f(w)\,dx = 0, \quad \|w\|^2 - \int_\Omega w^2 f'(w)\,dx < 0. \tag{10.45}$$

Eliminating the norm, we obtain

$$\int_{\Omega} w(f(w) - wf'(w))\,dx < 0. \tag{10.46}$$

Now, applying Theorem 10.1, we obtain equation (10.40), which implies that $\Lambda = 0$, and therefore w is a critical point of J. The lemma is proved. □

We further establish the existence of extremal (10.34) and the inequality $d > 0$ (see Theorem 10.3). First, we prove the following auxiliary assertion.

Lemma 10.5. *Assume that f satisfies conditions* (i) *and* (ii) *and also*

$$|f(u)| = O(|u|^p) \quad \text{for } p + 1 \le 2^*.$$

Then the functionals J and K are continuous on $\mathbb{H}_0^1(\Omega)$.

Proof. Indeed, it suffices to prove the continuity on $\mathbb{H}_0^1(\Omega)$ of the functionals

$$\int_{\Omega} F(u)\,dx, \quad \int_{\Omega} uf(u)\,dx. \tag{10.47}$$

The mean-value theorem implies that

$$F(u) - F(v) = \int_0^1 (u - v)f(u + \tau(v - u))\,d\tau; \tag{10.48}$$

therefore we have the following estimates:

$$\begin{aligned} \left|\int_{\Omega} (F(u) - F(v))\,dx\right| &\le \int_{\Omega}\int_0^1 |u - v||f(u + \tau(v - u))|\,d\tau\,dx \\ &\le \int_0^1 \left|\int_{\Omega} |f(u + \tau(v - u))|^r\,dx\right|^{1/r} \left|\int_{\Omega} |u - v|^s dx\right|^{1/s} d\tau \equiv \mathrm{I}. \end{aligned} \tag{10.49}$$

Since $|f(u)| = O(|u|^p)$, $p + 1 \le 2^*$, we set

$$s = p + 1, \quad r = \frac{p}{p+1}$$

and continue (10.49):

$$\mathrm{I} \le |u - v|_{p+1} \int_0^1 \left|\int_{\Omega} |v + \tau(u - v)|^{p+1}\,dx\right|^{p/(p+1)} d\tau$$

$$\leq S_{p+1}\|u-v\|\int_0^1 |v+\tau(u-v)|_{p+1}^p\,d\tau$$
$$\leq c(p,|u|_{p+1},|v|_{p+1})\|u-v\|. \tag{10.50}$$

Here c is a constant depending on p, $|u|_{p+1}$, and $|v|_{p+1}$. For the functional J, we have proved its Lipschitz continuity on $\mathbb{H}_0^1(\Omega)$. For the functional K, the proof is similar. □

Theorem 10.3. *Let f satisfy conditions* (i)–(iv). *Then $d > 0$. If $2 < \gamma < 2^*$, then there exists a unique extremal of the variational problem* (10.34). *If f is a convex functional, then there exists a unique positive extremal; in the case* (iii_a), *there exists two extremals: one positive and one negative.*

Proof. To prove the positiveness of d, we find the infimum of $J(u)$ under the conditions $\|u\| \neq 0$ and $K(u) = 0$. As we have seen in the proof of Theorem 10.1, it suffices to find the minimal value in the class of nonnegative functions in case (iii_a); in case (iii_b), in addition to the conditions for u in (10.34), we also assume that $u \geq 0$. Thus, we get the inequality

$$J(u) = \frac{1}{2}\|u\|^2 - \int_\Omega F(u)\,dx \geq \frac{1}{2}\|u\|^2 - \frac{1}{p+1}\int_\Omega uf(u)\,dx. \tag{10.51}$$

From $K(u) = 0$ we obtain

$$J(u) \geq \frac{p-1}{2(p+1)}\|u\|^2. \tag{10.52}$$

On the other hand, since

$$uf(u) \leq \gamma F(u) \leq c|u|^\gamma,$$

the Sobolev inequality implies

$$|u|_\gamma^2 \leq s_\gamma^2\|u\|^2 = s_\gamma^2\int_\Omega uf(u)\,dx \leq as_\gamma^2|u|_\gamma^\gamma, \tag{10.53}$$

where a is a some constant. From the last inequality we obtain the estimate

$$|u|_\gamma^{\gamma-2} \geq \frac{1}{as_\gamma^2}, \tag{10.54}$$

which implies the inequalities

$$\|u\| \geq \frac{1}{s\gamma}|u|_\gamma \geq \left(\frac{1}{as_\gamma^\gamma}\right)^{1/(\gamma-2)}. \tag{10.55}$$

Thus we get the following estimate for d:

$$d \geq \frac{1}{2}\frac{p-1}{p+1}\left(\frac{1}{as_\gamma^\gamma}\right)^{2/(\gamma-2)}. \tag{10.56}$$

Now we prove the existence of extremals (10.34). Let $\{u_n\}$ be a minimizing sequence:

$$d = \lim_{n\to+\infty} J(u_n), \quad u_n \in \mathbb{H}_0^1(\Omega), \quad \|u_n\| \neq 0, \quad K(u_n) = 0. \tag{10.57}$$

Note that this sequence is bounded in $\mathbb{H}_0^1(\Omega)$ due to inequality (10.52). Indeed, by (10.52) we see that

$$d_1 \geq J(u_n) \geq \frac{p-1}{2(p+1)}\|u_n\|^2,$$

where the first inequality follows from the convergence of the nonnegative numerical sequence $\{J(u_n)\}$. Therefore, it contains a subsequence $\{u_n\}$ that possesses the following two properties:

$$u_n \rightharpoonup w \quad \text{weakly in } \mathbb{H}_0^1(\Omega) \quad \text{as } n \to +\infty,$$
$$u_n \to w \quad \text{strongly in } \mathbb{L}^\gamma(\Omega) \quad \text{as } n \to +\infty$$

for $\gamma < 2^*$, where we have denoted the converging subsequence also by $\{u_n\}$. Thus, by the well-known theorem on weakly converging sequences we have

$$|u_n - w|_\gamma \to 0, \quad \|w\| \leq \liminf_{n\to\infty} \|u_n\|, \tag{10.58}$$

where the last limit inequality is a consequence of the weak sequential semicontinuity of the norm of a reflexive Banach space, in particular, the Hilbert space $\mathbb{H}_0^1(\Omega)$. We prove that

$$\liminf_{n\to\infty} \|u_n\| = \lim_{n\to\infty} \|u_n\|.$$

Indeed, by the condition we have

$$\frac{1}{2}\|u_n\|^2 = J(u_n) + \int_\Omega F(u_n)\,dx$$

and, moreover,

$$J(u_n) \to d, \quad \int_\Omega F(u_n)\,dx \to \int_\Omega F(u)\,dx$$

as $n \to \infty$. Therefore the numerical sequence $\{\|u_n\|\}$ converges.

From (10.55) we obtain that $|w|_\gamma \neq 0$:

$$|w|_\gamma = \lim_{n\to\infty} |u_n|_\gamma \geq s_\gamma \left(\frac{1}{as_\gamma^\gamma}\right)^{1/(\gamma-2)}. \tag{10.59}$$

By the limit inequality from (10.58) the following two situations are possible:

$$\|w\| = \lim_{n\to\infty} \|u_n\| \quad \text{and} \quad \|w\| < \lim_{n\to\infty} \|u_n\|.$$

Consider the first case. If

$$\|w\| = \lim_{n\to\infty} \|u_n\|,$$

then $K(w) = 0$ and $J(w) = d$, which proves the assertion of the theorem on the existence of an extremal.

Consider the second situation. If

$$\|w\| < \lim_{n\to\infty} \|u_n\|,$$

then $K(w) < 0$ and $J(w) < d$. By Lemma 10.2 there exists a constant $\bar\lambda < 1$ such that $K(\bar\lambda w) = 0$; then, for sufficiently small $\lambda > 0$, we have

$$K(\lambda u) > 0 \quad \text{for all } u \in \mathbb{H}_0^1(\Omega),\ u \neq 0.$$

For such λ, we have

$$\begin{aligned} J(\bar\lambda w) &= \frac{\bar\lambda^2}{2}\|w\|^2 - \int_\Omega F(\bar\lambda w)\,dx \\ &\leq \frac{\bar\lambda^2}{2}\lim_{n\to\infty}\|u_n\|^2 - \int_\Omega F(\bar\lambda w)\,dx \\ &= d + \int_\Omega \left(F(w) - F(\bar\lambda w) - \frac{1}{2}(1-\bar\lambda^2) w f(w)\right) dx, \end{aligned} \tag{10.60}$$

where we used the explicit form of the functionals $J(u_n)$ and $K(u_n)$, the fact $K(u_n) = 0$, and the following equalities:

$$\begin{aligned} \frac{\bar\lambda^2}{2}\|u_n\|^2 &= \frac{1}{2}\|u_n\|^2 - \int_\Omega F(u_n)\,dx + \frac{1}{2}(\bar\lambda^2 - 1)\|u_n\|^2 + \int_\Omega F(u_n)\,dx \\ &= \frac{1}{2}\|u_n\|^2 - \int_\Omega F(u_n)\,dx \\ &\quad + \frac{1}{2}(\bar\lambda^2 - 1)\int_\Omega u_n f(u_n)\,dx + \int_\Omega F(u_n)\,dx. \end{aligned} \tag{10.61}$$

Denote

$$I(\lambda) = \int_\Omega \left(F(w) - F(\lambda w) - \frac{1}{2}(1 - \lambda^2) w f(w) \right) dx \tag{10.62}$$

and consider nonnegative w for case (iii$_b$). By property (iv) we obtain the inequality

$$I(0) = \int_\Omega \left(F(w) - \frac{1}{2} w f(w) \right) dx < 0,$$

since

$$F(w) \le \frac{1}{p+1} w f(w) \quad \text{for } p > 1,$$

and hence

$$I(0) \le \int_\Omega \left(\frac{1}{p+1} w f(w) - \frac{1}{2} w f(w) \right) dx = -\frac{p-1}{2(p+1)} \int_\Omega w f(w)\, dx < 0,$$

since

$$0 \le \|w\|^2 < \lim_{n\to+\infty} \|u_n\|^2 = \lim_{n\to+\infty} \int_\Omega u_n f(u_n)\, dx = \int_\Omega w f(w)\, dx.$$

In addition, $I(1) = 0$.

Finally,

$$I'(\lambda) = \int_\Omega w(\lambda f(w) - f(\lambda w))\, dx > 0, \quad 0 < \lambda < 1, \tag{10.63}$$

since $f(0) = 0$ is convex down for $w > 0$, and therefore

$$f(w) > \frac{f(\lambda w)}{\lambda}, \quad \lambda \in (0, 1).$$

Thus, for $\lambda \in (0, 1)$ we have the inequality $I(\lambda) < 0$, which contradicts the definition of d:

$$J(\bar{\lambda} w) < d \quad \text{for } K(\bar{\lambda} w) = 0, \tag{10.64}$$

since, by construction, $J(\bar{\lambda} w) = d$.

Clearly, in case (iii$_a$), we must consider two variational problems, for positive and negative functions; therefore we obtain positive and negative extremals. The theorem is proved. □

It is easy to verify that the interior part of the potential well W is described by the inequalities

$$J(u) < d, \quad \frac{dJ(\lambda u)}{d\lambda} \geq 0, 0 < \lambda \leq 1.$$

As we have noted before, Sattinger studied the behavior of global solutions with initial states inside the potential well. Blow-up results are obtained in cases where initial data lie outside W.

Let

$$\Gamma \equiv \{u \in \mathbb{H}_0^1(\Omega),\ u \notin W\}.$$

In case (iii$_b$) the domain Γ is characterized by the inequalities

$$J(u) < d, \quad K(u) < 0. \tag{10.65}$$

In case (iii$_a$) the two exterior domains Γ_1 and Γ_2 are described by the inequalities

$$\Gamma_1 \equiv \{d_1 = \inf J(u),\ K(u) = 0,\ \|u\| \neq 0,\ u \geq 0\}, \tag{10.66}$$
$$\Gamma_2 \equiv \{d_2 = \inf J(u),\ K(u) = 0,\ \|u\| \neq 0,\ u \leq 0\}. \tag{10.67}$$

We also note that in both cases the exterior domains are unbounded.

10.3 Sobolev constant

This section does not directly relate to the theory of blow-up; we present here an exact description of the Sobolev constant, which plays a key role in reasoning on potential wells and so ultimately affects a better understanding of the method of saddle points of Sattinger and Payne.

Consider the following nonlinear problem for w:

$$\Delta w + |w|^{p-1} w = 0, \quad w = 0 \quad x \in \partial\Omega. \tag{10.68}$$

For all $\lambda > 0$, due to the homogeneity of the function $|u|^{p-1}u$, the family $(\lambda, \lambda^{1/(1-p)} w) = (\lambda, v(\lambda))$ is a one-parameter family of solutions to the problem

$$\Delta v + \lambda |v|^{p-1} v = 0, \quad v = 0 \text{ for } x \in \partial\Omega. \tag{10.69}$$

We define the Sobolev constant in the class $u \in \mathbb{H}_0^1(\Omega)$:

$$S_{p+1} = \inf \frac{\|u\|}{|u|_{p+1}}. \tag{10.70}$$

The Euler–Lagrange equation for this homogeneous variational problem is of the form

$$\Delta u + \lambda |u|^{p-1} u = 0$$

with Lagrange multipliers λ. Since the ratio $\|u\|/|u|_{p+1}$ is homogeneous, we can replace the extremal u by the scalar product of u and an arbitrary constant. Therefore

$$s_{p+1} = \frac{\|w\|}{|w|_{p+1}}. \tag{10.71}$$

On the other hand, (10.3) implies that

$$\|w\|^2 = (|w|_{p+1})^{p+1}$$

and

$$d = \frac{1}{2}\|w\|^2 - \frac{1}{p+1}(|w|_{p+1})^{p+1} = (|w|_{p+1})^{p+1}\left(\frac{p-1}{2(p+1)}\right).$$

Thus, the Sobolev constant is

$$s_{p+1} = \frac{\|w\|}{|w|_{p+1}} = |w|_{p+1}^{(p-1)/2} = \left(\frac{2(p+1)}{p-1}d\right)^{\frac{p-1}{2(p+1)}}. \tag{10.72}$$

10.4 Blow-up of solutions of hyperbolic equations

In this section, we obtain the following two results. First, we prove that a weak solution of problem (10.1) with initial data such that $u_0 \in \Gamma$ and $E(0) < d$ blows up on a finite time interval. Second, we show that if a solution starts inside the potential well and $E(0) < d$, then the kinetic energy

$$\int_\Omega \left(\frac{\partial u}{\partial t}\right)^2 dx$$

is bounded. Moreover, the function $u \in \mathbb{H}_0^1(\Omega)$ itself also remains bounded for all t. We denote by $(\cdot,\cdot)$ the scalar product in $\mathbb{L}^2(\Omega)$ and by $((\cdot,\cdot))$ in $\mathbb{H}_0^1(\Omega)$.

Definition 10.1. A weak solution $u(x,t)$ of problem (10.34) on an interval $[0, T_0)$ is a function satisfying the following conditions:

(1) $u(t)$ is a weakly continuous mapping of $[0, T_0)$ into $\mathbb{H}_0^1(\Omega)$, and the norms $\|u(t)\|$ and $|u(t)|_2$ are uniformly bounded on a compact subset of $[0, T_0)$;
(2) for all $0 \le t_1 < t_2 < T_0$ and $\varphi \in \mathbb{L}^2(\Omega)$, there exists a weakly continuous mapping of $[0, T)$ into $\mathbb{L}^2(\Omega)$:

$$(u,\varphi)\big|_{t_1}^{t_2} = \int_{t_1}^{t_2} (u_t, \varphi)\, ds;$$

(3) for any $\varphi : [0, T_0) \to \mathbb{H}_0^1(\Omega)$, we have

$$(u_t, \varphi)|_{t_1}^{t_2} = \int_{t_1}^{t_2} ((u_t, \varphi_t) - (\varphi, u) + (\varphi, f(u)))\, ds;$$

(4) for all $t_1 < t_2 < T_0$, the energy defined by the equality

$$E(t) = \frac{1}{2}(|u_t|_2^2 + \|u\|^2) - \int_\Omega F(u)\, dx$$

satisfies the inequality

$$E(t_2) \le E(t_1).$$

Remark 10.3. Note that this definition implies the weak absolute continuity of $u(t)$. Moreover, since $|u_t|$ is bounded, we conclude that $u(t)$ is weakly Lipschitz-continuous with respect to t and $(u(t), \varphi)$ is Lipschitz-continuous for all $\varphi \in \mathbb{L}^2(\Omega)$.

We further need the following result.

Lemma 10.6. *For a weak solution u of problem* (10.1), *define the integral*

$$M(t) = \int_\Omega u^2(x,t)\, dx.$$

Then, almost everywhere on the interval $[0, T_0)$, *there exists* $M''(t)$, *and* $M'(t)$ *is Lipschitz-continuous.*

Proof. Denote

$$Q(t,s) = (u(t), u(s)).$$

Since $u(t)$ and $u_t(t)$ are weakly continuous, we have

$$M'(t) = \left(\frac{\partial}{\partial t} Q(t,s) + \frac{\partial}{\partial s} Q(t,s)\right)_{s=t} = 2(u_t, u). \tag{10.73}$$

In property (3), we set $\varphi = u$; then

$$(u_t, u)|_{t_1}^{t_2} = \int_{t_1}^{t_2} (|u_t|^2 - \|u\|^2 + (u, f(u)))\, ds.$$

Therefore, for all $0 \le t_1 < t_2 < T$, we have

$$M'(t_2) - M'(t_1) = 2\int_{t_1}^{t_2} (|u_t|^2 - \|u\|^2 + (u, f(u)))\, ds. \tag{10.74}$$

Since all terms in the integrand are bounded on compact subsets of the set $[0, T_0)$, we conclude that $M'(t)$ is Lipschitz-continuous. Moreover, almost everywhere on $[0, T_0)$, there exists

$$M''(t) = 2(|u_t|_2^2 - \|u\|^2 + (u, f(u))). \tag{10.75}$$

The lemma is proved. □

The following assertion concerns the exterior domain Γ of the potential well under the condition that the initial total energy is less than the height of saddle points on the boundary of Γ.

Lemma 10.7. *Let $\Gamma \subset \mathbb{H}_0^1(\Omega)$ satisfy conditions* (10.65). *Then the solution* (10.1) *remains in Γ under the following condition*:

$$E(0) = \frac{1}{2}(\|u_0\|^2 + |v_0|_2^2) - \int_\Omega F(u_0)\,dx < J(w) = d. \tag{10.76}$$

Proof. Let $u(t)$ be a weak solution of problem (10.1) with the initial data $u(0) = u_0$ and $u_t(0) = v_0$. From the energy conservation law we conclude that

$$E(t) = E(0) \quad \Rightarrow \quad J(u(t)) < E(0).$$

Clearly, *if at a moment of time $t = t_0$, the solution $u(x,t)$ leaves the domain Γ, then* $K(u(t_0)) = 0$.

On the other hand, if $K(u(t_0)) = 0$, then the definition of d implies the inequality $J(u(t_0)) \geq d$. However, this contradicts the energy conservation law

$$d \leq J(u(t_0)) \leq E(t_0) = E(0) < d$$

and hence is impossible. The lemma is proved. □

Now we can state the main result on unbounded solutions.

Theorem 10.4. *Let the functional f satisfy conditions* (i)–(iv). *Denote by W the potential well corresponding to the potential energy of problem* (10.1). *If*

$$u_0 \in \Gamma \quad \text{and} \quad E(0) < d, \tag{10.77}$$

then the solution blows up in finite time: $|u|_2 \to +\infty$.

Proof. Let

$$M(t) = |u|_2^2, \quad M'(t) = 2(u_t, u). \tag{10.78}$$

From (10.75) we obtain the equality

$$M''(t) = 2|u_t|_2^2 + 2\int_\Omega uf(u)\,dx - 2\|u\|^2. \tag{10.79}$$

Since $u(x, t)$ lies in Γ for all (x, t) from the domain of existence, we have

$$\int_\Omega uf(u)\,dx - \|u\|^2 \geq 0 \quad \Rightarrow \quad M''(t) \geq 0. \tag{10.80}$$

Moreover, under condition (iv), we have

$$M'' \geq 2|u_t|_2^2 + 2(p+1)\int_\Omega F(u)\,dx - 2\|u\|^2. \tag{10.81}$$

The energy inequality of condition (iv) implies that

$$\int_\Omega F(u)\,dx \geq \frac{1}{2}(\|u\|^2 + |u_t|_2^2) - E(0). \tag{10.82}$$

From (10.81) and (10.82) we deduce the inequalities

$$\begin{aligned} M'' &\geq 2|u_t|_2^2 - 2\|u\|^2 + (p+1)(|u_t|_2^2 + \|u\|^2 - 2E(0)) \\ &\geq (p+3)|u_t|_2^2 + (p-1)\|u\|^2 - 2(p+1)E(0). \end{aligned} \tag{10.83}$$

From the variational inequality we obtain the estimate

$$\|u\|^2 \geq \lambda_1 |u|_2^2 = \lambda_1 M,$$

where λ_1 is the first eigenvalue of the Laplacian. Finally, from (10.83) we obtain

$$M''(t) \geq (p+3)|u_t|_2^2 + \lambda_1(p-1)M - 2(p+1)E(0). \tag{10.84}$$

By (10.80) the function M is convex with respect to time. Assume that there exists a time moment $t_1 > 0$ such that $M'(t_1) > 0$; then $M(t)$ increases for $t > t_1$. Therefore, $\lambda_1(p-1)M - 2(p+1)E(0)$ sooner or later becomes positive, and hence, for large values of time t, we have

$$M''(t) \geq (p+3)|u_t|_2^2. \tag{10.85}$$

For $\alpha = (p-1)/4$, we have the following inequalities:

$$\begin{aligned} (M^{-\alpha})'' &= -\frac{\alpha}{M^{\alpha+2}}\left(MM'' - \frac{p+3}{4}(M')^2\right) \\ &\leq -\frac{\alpha(p+3)}{M^{\alpha+2}}\left(|u|_2^2|u_t|_2^2 - \left|\int_\Omega uu_t\,dx\right|^2\right) \leq 0. \end{aligned} \tag{10.86}$$

Thus $M^{-\alpha}$ is concave for sufficiently large t, and there exists a finite value T of time for which $M^{-\alpha} \to 0$. In other words, the solution blows up in finite time:

$$\lim_{t\to T^-} M(t) = \infty. \tag{10.87}$$

Remark 10.4. At this point, we have in mind the following result. If the theorem on nonextendable solutions holds (see, e. g., [49, 79]), then the solution blows up in finite time. However, if we cannot apply this result, then the absence of global-in-time solvability and a lower estimate for the functional $M(t)$ on the interval $[0, T_0)$ can be proved. We also note that even in the case where the theorem on nonextendable solutions is valid, the situation can appear where the upper limit of the norm of solution is $+\infty$, whereas the limit itself does not exist. The corresponding example can be found in [80].

Proposition 10.7. *There exists a time moment $t \geq 0$ such that $M'(t) > 0$.*

Proof. Assume the contrary:

$$M'(t) \leq 0 \quad \text{for all } t \geq 0.$$

Since $M > 0$ and M is convex down, we conclude that M tends to a finite positive limit as $t \to \infty$. Indeed, M cannot tend to zero since, in this case, for large values of time, the solution lies inside the potential well, which is impossible.

Let us prove that there exist a constant $a > 0$ and a sequence $\{t_n\} \subset \mathbb{R}^1_+$ such that

$$M(t_n) \to a > 0, \quad M'(t_n) \to 0, \quad M''(t_n) \to 0$$

as $t_n \to \infty$. Indeed, assume that $M'(t) \leq 0$ for all t. Then, due to the positiveness and convexity ($M(t) > 0$, $M''(t) \geq 0$), the function $M(t)$ tends to a positive limit as $t \to \infty$, that is, there exists a sequence $\{t_n\}$ such that $M(t_n) \to a > 0$ as $t_n \to \infty$. Note that $M(t)$ cannot tend to zero or intersect zero since this means the hitting in the potential well, which is impossible. Thus,

$$M'(t) \to 0, \quad M''(t) \to 0;$$

these facts follow from the condition that $M'(t)$ and $M''(t)$ cannot be separated from zero by constants: otherwise, either the function $M(t)$ intersects the zero level in finite time, or $M'(t)$ becomes positive in finite time.

From (10.79) and (10.80) we obtain that

$$\lim_{t_n \to \infty} |u_t(t_n)|_2^2 = 0. \tag{10.88}$$

Then the energy inequality

$$\frac{1}{2}\|u(t_n)\|^2 + \frac{1}{2}|u_t(t_n)|_2^2 - \int_\Omega F(u(t_n))\, dx \leq E(0) \tag{10.89}$$

implies that, as $t_n \to \infty$,

$$\frac{1}{2}\|u(t_n)\|^2 - \int_\Omega F(u(t_n))\, dx \to b \leq E(0), \tag{10.90}$$

where b is some constant. On the other hand, from (10.79) we conclude that

$$\int_{\Omega} u(t_n)f(u(t_n))\,dx - \|u(t_n)\|^2 \to 0 \quad \text{as } t_n \to +\infty. \tag{10.91}$$

Lemma 10.8. *For any sequence $\{u_n\} \subset \mathbb{H}_0^1(\Omega)$ such that $\|u_n\| > 0$, $K(u_n) \le 0$, and $K(u_n) \to 0$, we have the following limit inequality:*

$$\lim_{n\to\infty} \inf J(u_n) \ge d. \tag{10.92}$$

Proof. First, we note that by inequality (10.52) the sequence $\{u_n\}$ is uniformly bounded with respect to $n \in \mathbb{N}$. Therefore we can choose a subsequence (we denote it also by $\{u_n\}$) such that

$$u_n \rightharpoonup w \quad \text{weakly in } \mathbb{H}_0^1(\Omega) \quad \text{as } n \to +\infty.$$

Moreover, due to the completely continuous embedding $\mathbb{H}_0^1(\Omega) \hookrightarrow \mathbb{L}^{p+1}(\Omega)$ for $p + 1 < 2^*$, we the following limit property for some subsequence of the sequence $\{u_n\}$:

$$u_n \to w \quad \text{strongly in } \mathbb{L}^{p+1}(\Omega) \quad \text{as } n \to +\infty.$$

Since $\|u_n\| > 0$, we have $\|w\| \neq 0$ and $K(w) \le 0$. Because of the weak lower semicontinuity of the norm of the Banach space, we have

$$\|w\| \le \liminf_{n\to+\infty} \|u_n\| \quad \text{as } n \to +\infty.$$

In addition,

$$\int_{\Omega} u_n f(u_n)\,dx \to \int_{\Omega} wf(w)\,dx \quad \text{as } n \to +\infty$$

and by the conditions of the lemma $K(u_n) \to 0$. Therefore

$$\liminf_{n\to+\infty} \|u_n\| = \lim_{n\to+\infty} \|u_n\|.$$

The following two cases are possible:

$$\lim_{n\to+\infty} \|u_n\| = \|w\| \quad \text{and} \quad \lim_{n\to+\infty} \|u_n\| > \|w\|.$$

In the first case,

$$K(w) = 0 \quad \text{and} \quad \lim_{n\to\infty} J(u_n) = J(w) \ge d.$$

In the second case, we have $K(w) < 0$. Then there exists $\bar{\lambda} > 0$ such that $K(\bar{\lambda}w) = 0$. Similarly to the theorem on extremals, we can show that

$$J(\bar{\lambda}w) < \liminf_{n\to\infty} J(u_n) + I(\bar{\lambda}),$$

where $I(\lambda)$ is defined by formula (10.62). In case (iii$_a$), we have $I(\bar{\lambda}) < 0$ and

$$\liminf_{n\to\infty} J(u_n) < d \quad \Rightarrow \quad J(\bar{\lambda}w) < d \quad \text{and} \quad K(\bar{\lambda}w) = 0,$$

which contradicts the definition of d.

In case (iii$_b$), we cannot assert that $w \geq 0$. We assume that f is convex and a sequence $\{u_n\}$ is such that

$$K(u_n) \to 0 \quad \text{and} \quad K(u_n) \leq 0.$$

As we have proved earlier, we have the inequalities

$$J(u_n) \geq J(|u_n|), \quad 0 \geq K(u_n) \geq K(|u_n|).$$

Since $K(|w|) < 0$, we can choose $\bar{\lambda} < 1$ such that $K(\bar{\lambda}|w|) = 0$. The reader can easily prove the following inequalities:

$$\begin{aligned}
J(\bar{\lambda}|w|) &= \frac{\bar{\lambda}^2}{2}\|w\|^2 - \int_\Omega F(\bar{\lambda}|w|)\,dx \\
&\leq \frac{\bar{\lambda}^2}{2}\liminf_{n\to\infty}\|u_n\|^2 - \int_\Omega F(\bar{\lambda}|w|)\,dx \\
&= \liminf_{n\to\infty}\frac{\|u_n\|^2}{2} + \frac{1-\bar{\lambda}^2}{2}\liminf_{n\to\infty}\|u_n\| - \int_\Omega F(\bar{\lambda}|w|)\,dx \\
&= \liminf_{n\to\infty}\Big(J(u_n) + \frac{1-\bar{\lambda}^2}{2}\|u_n\|^2 + \int_\Omega F(u_n)\,dx\Big) - \int_\Omega F(\bar{\lambda}|w|)\,dx \\
&\leq \liminf_{n\to\infty}\Big(J(u_n) + \frac{1-\bar{\lambda}^2}{2}\int_\Omega u_n f(u_n)\,dx + \int_\Omega F(u_n)\,dx\Big) - \int_\Omega F(\bar{\lambda}|w|)\,dx \\
&\leq \liminf_{n\to\infty} J(u_n) + \int_\Omega\Big(\frac{1-\bar{\lambda}^2}{2}wf(w) + F(w) - F(\bar{\lambda}|w|)\Big)\,dx \\
&\leq \liminf_{n\to\infty} J(u_n) + \int_\Omega\Big(\frac{1-\bar{\lambda}^2}{2}|w|f(|w|) + F(|w|) - F(\bar{\lambda}|w|)\Big)\,dx \\
&\leq \liminf_{n\to\infty} J(u_n).
\end{aligned} \tag{10.93}$$

Now, if

$$\liminf_{n\to\infty} J(u_n) < d,$$

then there exists a function $\bar{\lambda}|w|$ for which $K(\bar{\lambda}|w|) = 0$ and $J(\bar{\lambda}|w|) < d$, which is impossible. This contradiction proves (10.92). The lemma is proved. □

Now we can apply Lemma 10.8 to the sequence $\{u(t_n)\}$ and obtain

$$\liminf_{t_n\to+\infty} J(u(t_n)) \geq d > E(0) = \liminf_{t_n\to+\infty} E(t_n) \geq \liminf_{t_n\to+\infty} J(u(t_n)),$$

where we have applied the conservation law $E(t_n) = E(0)$, which is valid for any sequence $\{t_n\}$. The contradiction obtained proves Proposition 10.7. □

Proposition 10.7 leads us to the main theorem on the blow-up. Thus, Theorem 10.4 is proved. □

10.5 Blow-up of solutions of a parabolic equation

The method of saddle points can be also applied to problems for parabolic equations. In a bounded domain $\Omega \subset \mathbb{R}^N$ with sufficiently smooth boundary $\partial\Omega$, we consider the following problem:

$$u_t = \Delta u + f(u), \quad u(x,0) = u_0(x), \quad x \in \Omega,\ t > 0, \tag{10.94}$$

$$u = 0 \quad x \in \partial\Omega. \tag{10.95}$$

In the definition of a weak solution (Definition 10.1), items (3) and (4) must be replaced by the following conditions:
(3′) for all $\varphi : [0,T) \to \mathbb{H}_0^1(\Omega)$,

$$(\varphi, u_t) + ((\varphi, u)) - (\varphi, f(u)) = 0;$$

(4′) the functional $J(u)$ satisfies the inequality

$$\int_0^t |u_s|^2\,ds + J(u) \leq J(u_0).$$

As in the previous sections, we can prove the following results.

Lemma 10.9. *For the integral*

$$M_1(t) = \int_0^t |u|_2^2\,ds, \tag{10.96}$$

the second derivative M_1'' exists almost everywhere on $[0,t)$, and the first derivative $M_1'(t)$ is Lipschitz-continuous.

Lemma 10.10. *Let Γ be a subdomain of $\mathbb{H}_0^1(\Omega)$ satisfying the conditions* (10.65). *Then the solution of* (10.94) *does not leave Γ if*

$$J(u_0) < J(w) = d. \tag{10.97}$$

The main blow-up result is the following theorem.

Theorem 10.5. *Let $u_0 \in \Gamma$ and $J(u_0) < d$. Then the weak solution of problem* (10.94) *blows up in finite time*: $|u|_2 \to \infty$ *as* $t \to T^-$.

Proof. Assume the contrary. From (10.96) we obtain that

$$M_1'(t) = |u|_2^2 = |u_0|_2^2 + 2\int_0^t (u, u_t)\, ds. \tag{10.98}$$

Using property (3′), we rewrite this expression in the form

$$M_1'(t) = |u_0|_2^2 - 2\int_0^t (\|u\|^2 - (u, f(u)))\, ds; \tag{10.99}$$

then the second derivative is

$$M_1''(t) = 2((u, f(u)) - \|u\|^2) = -2K(u). \tag{10.100}$$

Since $u_0 \in \Gamma$, $K(u)$ remains negative for all t from the domain of the solution. Therefore,

$$M_1'(t) > 0 \quad \text{and} \quad M_1''(t) \geq 0, \quad K(u) < 0$$

for all $t \leq T$. We prove that the sequence

$$\{K(u_n) = K(u(\cdot, t_n))\}$$

does not tend to zero as $t_n \to \infty$; to this end, we apply the result proved for estimate (10.92):

$$\liminf_{n\to\infty} J(u_n) \geq d > J(u_0).$$

This contradicts condition (4′); therefore, $K(u) < 0$ on the whole interval of existence, and the function $M'(t)$ becomes greater than an arbitrary predetermined constant as t increases.

Condition (iv) yields the estimate

$$M_1''(t) \geq 2(p+1)\int_\Omega F(u)\, dx - 2\|u\|^2 = \mathrm{I}, \tag{10.101}$$

which can be continued by using the definition of $J(u)$ and property (4′):

$$\mathrm{I} \geq 2(p+1)\int_0^t |u_\eta|_2^2\, d\eta + (p-1)\|u\|^2 - 2(p+1)J(u_0)$$

$$\geq 2(p+1)\int_0^t |u_\eta|_2^2\,d\eta + (p-1)\lambda_1 M_1'(t) - 2(p+1)J(u_0). \tag{10.102}$$

Consider the following expression:

$$\begin{aligned} &M_1 M_1'' - \frac{p+1}{2}(M_1')^2 \\ &\quad = 2(p+1)\left(\int_0^t |u|_2^2\,d\eta \int_0^t |u_\eta|_2^2\,d\eta - \left|\int_0^t\int_\Omega uu_\eta\,dx\,d\eta\right|^2\right) \\ &\quad + (p-1)\lambda_1 M_1 M_1' - (p+1)|u_0|_2^2 M_1' \\ &\quad - 2(p+1)J(u_0)M_1 + \frac{p+1}{2}|u_0|_2^4. \end{aligned} \tag{10.103}$$

The first term is nonnegative by the Cauchy–Bunyakovsky inequality, and the second term grows faster than all other terms. Therefore, for sufficiently large values of time $t > t^*$, the right-hand side is positive. As in the case of hyperbolic problems, this leads to the blow-up in finite time:

$$M_1^{-(p-1)/2} \leq M_1^{-(p-1)/2}(t^*)\left(1 - \frac{p-1}{2}\frac{M_1'(t^*)}{M_1(t^*)}(t-t^*)\right). \tag{10.104}$$

The theorem is proved. □

10.6 Bibliographical notes

The material of this chapter is taken from [82]. Note that this method was successfully applied to problems for the nonlinear Klein–Gordon equation (see [106]). Moreover, this method can also be used for Sobolev equations of the form

$$\frac{\partial}{\partial t}(\Delta u - u) + \Delta u + f(x,u) = 0.$$

11 Breaking of solutions of wave equations

In this chapter, we consider the breaking of solutions of wave equation. First, we formulate the notion of breaking. We say that at a point $x_0 \in \mathbb{R}^1$ and time instant $t_0 \in \mathbb{R}^1_+$, a solution $u = u(x,t)$ of a certain nonlinear wave equation *breaks* (or that the breaking of a solution occurs) if the solution itself remains bounded everywhere in $\mathbb{R}^2_+$, whereas the derivative $u_x(x_0, t_0)$ becomes infinite. From the definition of the notion of breaking we see that methods developed in the previous sections are inapplicable. Moreover, as a rule, in cases where breaking occurs, weak solutions of the corresponding nonlinear equations exist globally in time. Note that this situation is observed even for simple ordinary differential equations of order 1, for example, for the problem

$$\frac{dy}{dt} = \frac{1}{1 - y(t)}, \quad y(0) = 0 \quad \Rightarrow \quad y(t) = 1 \pm \sqrt{1 - 2t}, \quad t \in [0, 1/2),$$

and also for complicated equations of hydrodynamical type. We consider certain classical examples of the proof of the breaking of solution.

11.1 Camassa–Holm equation

The Cauchy problem for the Camassa–Holm equation is of the following form:

$$\begin{gathered} u_t - u_{xxt} + 3uu_x = 2u_x u_{xx} + uu_{xxx}, \quad x \in \mathbb{R}^1,\ t > 0, \\ u(x,0) = u_0(x). \end{gathered} \tag{11.1}$$

Theorem 11.1 (see [12]). *If $u_0(x) \in \mathbb{H}^3(\mathbb{R}^1)$, then there exists a unique, local in time $t \in [0,T)$ solution $u(x,t)$ of the Cauchy problem* (11.1) *of the class*

$$u \in \mathbb{C}([0,T); \mathbb{H}^3(\mathbb{R}^1)) \cap \mathbb{C}^{(1)}([0,T); \mathbb{H}^2(\mathbb{R}^1)).$$

Note that equation (11.1) can be rewritten in the form

$$(\mathbb{I} - \partial_x^2)(u_t + uu_x) = -2uu_x - u_x u_{xx} = -\partial_x\left(u^2 + \frac{1}{2}u_x^2\right). \tag{11.2}$$

The operator

$$(\mathbb{I} - \partial_x^2)^{-1}$$

can be represented in the convolution form:

$$(\mathbb{I} - \partial_x^2)^{-1} f = p * f, \quad p(x) \equiv \frac{1}{2} e^{-|x|}, \quad f \in \mathbb{L}^2(\mathbb{R}^1).$$

https://doi.org/10.1515/9783110602074-011

Therefore

$$u_t + uu_x = -\partial_x\left(p * \left(u^2 + \frac{1}{2}u_x^2\right)\right) \quad \text{in } \mathbb{C}([0,T); \mathbb{H}^1(\mathbb{R}^1)), \tag{11.3}$$

where $*$ denotes the convolution with respect to the spatial variable. Differentiating (11.3) with respect to $x \in \mathbb{R}^1$, we obtain the following equalities:

$$\begin{aligned} u_{tx} + u_x^2 + uu_{xx} &= -\partial_x^2\left(p * \left(u^2 + \frac{1}{2}u_x^2\right)\right) \\ &= (\mathbb{I} - \partial_x^2 - \mathbb{I})\left(p * \left(u^2 + \frac{1}{2}u_x^2\right)\right) \\ &= u^2 + \frac{1}{2}u_x^2 - p * \left(u^2 + \frac{1}{2}u_x^2\right); \end{aligned} \tag{11.4}$$

therefore

$$u_{tx} + uu_{xx} = u^2 - \frac{1}{2}u_x^2 - p * \left(u^2 + \frac{1}{2}u_x^2\right) \quad \text{in } \mathbb{C}([0,T); \mathbb{H}^1(\mathbb{R}^1)). \tag{11.5}$$

Theorem 11.2. *Let $u_0 \in \mathbb{H}^3(\mathbb{R}^1)$ be an odd function such that $u_0'(0) < 0$. Then the corresponding solution of the Cauchy problem* (11.1) *breaks in finite time: $0 < t_0 \leq (2|u_0'(0)|)^{-1}$ at the point $x = 0$.*

Proof. Let $[0, T)$ be the maximal interval of the existence to the corresponding solution of problem (11.1) of the class

$$u \in \mathbb{C}([0,T); \mathbb{H}^3(\mathbb{R}^1)) \cap \mathbb{C}^{(1)}([0,T); \mathbb{H}^2(\mathbb{R}^1))$$

with the initial function $u_0(x) \in \mathbb{H}^3(\mathbb{R}^1)$. Note that

$$v(x,t) \equiv -u(-x,t), \quad t \in [0,T),\ x \in \mathbb{R}^1,$$

is also a solution of problem (11.1) of the same class as $u(x,t)$. Due to the uniqueness of a solution, we conclude that $v \equiv u$, and therefore $u(\cdot, t)$ is an odd function with respect to $x \in \mathbb{R}^1$ for all $t \in [0,T)$. In particular, we have

$$u(0,t) = u_{xx}(0,t) = 0 \quad \text{for all } t \in [0,T). \tag{11.6}$$

Introduce the new function $g(t) \equiv u_x(0,t)$ for $t \in [0,T)$ and note that $g \in \mathbb{C}^{(1)}([0,T); \mathbb{R}^1)$. From (11.5) and (11.6) we obtain

$$\frac{dg}{dt} = -\frac{1}{2}g^2(t) - \frac{1}{2}\int_{\mathbb{R}^1} e^{-|y|}\left(u^2 + \frac{1}{2}u_x^2\right) dy \leq -\frac{1}{2}g^2(t), \quad t \in [0,T).$$

Therefore

$$\frac{1}{g(t)} \geq \frac{1}{g_0} + \frac{t}{2}, \quad t \in [0,T),$$

and hence $T < -2/g_0$. The theorem is proved. □

11.2 Whitham equation

In this section, we consider the question of the breaking of solutions of the Whitham equation that have the form of solitary waves. Results are taken from [24]. The Whitham equation is the following nonlinear nonlocal equation (see [76]):

$$u_t + uu_x + \int_{-\infty}^{+\infty} \mathrm{K}(x-s)u_s(s,t)\,ds = 0, \quad x \in \mathbb{R}^1,\ t > 0. \tag{11.7}$$

Assume that at some time moment (we set $t = 0$), a solution of the Whitham equation (11.7) has the form of a solitary wave, whose existence was proved by Gabov [24]. We also assume that the amplitude of this function decreases on both sides of its maximum and $u(x,t) \to 0$ as $x \to \pm\infty$. Let $u(x,t)$ be twice differentiable on some time interval $t \in [0, T_0)$ with respect to $x \in \mathbb{R}^1$.

Let functions $x_1(t)$ and $x_2(t)$ be such that

$$u_x(x_1(t),t) = \min_{x\in\mathbb{R}^1} u_x(x,t) = v(t) < 0, \quad u_x(x_2(t),t) = \max_{x\in\mathbb{R}^1} u_x(x,t) = w(t) > 0.$$

We assume that $x_i(t) \in \mathbb{C}^{(1)}[0,T_0)$ on the interval $[0,T_0)$. Clearly, $x_1(t) > x_2(t)$.

Assume that the function $\mathrm{K}(s)$ is differentiable and its derivative is absolutely integrable. Then, differentiating equation (11.7), we obtain

$$u_{xt} + uu_{xx} + u_x^2 + \int_{-\infty}^{+\infty} \mathrm{K}'(x-s)u_s(s,t)\,ds = 0.$$

Substituting the functions $x_i = x_i(t)$ into this equation, for the functions w and v, we obtain

$$\frac{dw}{dt} + w^2 + \int_{-\infty}^{+\infty} \frac{d}{dx}\mathrm{K}(x_2(t)-s)u_s(s,t)\,ds = 0, \tag{11.8}$$

$$\frac{dv}{dt} + v^2 + \int_{-\infty}^{+\infty} \frac{d}{dx}\mathrm{K}(x_1(t)-s)u_s(s,t)\,ds = 0; \tag{11.9}$$

here we take into account the fact that $u_{xx}(x_i(t),t) = 0$ at extremum points of the function $u_x(x,t)$. We estimate the integral terms in (11.8) and (11.9):

$$\left|\int_{-\infty}^{+\infty} \frac{d}{dx}\mathrm{K}(x_i(t)-s)u_s(s,t)\,ds\right| \le \alpha \max_{x\in\mathbb{R}^1} |u_x(x,t)|, \tag{11.10}$$

where

$$\alpha = \int_{-\infty}^{+\infty} |\mathrm{K}'(x)|\,dx.$$

Using the definition of the functions $v(t)$ and $w(t)$, we obtain the estimates

$$\max_{x\in\mathbb{R}^1} |u_x(x,t)| \le w(t) - v(t), \tag{11.11}$$

$$\max_{x\in\mathbb{R}^1} |u_x(x,t)| \le -v(t) \tag{11.12}$$

under the condition $w(t) \le |v(t)|$ or $w(t) + v(t) \le 0$. From equations (11.8) and (11.9) and from estimates (11.10) and (11.11) we deduce the inequalities

$$\frac{dw}{dt} + w^2 \le \alpha\big(w(t) - v(t)\big), \tag{11.13}$$

$$\frac{dv}{dt} + v^2 \le \alpha\big(w(t) - v(t)\big). \tag{11.14}$$

Similarly, from (11.9) and (11.12) we obtain the following "conditional" inequality: if $w(t) + v(t) \le 0$, then

$$\alpha v(t) \le \frac{dv}{dt} + v^2 \le -\alpha v(t). \tag{11.15}$$

Thus, the functions $w(t)$ and $v(t)$ satisfy system (11.13)–(11.15) of differential and "conditionally" differential inequalities.

If we show that a solution of the system of inequalities (11.13)–(11.15) under certain condition possesses the property

$v(t)$ monotonically decreases and tends to $-\infty$ as $t \to T_0$, where $T_0 < +\infty$,

then the breaking of this solution will be proved.

Lemma 11.1. *If $w(t) + v(t) \le 0$ and $v(0) < -\alpha$, then $v(t)$ decreases and tends to $-\infty$. Moreover, we have the following inequalities:*

$$|v(t)| \ge \frac{\alpha}{1 - c_1 e^{\alpha t}}, \quad t < T^*, \tag{11.16}$$

$$|v(t)| \le \frac{\alpha c_2 e^{\alpha t}}{1 - c_2 e^{\alpha t}}, \quad t < T_*, \tag{11.17}$$

where

$$c_1 = \frac{|v(0)| - \alpha}{|v(0)|} < c_2^{-1} = \frac{|v(0)|}{|v(0)| + \alpha},$$

$$T_* = \frac{1}{\alpha}\ln\left(\frac{1}{c_2}\right) < T^* = \frac{1}{\alpha}\ln\left(\frac{1}{c_1}\right).$$

Proof. We set

$$m(t) = |v(t)| = -v(t) > 0.$$

Then from (11.15) we have

$$m(m+\alpha) \geq \frac{dm}{dt} \geq m(m-\alpha), \tag{11.18}$$

which, for $m(0) = -v(0) > \alpha$, implies the increasing of $m(t)$ or decreasing of $v(t)$. Due to the increasing of $m(t)$, the right-hand side of (11.18) is positive for all t. We rewrite the second inequality in the form

$$\frac{1}{m(m-\alpha)}\frac{dm}{dt} \geq 1,$$

integrate it from 0 to t, and obtain

$$\ln\left(\frac{m(t)-\alpha}{m(t)}\right) - \ln\left(\frac{m(0)-\alpha}{m(0)}\right) \geq \alpha t.$$

Solving this elementary inequality, we arrive at estimate (11.16) for $t < T^*$, where T^* is defined from the equality

$$1 - c_1 e^{\alpha t} = 0.$$

Similarly, from the left-hand side of inequality (11.18) we can deduce estimate (11.17). Thus we have proved the breaking of the solution of problem (11.7). The lemma is proved. □

11.3 Bibliographical notes

The material of this chapter is taken from [12, 24, 76].

Note that the twice continuous differentiability of the solution $u(x,t)$ with respect to the variable x for all $t \in [0,T]$ and the continuous differentiability of the functions $x_i(t)$ are essential restrictions for application of the method described in this section.

However, the Whitham equation, as an equation with specific nonlinearity $(u^2)_x$, admits the method of nonlinear capacity of Pokhozhaev and the method of test functions (see Chapter 1).

A Auxiliary and additional results

A.1 Differential inequality I

Consider the following ordinary differential inequality, which plays an important role in our studies:

$$\Phi\Phi'' - \alpha(\Phi')^2 + \gamma\Phi'\Phi + \beta\Phi \geq 0, \quad \alpha > 1,\ \beta \geq 0,\ \gamma \geq 0, \tag{A.1}$$

where

$$\Phi(t) \in \mathbb{C}^{(2)}([0,T]), \quad \Phi(t) \geq 0,\ \Phi(0) > 0.$$

We divide both sides of inequality (A.1) by $\Phi^{1+\alpha}$ and after simple transformations obtain

$$\left(\frac{\Phi'}{\Phi^\alpha}\right)' + \gamma\frac{\Phi'}{\Phi^\alpha} + \beta\Phi^{-\alpha} \geq 0.$$

This implies

$$\frac{1}{1-\alpha}(\Phi^{1-\alpha})'' + \frac{\gamma}{1-\alpha}(\Phi^{1-\alpha})' + \beta\Phi^{-\alpha} \geq 0. \tag{A.2}$$

Introducing the notation

$$Z(t) = \Phi^{1-\alpha}(t), \tag{A.3}$$

from (A.2) we obtain

$$Z'' + \gamma Z' - \beta(\alpha-1)Z^{\alpha_1} \leq 0, \quad \alpha_1 = \frac{\alpha}{\alpha-1}. \tag{A.4}$$

Introducing the notation

$$Y(t) = e^{\gamma t}Z(t), \tag{A.5}$$

from (A.4) we obtain

$$Y'' - \gamma Y' - \beta(\alpha-1)e^{-\delta t}Y^{\alpha_1} \leq 0, \quad \delta = \frac{\gamma}{\alpha-1}. \tag{A.6}$$

We have the following chain of equalities:

$$Y' = (\Phi^{1-\alpha}e^{\gamma t})' = \Phi^{-\alpha}(\alpha-1)e^{\gamma t}\left[-\Phi'(t) + \frac{\gamma}{\alpha-1}\Phi(t)\right]. \tag{A.7}$$

Assume that the initial condition

$$\Phi'(0) > \frac{\gamma}{\alpha-1}\Phi(0) \tag{A.8}$$

https://doi.org/10.1515/9783110602074-012

is fulfilled; then there exists $t_0 > 0$ such that

$$\Phi'(t) > \frac{\gamma}{\alpha - 1}\Phi(t) \quad \text{for } t \in [0, t_0). \tag{A.9}$$

Then from inequality (A.9) and expressions (A.7) we obtain

$$Y'(t) < 0 \quad \text{for } t \in [0, t_0).$$

Since

$$-\gamma Y'(t) \geq 0 \quad \text{for } t \in [0, t_0),$$

from inequality (A.6) we obtain the estimate

$$Y'' - \beta(\alpha - 1)e^{-\delta t} Y^{\alpha_1} \leq 0 \quad \text{for } t \in [0, t_0), \tag{A.10}$$

where $\delta = \gamma/(\alpha - 1)$. Multiplying both sides of inequality (A.10) by Y', we obtain the inequality

$$Y'Y'' - \beta(\alpha - 1)e^{-\delta t} Y^{\alpha_1} Y' \geq 0 \quad \text{for } t \in [0, t_0). \tag{A.11}$$

Note the following equality:

$$e^{-\delta t} Y^{\alpha_1} Y' = \frac{d}{dt}[e^{-\delta t} Y^{1+\alpha_1}] + \delta e^{-\delta t} Y^{1+\alpha_1} - \alpha_1 e^{-\delta t} Y^{\alpha_1} Y'.$$

Therefore

$$e^{-\delta t} Y^{\alpha_1} Y' = \frac{1}{1 + \alpha_1}\frac{d}{dt}[e^{-\delta t} Y^{1+\alpha_1}] + \frac{1}{1 + \alpha_1}\delta e^{-\delta t} Y^{1+\alpha_1}. \tag{A.12}$$

Substituting (A.12) into (A.11), we obtain the inequality

$$Y'Y'' - \frac{\beta(\alpha - 1)}{1 + \alpha_1}\frac{d}{dt}[e^{-\delta t} Y^{1+\alpha_1}] - \frac{\beta(\alpha - 1)\delta}{1 + \alpha_1} e^{-\delta t} Y^{1+\alpha_1} \geq 0$$

for $t \in [0, t_0)$, which immediately implies the inequality

$$Y'Y'' - \frac{\beta(\alpha - 1)}{1 + \alpha_1}\frac{d}{dt}[e^{-\delta t} Y^{1+\alpha_1}] \geq 0 \quad \text{for } t \in [0, t_0). \tag{A.13}$$

Integrating the last inequality, we arrive at the estimate

$$(Y')^2 \geq A^2 + \frac{2\beta(\alpha - 1)^2}{2\alpha - 1} e^{-\delta t} Y^{1+\alpha_1} \geq A^2, \tag{A.14}$$

where

$$A^2 \equiv (Y'(0))^2 - \frac{2\beta(\alpha - 1)^2}{2\alpha - 1} Y^{1+\alpha_1}(0). \tag{A.15}$$

Now we assume that the condition

$$A^2 > 0$$

is fulfilled; by simple transformations it can be reduced to the form

$$A^2 = (\alpha - 1)^2 \Phi^{-2\alpha}(0)\left[\left(\Phi'(0) - \frac{\gamma}{\alpha - 1}\Phi(0)\right)^2 - \frac{2\beta}{2\alpha - 1}\Phi(0)\right] > 0. \tag{A.16}$$

Therefore, the condition $A^2 > 0$ is equivalent to the condition

$$\left(\Phi'(0) - \frac{\gamma}{\alpha - 1}\Phi(0)\right)^2 > \frac{2\beta}{2\alpha - 1}\Phi(0). \tag{A.17}$$

Thus, from inequalities (A.14) and (A.16) we conclude that

$$Y'(t) \le -A < 0 \quad \Rightarrow \quad \Phi'(t_0) > \frac{\gamma}{\alpha - 1}\Phi(t_0).$$

Therefore $Y'(t_0) < 0$. Applying the algorithm of extension of solutions in time, we obtain

$$Y'(t) < 0 \quad \text{for all } t \in [0, T].$$

Therefore, we have the following inequalities:

$$\begin{aligned} |Y'| \ge A > 0 \quad &\Rightarrow \quad Y'(t) \le -A \quad \Rightarrow \quad Y(t) \le Y(0) - At \quad \Rightarrow \\ &\Rightarrow \quad \Phi^{1-\alpha}(t) \le e^{-\gamma t}[\Phi^{1-\alpha}(0) - At] \quad \Rightarrow \\ &\Rightarrow \quad \Phi(t) \ge \frac{e^{\gamma t/(\alpha-1)}}{[\Phi^{1-\alpha}(0) - At]^{1/(\alpha-1)}}. \end{aligned} \tag{A.18}$$

Thus, the following theorem is proved.

Theorem A.1. *Let $\Phi(t) \in \mathbb{C}^{(2)}([0, T])$, and let the following conditions be fulfilled:*

$$\Phi'(0) > \frac{\gamma}{\alpha - 1}\Phi(0), \tag{A.19}$$

$$\left(\Phi'(0) - \frac{\gamma}{\alpha - 1}\Phi(0)\right)^2 > \frac{2\beta}{2\alpha - 1}\Phi(0), \tag{A.20}$$

where $\Phi(t) \ge 0$ and $\Phi(0) > 0$. Then the time $T > 0$ cannot be infinitely large; namely, we have the following inequality:

$$T \le T_\infty \le \Phi^{1-\alpha}(0)A^{-1},$$

where

$$A^2 \equiv (\alpha - 1)^2 \Phi^{-2\alpha}(0)\left[\left(\Phi'(0) - \frac{\gamma}{\alpha - 1}\Phi(0)\right)^2 - \frac{2\beta}{2\alpha - 1}\Phi(0)\right]$$

and

$$\Phi(t) \ge \frac{e^{\gamma t/(\alpha-1)}}{[\Phi^{1-\alpha}(0) - At]^{1/(\alpha-1)}}.$$

A.2 Differential inequality II

Let a function

$$\Phi(t) \in \mathbb{C}^{(2)}([0,T]), \quad \Phi(t) \geq 0 \quad \text{for } t \in [0,T],\ T > 0,$$

satisfy the ordinary integro-differential inequality

$$\Phi\Phi'' - \alpha(\Phi')^2 + \beta\Phi\Phi' + \gamma_1\Phi^2 + \gamma_2 \int_0^t ds\, h(t-s)\Phi(s)\Phi(t) \geq 0 \tag{A.21}$$

for $\alpha > 1$, $\beta, \gamma_1, \gamma_2 \geq 0$, and all $t \in [0,T]$. Assume that the function $h(t)$ belongs to the class $\mathbb{C}(\mathbb{R}^1_+)$ and satisfies the inequality

$$|h(t)| \leq h_0 e^{-at}, \quad h_0 > 0,\ a > 0. \tag{A.22}$$

We obtain sufficient conditions of the appearance of a second-type discontinuity of the function $\Phi(t)$ satisfying the conditions specified before and inequality (A.21).

First, we note that inequalities (A.21) and (A.22) imply the inequality

$$\Phi\Phi'' - \alpha(\Phi')^2 + \beta\Phi\Phi' + \gamma_1\Phi^2 + \gamma_3 \int_0^t ds\, e^{-a(t-s)}\Phi(s)\Phi(t) \geq 0, \tag{A.23}$$

where $\gamma_3 = h_0\gamma_2$. Since $\Phi(t) \in \mathbb{C}^{(2)}([0,T])$ for certain $T > 0$, we have the following integration-by-parts formula:

$$\begin{aligned}\int_0^t ds\, e^{-a(t-s)}\Phi(s) &= e^{-at}\int_0^t ds\, e^{as}\Phi(s)\\ &= e^{-at}\left[\frac{1}{a}e^{as}\Phi(s)\Big|_{s=0}^{s=t} - \frac{1}{a}\int_0^t ds\, e^{as}\Phi'(s)\right]\\ &= \frac{1}{a}\Phi(t) - \frac{1}{a}e^{-at}\Phi(0) - \frac{1}{a}\int_0^t ds\, e^{-a(t-s)}\Phi'(s).\end{aligned} \tag{A.24}$$

Since $\Phi(t) \geq 0$, the chain of equalities (A.24) implies the upper estimate

$$\int_0^t ds\, e^{-a(t-s)}\Phi(s) \leq \frac{1}{a}\Phi(t) - \frac{1}{a}\int_0^t ds\, e^{-a(t-s)}\Phi'(s) \tag{A.25}$$

for all $t \in [0,T]$. Since $\Phi(t) \in \mathbb{C}^{(1)}([0,T])$, we conclude that, under the initial condition

$$\Phi'(0) > 0, \tag{A.26}$$

there exists a time moment $t_1 \in (0, T]$ such that

$$\Phi'(t) \geq 0 \quad \text{for all } t \in [0, t_1]. \tag{A.27}$$

Then for all $t \in [0, t_1]$, from inequalities (A.25) and (A.27) we obtain the following inequality:

$$\int_0^t ds\, e^{-a(t-s)}\Phi(s) \leq \frac{1}{a}\Phi(t). \tag{A.28}$$

Therefore, the integro-differential inequality (A.23) implies the ordinary differential inequality

$$\Phi\Phi'' - \alpha(\Phi')^2 + \beta\Phi\Phi' + \gamma\Phi^2 \geq 0 \quad \text{for } t \in [0, t_1],\ t_1 \in (0, T], \tag{A.29}$$

where $\alpha > 1$, $\beta \geq 0$, $\gamma = \gamma_1 + \gamma_3 a^{-1} \geq 0$.

Now we examine inequality (A.29). We have

$$\begin{aligned} &\frac{\Phi''}{\Phi^\alpha} - \alpha\frac{(\Phi')^2}{\Phi^{1+\alpha}} + \beta\frac{\Phi'}{\Phi^\alpha} + \gamma\Phi^{1-\alpha} \geq 0 \quad \Rightarrow \\ &\Rightarrow \quad \left[\frac{\Phi'}{\Phi^\alpha}\right]' + \beta\frac{\Phi'}{\Phi^\alpha} + \gamma\Phi^{1-\alpha} \geq 0 \quad \Rightarrow \\ &\Rightarrow \quad \frac{1}{1-\alpha}Z'' + \frac{\beta}{1-\alpha}Z' + \gamma Z \geq 0, \end{aligned} \tag{A.30}$$

where

$$Z(t) \equiv \Phi^{1-\alpha}(t). \tag{A.31}$$

From (A.30) we obtain the inequality

$$Z'' + \beta Z - \delta Z \leq 0, \quad \delta \equiv \gamma(\alpha - 1). \tag{A.32}$$

Introduce the new function

$$Y(t) = Ze^{\beta t}. \tag{A.33}$$

We have the following equalities:

$$\begin{aligned} Z'' + \beta Z' &= [e^{\beta t}Z']'e^{-\beta t} = [(e^{\beta t}Z)' - \beta e^{\beta t}Z]'e^{-\beta t} \\ &= [(e^{\beta t}Z)'' - \beta(e^{\beta t}Z)']e^{-\beta t} = [Y'' - \beta Y']e^{-\beta t}. \end{aligned} \tag{A.34}$$

Taking into account notation (A.33), from equations (A.34) and inequality (A.32) we obtain the inequality

$$Y'' - \beta Y' - \delta Y \leq 0 \quad \text{for } t \in [0, t_1], \quad \delta = (\alpha - 1)\gamma. \tag{A.35}$$

We separately consider the expression for $Y'(t)$:

$$Y' = (Z' + \beta Z)e^{\beta t} = ((1-\alpha)\Phi^{-\alpha}\Phi' + \beta\Phi^{1-\alpha})e^{\beta t}$$
$$= (\alpha - 1)e^{\beta t}\Phi^{-\alpha}\left(\frac{\beta}{\alpha-1}\Phi - \Phi'\right). \tag{A.36}$$

Assume that at the initial time moment the following condition is fulfilled:

$$\Phi'(0) > \frac{\beta}{\alpha-1}\Phi(0). \tag{A.37}$$

Since $\Phi(t) \in \mathbb{C}^{(1)}([0,T])$, there exists a time moment $t_2 \in (0, t_1]$ such that

$$\Phi'(t) \geq \frac{\beta}{\alpha-1}\Phi(t) \tag{A.38}$$

for $t \in [0, t_2]$. If inequality (A.38) is valid, then from (A.36) we obtain

$$Y'(t) \leq 0 \quad \text{for all } t \in [0, t_2]. \tag{A.39}$$

This and inequality (A.35) imply the inequality

$$Y'' - \delta Y \leq 0 \quad \text{for } t \in [0, t_2], \quad \delta = (\alpha - 1)\gamma. \tag{A.40}$$

Multiplying both sides of inequality (A.40) by Y', we obtain the inequality

$$\frac{1}{2}\frac{d}{dt}(Y'(t))^2 \geq \frac{\delta}{2}\frac{d}{dt}(Y(t))^2 \quad \text{for } t \in [0, t_2]. \tag{A.41}$$

Integrating (A.41) and taking into account the initial condition, we obtain

$$(Y'(t))^2 \geq A^2 + \delta(Y(t))^2, \tag{A.42}$$

where

$$A^2 \equiv (Y'(0))^2 - \delta(Y(0))^2. \tag{A.43}$$

We have

$$Y'(0) = (\alpha - 1)\Phi^{-\alpha}(0)\left[\frac{\beta}{\alpha-1}\Phi(0) - \Phi'(0)\right], \quad Y(0) = \Phi^{1-\alpha}(0).$$

Then

$$A^2 = (\alpha-1)^2\Phi^{-2\alpha}(0)\left[\left(\Phi'(0) - \frac{\beta}{\alpha-1}\Phi(0)\right)^2 - \frac{\gamma}{\alpha-1}(\Phi(0))^2\right]. \tag{A.44}$$

Assume that $A^2 > 0$; taking into account (A.44), we obtain the condition for the initial data of the problem:

$$\Phi'(0) > \frac{\beta}{\alpha - 1}\Phi(0) + \left(\frac{\gamma}{\alpha - 1}\right)^{1/2}\Phi(0). \tag{A.45}$$

Under condition (A.45), we obtain from (A.42) the following chain of inequalities:

$$\begin{aligned}(Y'(t))^2 \geq A^2 + \delta(Y(t))^2 \quad &\Rightarrow \quad (Y'(t))^2 \geq A^2 \quad \Rightarrow \\ \Rightarrow \quad -Y'(t) \geq A \quad &\Rightarrow \quad Y'(t) \leq -A \quad \text{for all } t \in [0, t_2].\end{aligned} \tag{A.46}$$

From the last inequality of this chain we obtain

$$\begin{aligned}(\alpha - 1)e^{-\beta t}\Phi^{-\alpha}(t)\left[\frac{\beta}{\alpha - 1}\Phi(t) - \Phi'(t)\right] \leq -A \quad &\Rightarrow \\ \Rightarrow \quad \Phi'(t) - \frac{\beta}{\alpha - 1}\Phi(t) \geq \frac{A}{\alpha - 1}e^{\beta t}\Phi^{\alpha}(t) > 0&\end{aligned} \tag{A.47}$$

for all $t \in [0, t_2]$, where the last inequality is a consequence of the following facts: first, $\Phi(0) > 0$, and second, $\Phi'(t) \geq 0$ for $t \in [0, t_2]$. Therefore, if $t_2 < T$, then, for $t = t_2$, we have

$$\Phi'(t_2) > \frac{\beta}{\alpha - 1}\Phi(t_2),$$

which, in particular, implies

$$\Phi'(t_2) > 0.$$

We choose $t = t_2$ as the initial time moment; then due to the fact that $\Phi(t) \in \mathbb{C}^{(1)}([0, T])$, there exists a moment $t_3 > t_2$ such that

$$\Phi'(t) \geq \frac{\beta}{\alpha - 1}\Phi(t) \quad \text{for all } t \in [0, t_3];$$

in particular, $\Phi'(t) \geq 0$ for all $t \in [0, t_3]$. Continuing this process of extending solutions in time, we arrive at the inequality

$$Y'(t) \leq -A \quad \text{for all } t \in [0, T]. \tag{A.48}$$

Integrating this inequality by time, we obtain

$$Y(t) \leq Y(0) - At \quad \text{for all } t \in [0, T]. \tag{A.49}$$

We have

$$e^{\beta t}\Phi^{1-\alpha}(t) \leq \Phi_0^{1-\alpha} - At \quad \Rightarrow \quad \Phi^{1-\alpha}(t) \leq e^{-\beta t}[\Phi_0^{1-\alpha} - At] \quad \Rightarrow$$

$$\Rightarrow \quad \Phi^{\alpha-1} \geq \frac{e^{\beta t}}{\Phi_0^{1-\alpha} - At} \quad \Rightarrow$$

$$\Rightarrow \quad \Phi(t) \geq \frac{\exp\{\beta/(\alpha-1)t\}}{[\Phi_0^{1-\alpha} - At]^{1/(\alpha-1)}} \quad \text{for all } t \in [0, T]. \qquad \text{(A.50)}$$

Thus, we have proved the following theorem.

Theorem A.2. *Let a nonnegative function $\Phi(t) \in \mathbb{C}^{(2)}([0,T])$ for certain $T > 0$ satisfy the integro-differential inequality* (A.21), *let the function $h(t)$ satisfy the conditions specified above, and let the following initial conditions be fulfilled*:

$$\Phi'(0) > \frac{\beta}{\alpha-1}\Phi(0) + \left(\frac{\gamma}{\alpha-1}\right)^{1/2}\Phi(0), \quad \Phi(0) > 0. \qquad \text{(A.51)}$$

Then $T \neq +\infty$. Moreover,

$$\Phi(t) \geq \frac{\exp\{\beta/(\alpha-1)t\}}{[\Phi_0^{1-\alpha} - At]^{1/(\alpha-1)}}, \qquad \text{(A.52)}$$

where

$$\alpha > 1, \quad \beta \geq 0, \quad \gamma = \gamma_1 + \frac{h_0\gamma_2}{a}, \quad \gamma_1 \geq 0, \quad \gamma_2 \geq 0,$$

$$A \equiv (\alpha-1)\Phi_0^{-\alpha}\left[\left(\Phi'(0) - \frac{\beta}{\alpha-1}\Phi_0\right)^2 - \frac{\gamma}{\alpha-1}\Phi_0^2\right] > 0, \quad \Phi_0 = \Phi(0).$$

Remark A.1. We consider a generalization of the conditions for the function $h(t)$. Assume that there exists a function

$$\chi(t) \in \mathbb{C}^{(1)}(\mathbb{R}^1_+)$$

such that

$$|h(t)| \leq \chi'(t) \quad \text{for all } t \in \mathbb{R}^1_+ \qquad \text{(A.53)}$$

and, moreover,

$$\chi(t) \leq 0 \quad \text{for } t \geq 0. \qquad \text{(A.54)}$$

Taking into account inequality (A.53), we obtain the following chain of relations:

$$\int_0^t ds\, h(t-s)\Phi(s) \leq \int_0^t ds\, |h(t-s)|\Phi(s)$$

$$\leq \int_0^t ds\chi'(t-s)\Phi(s) = -\int_0^t d\chi(t-s)\Phi(s)$$

$$= -\chi(t-s)\Phi(s)\Big|_{s=0}^{s=t} + \int_0^t ds\, \chi(t-s)\Phi'(s)$$

$$= -\chi(0)\Phi(t) + \chi(t)\Phi(0) + \int_0^t ds\, \chi(t-s)\Phi'(s). \qquad \text{(A.55)}$$

Repeating the reasoning for the function

$$\chi(t) = -\frac{h_0}{a}e^{-at}, \quad a > 0,\ h_0 > 0,$$

we obtain an analog of Theorem A.2, in which we must replace

$$\gamma = \gamma_1 + \frac{h_0\gamma_2}{a}$$

by

$$\gamma = \gamma_1 + \gamma_2|\chi(0)|. \qquad \text{(A.56)}$$

Now we consider the following integro-differential inequality:

$$\Phi\Phi'' - \alpha(\Phi')^2 + \gamma_1\Phi^2 + \gamma_3\Phi^{1+\lambda} + \gamma_2\int_0^t ds\, h(t-s)\Phi(s)\Phi(t) \geq 0, \qquad \text{(A.57)}$$

where $\alpha > 1$ and $\gamma_1, \gamma_2, \gamma_3, \lambda \geq 0$. Arguing as in the proof of the differential inequality (A.29), we obtain from (A.57) the inequality

$$\Phi\Phi'' - \alpha(\Phi')^2 + \gamma_4\Phi^2 + \gamma_3\Phi^{1+\lambda} \geq 0, \quad t \in [0, t_1],\ t_1 \in (0, T], \qquad \text{(A.58)}$$

where

$$\gamma_4 = \gamma_1 + \gamma_2|\chi(0)|.$$

Moreover, if $\Phi'(0) > 0$, then

$$\Phi'(t) \geq 0 \quad \text{for } t \in [0, t_1]. \qquad \text{(A.59)}$$

In addition, we assume that $\Phi(0) > 0$. Then inequality (A.58) implies the following chain of inequalities:

$$\frac{\Phi''}{\Phi^\alpha} - \alpha\frac{(\Phi')^2}{\Phi^{1+\alpha}} + \gamma_4\Phi^{1-\alpha} + \gamma_3\Phi^{\lambda-\alpha} \geq 0 \quad \Rightarrow$$

$$\Rightarrow \quad \left[\frac{\Phi'}{\Phi^\alpha}\right]' + \gamma_4\Phi^{1-\alpha} + \gamma_3\Phi^{\lambda-\alpha} \geq 0 \quad \Rightarrow$$

$$\begin{aligned}
&\Rightarrow \quad \frac{1}{1-\alpha}(\Phi^{1-\alpha})'' + \gamma_4\Phi^{1-\alpha} + \gamma_3\Phi^{\lambda-\alpha} \geq 0 \quad \Rightarrow \\
&\Rightarrow \quad Z(t) \equiv \Phi^{1-\alpha}, \quad -\frac{1}{\alpha-1}Z'' + \gamma_4 Z + \gamma_3 Z^{(\lambda-\alpha)/(1-\alpha)} \geq 0 \quad \Rightarrow \\
&\Rightarrow \quad Z'' \leq (\alpha-1)\gamma_4 Z + (\alpha-1)\gamma_3 Z^{(\lambda-\alpha)/(1-\alpha)}.
\end{aligned} \tag{A.60}$$

Note that

$$Z' = (1-\alpha)\Phi^{-\alpha}\Phi' \leq 0 \quad \text{for } t \in [0, t_1].$$

Therefore, from the final inequality in (A.60) we obtain

$$Z'Z'' \geq (\alpha-1)\gamma_4 ZZ' + (\alpha-1)\gamma_3 Z'Z^{(\lambda-\alpha)/(1-\alpha)}.$$

Thus, we conclude that

$$\frac{1}{2}\frac{d}{dt}(Z')^2 \geq \frac{(\alpha-1)\gamma_4}{2}\frac{d}{dt}Z^2 + \frac{(\alpha-1)\gamma_3}{\delta}\frac{d}{dt}Z^\delta, \tag{A.61}$$

where

$$\delta = 1 + \frac{\lambda-\alpha}{1-\alpha}. \tag{A.62}$$

Assuming that $\delta > 0$, we have

$$\begin{aligned}
\delta > 0 \quad &\Rightarrow \quad 1 + \frac{\lambda-\alpha}{1-\alpha} > 0 \quad \Rightarrow \quad \frac{1+\lambda-2\alpha}{1-\alpha} > 0 \quad \Rightarrow \\
&\Rightarrow \quad \frac{2\alpha-1-\lambda}{\alpha-1} > 0 \quad \Rightarrow \quad \lambda < 2\alpha - 1.
\end{aligned}$$

Moreover, assume that

$$\lambda < 2\alpha - 1. \tag{A.63}$$

Then, integrating inequality (A.61) by $t \in [0, t_1]$, we obtain

$$(Z')^2 \geq (\alpha-1)\gamma_4 Z^2 + \frac{2(\alpha-1)\gamma_3}{\delta}Z^\delta + A^2, \tag{A.64}$$

where

$$A^2 \equiv (Z'(0))^2 - (\alpha-1)\gamma_4(Z(0))^2 - \frac{2(\alpha-1)\gamma_3}{\delta}(Z(0))^\delta, \quad \delta = 1 + \frac{\lambda-\alpha}{1-\alpha}.$$

Assume that at the initial time moment the function $\Phi(t)$ satisfies the condition $A^2 > 0$:

$$\begin{aligned}
A^2 &= (\alpha-1)^2\Phi^{-2\alpha}(0)(\Phi'(0))^2 - (\alpha-1)\gamma_4(\Phi(0))^{2-2\alpha} \\
&\quad - \frac{2(\alpha-1)\gamma_3}{\delta}(\Phi(0))^{\delta-\delta\alpha} = (\alpha-1)^2\Phi^{-2\alpha}(0)
\end{aligned}$$

$$\times\left[(\Phi'(0))^2 - \frac{\gamma_4}{\alpha-1}(\Phi(0))^2 - \frac{2\gamma_3}{(\alpha-1)\delta}(\Phi(0))^{\delta(1-\alpha)+2\alpha}\right]. \quad (A.65)$$

Since

$$\delta(1-\alpha) + 2\alpha = 1 - \alpha + \lambda - \alpha - 2\alpha = 1 + \lambda,$$

from (A.65) we obtain the following condition, which is equivalent to $A^2 > 0$:

$$\Phi'(0) > \left(\frac{\gamma_4}{\alpha-1}(\Phi(0))^2 + \frac{2\gamma_3}{(\alpha-1)\delta}(\Phi(0))^{1+\lambda}\right)^{1/2}. \quad (A.66)$$

If (A.66) holds, then from inequality (A.64) we obtain

$$\begin{aligned}(Z')^2 \geq A^2 \quad &\Rightarrow \quad |Z'| \geq A \quad \Rightarrow \quad -Z' \geq A \quad \Rightarrow \quad Z' \leq -A \quad \Rightarrow \\ &\Rightarrow \quad (1-\alpha)\Phi^{-\alpha}\Phi' \leq -A \quad \Rightarrow \quad \Phi' \geq \frac{A}{\alpha-1}\Phi^{\alpha}(t) > 0\end{aligned} \quad (A.67)$$

for $t \in [0, t_1]$ since

$$\Phi(0) > 0, \quad \Phi'(t) \geq 0 \quad \text{for } t \in [0, t_1].$$

Continuing similar reasonings, we obtain the estimate

$$\Phi'(t) \geq 0 \quad \text{for all } t \in [0, T];$$

then (A.67) implies the following chain of inequalities:

$$\begin{aligned}Z(t) \leq Z(0) - At \quad &\Rightarrow \quad \Phi^{1-\alpha}(t) \leq \Phi_0^{1-\alpha} - At \quad \Rightarrow \\ &\Rightarrow \quad \Phi^{\alpha-1}(t) \geq \frac{1}{\Phi_0^{1-\alpha} - At} \quad \Rightarrow \quad \Phi(t) \geq \frac{1}{[\Phi_0^{1-\alpha} - At]^{1/(\alpha-1)}}.\end{aligned}$$

This implies that $T \neq +\infty$.

Theorem A.3. *Assume that, for certain $T > 0$, a nonnegative function $\Phi(t) \in \mathbb{C}^{(2)}([0, T])$ satisfies the integro-differential inequality* (A.57), *conditions* (A.53)–(A.54) *for the functions $h(t)$ are fulfilled, and the initial conditions*

$$\Phi'(0) > \left(\frac{\gamma_4}{\alpha-1}(\Phi(0))^2 + \frac{2\gamma_3}{(\alpha-1)\delta}(\Phi(0))^{1+\lambda}\right)^{1/2}, \quad \Phi(0) > 0, \quad (A.68)$$

and the condition

$$0 \leq \lambda < 2\alpha - 1 \quad (A.69)$$

hold. Then $T \neq +\infty$. Moreover,

$$\Phi(t) \geq \frac{1}{[\Phi_0^{1-\alpha} - At]^{1/(\alpha-1)}}, \quad (A.70)$$

where

$$\alpha > 1, \quad \gamma_3 \geq 0, \quad \gamma_4 = \gamma_1 + \gamma_2 |\chi(0)|,$$
$$\gamma_1 \geq 0, \quad \gamma_2 \geq 0, \quad \delta = 1 + \frac{\lambda - \alpha}{1 - \alpha},$$

and

$$A^2 = (\alpha - 1)^2 \Phi^{-2\alpha}(0) \times \left[(\Phi'(0))^2 - \frac{\gamma_4}{\alpha - 1} (\Phi(0))^2 - \frac{2\gamma_3}{(\alpha - 1)\delta} (\Phi(0))^{\delta(1-\alpha)+2\alpha} \right].$$

A.3 Differential inequality III

Consider the following ordinary integro-differential inequality:

$$\Phi\Phi'' - \alpha(\Phi')^2 + \beta\Phi^2 + \gamma T_1 \int_0^t ds\, \Phi(s)\Phi(t) \geq 0, \quad t \in [0, T_1], \tag{A.71}$$

where

$$\alpha > 1, \quad \beta \geq 0, \quad \gamma \geq 0.$$

Let the function $\Phi(t)$ belong to the class $\mathbb{C}^{(2)}([0, T_\infty))$, and let $\Phi(t)$ be a nonextendable solution on the maximal time interval $T_\infty > 0$, that is, either $T_\infty = +\infty$ or $T_\infty < +\infty$, and in the latter case,

$$\limsup_{t \to T_\infty - 0} \Phi(t) = +\infty.$$

Now we obtain sufficient conditions for $\Phi(0)$ and $\Phi'(0)$ under which, for any fixed $T_1 > 0$, the following limit relation holds:

$$\limsup_{t \uparrow T_\infty} \Phi(t) = +\infty, \quad T_\infty \in (0, T_1].$$

We have the following integration-by-parts formula:

$$\int_0^t ds\, \Phi(s) = s\Phi(s)\Big|_{s=0}^{s=t} - \int_0^t ds\, s\Phi'(s). \tag{A.72}$$

Then, under the initial condition $\Phi'(0) > 0$, there exists a time moment $t_1 \in (0, T_\infty)$ such that

$$\Phi'(t) \geq 0 \quad \text{for } t \in [0, t_1]; \tag{A.73}$$

therefore, taking into account (A.72), from inequality (A.71) we obtain the inequality

$$\Phi\Phi'' - \alpha(\Phi')^2 + [\beta + \gamma T_1^2]\Phi^2 \geq 0, \quad t \in [0, t_1]. \tag{A.74}$$

Dividing both sides of inequality (A.74) by $\Phi^{1+\alpha}$ and taking into account the condition $\alpha > 1$, we obtain

$$\begin{aligned} &\frac{\Phi''}{\Phi^\alpha} - \alpha\frac{(\Phi')^2}{\Phi^{1+\alpha}} + (\beta + \gamma T_1^2)\Phi^{1-\alpha} \geq 0 \quad \Rightarrow \\ &\Rightarrow \quad \left(\frac{\Phi'}{\Phi^\alpha}\right)' + (\beta + \gamma T_1^2)\Phi^{1-\alpha} \geq 0 \quad \Rightarrow \\ &\Rightarrow \quad \frac{1}{1-\alpha}(\Phi^{1-\alpha})'' + (\beta + \gamma T_1^2)\Phi^{1-\alpha} \geq 0. \end{aligned} \tag{A.75}$$

Introduce the notation

$$Z(t) = \Phi^{1-\alpha} \quad \text{for } \alpha > 1; \tag{A.76}$$

then from (A.75) we obtain the inequality

$$Z''(t) \leq (\alpha - 1)(\beta + \gamma T_1^2)Z(t) \quad \text{for } t \in [0, t_1]. \tag{A.77}$$

Note that

$$Z'(t) = (1 - \alpha)\Phi^{-\alpha}(t)\Phi'(t) \leq 0 \quad \text{for } t \in [0, t_1], \tag{A.78}$$

since $\alpha > 1$ and inequality (A.73) holds. Multiplying both sides of inequality (A.77) by $Z'(t)$, we obtain the following chain of inequalities:

$$\begin{aligned} &Z'Z'' \geq (\alpha - 1)(\beta + \gamma T_1^2)ZZ' \quad \Rightarrow \\ &\Rightarrow \quad \frac{1}{2}\frac{d}{dt}(Z')^2 \geq \frac{(\alpha - 1)(\beta + \gamma T_1^2)}{2}\frac{d}{dt}Z^2 \quad \Rightarrow \\ &\Rightarrow \quad (Z')^2 \geq A^2 + Z^2 \geq A^2, \end{aligned} \tag{A.79}$$

where

$$A^2 \equiv (Z'(0))^2 - (\alpha - 1)(\beta + \gamma T_1^2)(Z(0))^2. \tag{A.80}$$

Now we assume that, for certain fixed $T_1 > 0$, the inequality $A^2 > 0$ holds. Then we arrive at the inequality

$$\begin{aligned} A^2 \equiv {} & (\alpha - 1)^2\Phi(0)^{-2\alpha}(\Phi'(0))^2 \\ & - (\alpha - 1)(\beta + \gamma T_1^2)\Phi(0)^{-2\alpha}(\Phi(0))^2. \end{aligned} \tag{A.81}$$

Assume that $\Phi(0) > 0$; then from (A.81) we obtain the inequalities

$$\Phi'(0) > \left(\frac{\beta + \gamma T_1^2}{\alpha - 1}\right)^{1/2} \Phi(0), \quad \Phi(0) > 0. \tag{A.82}$$

Due to (A.79), we have the following inequalities:

$$\begin{aligned} (Z')^2 \geq A^2 \quad &\Rightarrow \quad |Z'| \geq A \quad \Rightarrow \\ &\Rightarrow \quad Z' \leq -A \quad \Rightarrow \quad (1-\alpha)\Phi^{-\alpha}(t)\Phi'(t) \leq -A < 0. \end{aligned} \tag{A.83}$$

Since $\Phi'(t) \geq 0$ for $t \in [0, t_1]$, we conclude that $\Phi(t) \geq \Phi(0) > 0$, and (A.83) implies the inequality

$$\Phi'(t) \geq \frac{A}{\alpha - 1}\Phi^{\alpha}(t) \geq \frac{A}{\alpha - 1}\Phi^{\alpha}(0) > 0 \quad \text{for } t \in [0, t_1].$$

Therefore

$$\Phi'(t_1) > 0.$$

Using the same procedure, we obtain

$$\Phi'(t) > 0 \quad \text{for } t \in [0, T_\infty).$$

Then from inequality (A.79) we deduce

$$Z(t) \leq Z(0) - At \quad \text{for } t \in [0, T_\infty).$$

Taking into account definition (A.76), for the function $Z(t)$, we obtain the following inequality:

$$\Phi(t) \geq \frac{1}{[\Phi^{1-\alpha}(0) - At]^{1/(\alpha-1)}} \quad \text{for } t \in [0, T_\infty), \tag{A.84}$$

which, obviously, implies the inequality

$$T_\infty \in \left(0, \frac{\Phi^{1-\alpha}(0)}{A}\right], \tag{A.85}$$

and, moreover,

$$\limsup_{t \uparrow T_\infty} \Phi(t) = +\infty.$$

Under the condition

$$\frac{\Phi^{1-\alpha}(0)}{A} \leq T_1,$$

we obtain the following chain of inequalities:

$$\begin{aligned}
&\Phi^{2-2\alpha}(0)T_1^{-2} \leq A^2 \quad \Rightarrow \\
&\Rightarrow \quad \Phi^{2-2\alpha}(0)T_1^{-2} \leq (\alpha-1)^2\Phi(0)^{-2\alpha}(\Phi'(0))^2 \\
&\qquad - (\alpha-1)(\beta+\gamma T_1^2)\Phi(0)^{-2\alpha}(\Phi(0))^2 \quad \Rightarrow \\
&\Rightarrow \quad (\Phi'(0))^2 \geq \frac{\beta+\gamma T_1^2}{\alpha-1}(\Phi(0))^2 + \frac{1}{T_1^2(\alpha-1)^2}(\Phi(0))^2,
\end{aligned} \tag{A.86}$$

which implies the required conditions for the initial data of the problem:

$$\Phi'(0) \geq \left(\frac{\beta+\gamma T_1^2}{\alpha-1} + \frac{1}{T_1^2(\alpha-1)^2}\right)^{1/2}\Phi(0), \quad \Phi(0) > 0. \tag{A.87}$$

Thus, we conclude that the lifetime of a strong generalized solution T_∞ satisfies the inequality

$$T_\infty \leq T_1.$$

We have proved the following theorem.

Theorem A.4. *Let $\Phi(t) \in \mathbb{C}^{(2)}[0, T_\infty)$ be a nonextendable solution of the integro-differential inequality* (A.71) *for certain maximal $T_\infty > 0$, and, moreover, let the following conditions be fulfilled:*

$$\Phi'(0) \geq \left(\frac{\beta+\gamma T_1^2}{\alpha-1} + \frac{1}{T_1^2(\alpha-1)^2}\right)^{1/2}\Phi(0), \quad \Phi(0) > 0,\ T_1 > 0. \tag{A.88}$$

Then the function $\Phi(t)$ is discontinuous at the point $T_\infty \in (0, T_1]$:

$$\limsup_{t\uparrow T_\infty} \Phi(t) = +\infty. \tag{A.89}$$

A.4 Auxiliary results of the theory of vector fields

Let $\Omega \subset \mathbb{R}^3$ be a bounded Lipschitzian domain with smooth boundary $\partial\Omega \in \mathbb{C}^{2,\delta}$, $\delta \in (0,1]$. Denote by $\mathbb{L}^p(\Omega)$, $1 < p < +\infty$ (respectively, by $\mathbb{L}^\infty(\Omega)$) the space of real-valued functions defined on Ω and absolutely integrable with power p (respectively, essentially bounded) with respect to the Lebesgue measure $dx = dx_1dx_2dx_3$. These spaces with the norm

$$\|u\|_p = \left(\int_\Omega |u(x)|^p dx\right)^{1/p}$$

and

$$\|u\|_\infty = \operatorname{ess.sup}_{\Omega} |u(x)|,$$

respectively, are Banach spaces. Introduce the Banach spaces of vector-valued functions:

$$\mathbf{u} \in \mathbf{L}^q(\Omega) = \mathbb{L}^q(\Omega) \times \mathbb{L}^q(\Omega) \times \mathbb{L}^q(\Omega), \quad q \in [1, +\infty],$$

where

$$|\mathbf{u}| = (\mathbf{u}, \mathbf{u})^{1/2}, \quad (\mathbf{u}, \mathbf{v}) = u_1 v_1 + u_2 v_2 + u_3 v_3,$$
$$\mathbf{u} = (u_1, u_2, u_3), \quad \mathbf{v} = (v_1, v_2, v_3).$$

We denote by $(\mathbf{u}, \mathbf{v})_2$ the scalar product in $\mathbf{L}^2(\Omega)$. Let $\mathbf{H}_0^1(\Omega)$ be the space of vector-valued functions $\mathbf{u} = (u_1, u_2, u_3)$ such that $u_k \in \mathbb{H}_0^1(\Omega)$, $k = \overline{1, 3}$. We can prove that

$$\mathbf{H}_0^1(\Omega) = \mathbb{H}_0^1(\Omega) \times \mathbb{H}_0^1(\Omega) \times \mathbb{H}_0^1(\Omega),$$

so that $\mathbf{H}_0^1(\Omega)$ is a Hilbert space with respect to the scalar product

$$((\mathbf{u}, \mathbf{v})) = \sum_{i=1}^{3} (D_i \mathbf{u}, D_i \mathbf{v})_2, \quad D_i = \frac{\partial}{\partial x_i},$$

with the norm

$$\|\mathbf{u}\|_{+1} = \left(\sum_{j=1}^{3} \|D_j \mathbf{u}\|_2^2 \right)^{1/2}.$$

We denote by $\mathbf{H}^{-1}(\Omega)$ the dual space of the Hilbert space $\mathbf{H}_0^1(\Omega)$:

$$(\mathbf{H}_0^1(\Omega))' = \mathbf{H}^{-1}(\Omega);$$

its norm is denoted by $\|\mathbf{u}\|_{-1}$. Moreover, we will also use the space

$$\mathbf{D}(\Omega) = \mathcal{D}(\Omega) \times \mathcal{D}(\Omega) \times \mathcal{D}(\Omega),$$

where $\mathcal{D}(\Omega)$ is the space of compactly supported test functions: $\mathbf{u} \in \mathbf{D}(\Omega)$ if $u_k \in \mathcal{D}(\Omega)$, $k = 1, 2, 3$, $\mathbf{u} = (u_1, u_2, u_3)$. Introduce the space

$$\mathbf{V}(\Omega) = (\mathbf{u} \in \mathbf{D}(\Omega) : \operatorname{div} \mathbf{u} = 0).$$

We denote the closure of the space $\mathbf{V}(\Omega)$ in the space $\mathbf{H}_0^1(\Omega)$ by $\mathbf{H}(\Omega)$ and the closure of the space $\mathbf{V}(\Omega)$ in the space $\mathbf{L}^2(\Omega)$ by $\mathbf{X}(\Omega)$. The space $\mathbf{H}(\Omega)$ is densely embedded in $\mathbf{X}(\Omega)$; moreover, the embedding is continuous.

We denote by $\mathbf{H}'(\Omega)$ the dual space of the Hilbert space $\mathbf{H}(\Omega)$.

Recall the following important fact (see, e. g., [54]).

Theorem A.5. *Let Ω be an open bounded Lipschitzian set. Then*

$$\mathbf{H}(\Omega) = (\mathbf{u} \in \mathbf{H}_0^1(\Omega), \operatorname{div} \mathbf{u} = 0).$$

A.5 Theorem on nonextendable solution

Consider the abstract Cauchy problem

$$\frac{dy}{dt} = A(y), \quad y(0) = y_0, \tag{A.90}$$

in a Banach space $\mathbb{B}$ with norm $\|\cdot\|$, $y_0 \in \mathbb{B}$. A solution of the Cauchy problem (A.90) is a function

$$y \in \mathbb{C}^{(1)}([0, T_0), \mathbb{B}), \quad T_0 > 0.$$

We treat the equation in the sense of strong differentiation in the Banach space $\mathbb{B}$ for all $t \in [0, T_0)$. The following important theorem was proved in [79].

Theorem A.6. *Let an operator $A(y)$ be locally Lipschitz-continuous, that is, there exists a function $\mu : [0, +\infty) \to [0, +\infty)$, which is bounded on any bounded subset of the ray $[0, +\infty)$ and is such that, for any $z_1, z_2 \in \mathbb{B}$,*

$$\|A(z_1) - A(z_2)\| \le \mu(\max(\|z_1\|, \|z_2\|))\|z_1 - z_2\|. \tag{A.91}$$

Then there exists a unique solution such that either $T_0 = +\infty$ or $T_0 < +\infty$, and in the latter case,

$$\lim_{t \uparrow T_0} \|y(t)\| = +\infty. \tag{A.92}$$

Such a solution is said to be nonextendable; each solution can be extended to a solution satisfying this condition. In other words, any solution is a restriction of a certain nonextendable solution to a narrower time interval.

A.6 Examples of dispersion blow-up in linear equations

The examples presented in this section are taken from [5–8, 65, 102].

Example A.1 (inversely parabolic equation). Consider the following Cauchy problem for the linear inversely parabolic equation:

$$\frac{\partial u}{\partial t} + \Delta u = 0, \quad u(x, 0) = u_0(x), \quad t \in [0, T),\ x \in \mathbb{R}^N,\ T > 0. \tag{A.93}$$

We take the initial function

$$u_0(x) = \frac{1}{T^{N/2}} \exp(-|x|^2/4T);$$

it is easy to verify that the function

$$u(x, t) = \frac{1}{(T - t)^{N/2}} \exp(-|x|^2/[4(T - t)]) \tag{A.94}$$

is the solution of the Cauchy problem (A.93). From the explicit form of the function (A.94) we conclude that

$$\lim_{t\to T^-} u(0,t) = +\infty,$$

that is, a solution of the Cauchy problem (A.93) blows up for a finite time $T > 0$.

Example A.2 (parabolic equation). Consider the following Cauchy problem for the linear parabolic equation:

$$\frac{\partial u}{\partial t} = \Delta u, \quad u(x,0) = u_0(x), \quad x \in \mathbb{R}^N,\ t \in [0,T), \tag{A.95}$$

whose initial function *does not belong* to the Tikhonov–Teclind class (see, e. g., [54]):

$$u_0(x) \geq A\exp(\gamma|x|^2), \quad \gamma > 0,\ A > 0. \tag{A.96}$$

We assume that

$$T < T_0 = \frac{1}{4\gamma}.$$

Introduce the following test function:

$$\varphi(t,x) = \varphi_T(t,x) = \frac{1}{(T-t)^{N/2}}\exp\left(-\frac{|x|^2}{4(T-t)}\right); \tag{A.97}$$

it satisfies the linear, inversely parabolic equation (A.93) on the interval $[0,T)$. Introduce the functional

$$\psi(t,T) \equiv \int_{\mathbb{R}^N} u(x,t)\varphi(t,x)\,dx, \quad t \in [0,T). \tag{A.98}$$

First, we prove that this functional is well defined for $t \in [0,T]$ under the condition that there exists a solution $u(x,t)$ of problem (A.95) for $t \in [0,T]$, which is bounded in a small semineighborhood of the point $(x,t) = (0,T)$. Indeed, after the change of variables

$$z = \frac{x}{(T-t)^{1/2}}, \quad t \in [0,T],$$

we obtain the equality

$$\psi(t,T) = \int_{\mathbb{R}^N} u(z(T-t)^{1/2},t)\exp(-|z|^2)\,dz;$$

this function is bounded on the segment $t \in [0,T]$ if the function $u(x,t)$ is bounded in a small semineighborhood of the point $(x,t) = (0,T)$.

We differentiate the function $\psi(t,T)$ with respect to time $t \in (0,T)$:

$$\begin{aligned}\frac{d\psi(t,T)}{dt} &= \int_{\mathbb{R}^N} u_t(x,t)\varphi(t,x)\,dx + \int_{\mathbb{R}^N} u(x,t)\varphi_t(t,x)\,dx\\ &= \int_{\mathbb{R}^N} \Delta u(x,t)\varphi(t,x)\,dx - \int_{\mathbb{R}^N} u(x,t)\Delta\varphi(t,x)\,dx\\ &= \int_{\mathbb{R}^N} u(x,t)\Delta\varphi(t,x)\,dx - \int_{\mathbb{R}^N} u(x,t)\Delta\varphi(t,x)\,dx = 0. \end{aligned} \tag{A.99}$$

Therefore, we arrive at the well-known identity (see, e. g., [23]):

$$\begin{aligned}\psi(t,T) = \psi(0,T) \quad \Rightarrow \quad \int_{\mathbb{R}^N} u(x,t)\varphi(t,x)\,dx &= \int_{\mathbb{R}^N} u_0(x)\varphi(0,x)\,dx\\ &\geq A\int_{\mathbb{R}^N} \exp(\gamma|x|^2 - |x|^2/(4T))\,dx\\ &= A\left(\frac{4T}{1-4T\gamma}\right)^{N/2}\int_{\mathbb{R}^N} e^{-|z|^2}\,dz. \end{aligned} \tag{A.100}$$

The last equality implies

$$\lim_{T\to T_1^-} \psi(0,T) = +\infty, \quad T_1 \leq T_0 = \frac{1}{4\gamma}.$$

Therefore, the solution of the linear problem (A.95)–(A.96) is undefined for $T \geq T_0$. On the other hand, if there exists a nonextendable solution of problem (A.95)–(A.96) on the maximal interval $[0,T_2)$, then there exists a sequence $(x_n,t_n) \to (0,T_2^-)$ such that the following limit property holds:

$$\lim_{n\to+\infty} |u(x_n,t_n)| = +\infty \quad \text{for } T_2 \leq T_1 \leq T_0.$$

Remark A.2. In the proof of nonexistence of a global solution to problem (A.95)–(A.96), we in fact used the method of test functions with a specific choice of a test function, as Fujita did (see Sec. 3.2). However, this result can be obtained by a direct calculation. Indeed, we have the following formal chain of inequalities:

$$\begin{aligned}u(0,t) &= \frac{1}{(4\pi t)^{N/2}}\int_{\mathbb{R}^N} \exp\left(-\frac{1}{4t}|y|^2\right)u_0(y)\,dy\\ &\geq A\frac{1}{(4\pi t)^{N/2}}\int_{\mathbb{R}^N} \exp(-[1/(4t)-\gamma]|y|^2)\,dy = \frac{c_0}{(1-4\gamma t)^{N/2}}.\end{aligned}$$

This immediately implies the nonexistence of solutions of the problem for $t \geq (4\gamma)^{-1}$.

Example A.3 (linear Korteweg–de Vries equation). We recall the well-known result of Benjamin, Bona, and Mahony. Consider the Cauchy problem

$$\frac{\partial u}{\partial t} + \frac{\partial^3 u}{\partial x^3} = 0, \quad u(x,0) = u_0(x), \quad x \in \mathbb{R}^N,\ t > 0, \tag{A.101}$$

$$u_0(x) = \frac{\mathrm{Ai}(-x)}{(1+|x|^2)^m}, \quad \frac{1}{8} < m < \frac{1}{4}. \tag{A.102}$$

Here $\mathrm{Ai}(z)$ is the Airy function

$$\mathrm{Ai}(z) \equiv \frac{1}{\pi} \int_0^{+\infty} \cos\left(\frac{t^3}{3} + zt\right) dt.$$

We can verify that $u_0(x) \in \mathbb{L}^\infty(\mathbb{R}^1) \cap \mathbb{L}^2(\mathbb{R}^1) \cap \mathbb{C}^\infty(\mathbb{R}^1)$. Then the solution of the Cauchy problem (A.101)–(A.102) has the form

$$u(x,t) = \frac{c_0}{t^{1/3}} \int_{\mathbb{R}^1} \mathrm{Ai}\left(\frac{x-y}{t^{1/3}}\right) \frac{\mathrm{Ai}(-y)}{(1+y^2)^m}\, dy. \tag{A.103}$$

Using properties of the Airy function, we obtain

$$\lim_{(x,t)\to(0,1)} u(x,t) = c_0 \int_{\mathbb{R}^1} \frac{\mathrm{Ai}^2(-y)}{(1+y^2)^m}\, dy = +\infty,$$

that is, the solution of the Cauchy problem (A.101)–(A.102) blows up for a finite time.

Example A.4 (linear Schrödinger equation). Consider the following Cauchy problem:

$$i\frac{\partial u}{\partial t} + \Delta u = 0, \quad u(x,0) = u_0(x), \quad x \in \mathbb{R}^N,\ t > 0, \tag{A.104}$$

where

$$u_0(x) = \frac{e^{-i|x|^2}}{(1+|x|^2)^m}, \quad \frac{N}{4} < m \leq \frac{N}{2}. \tag{A.105}$$

Clearly, $u_0(x) \in \mathbb{C}^{(\infty)}(\mathbb{R}^N) \cap \mathbb{L}^\infty(\mathbb{R}^N) \cap \mathbb{L}^2(\mathbb{R}^N)$. Then the unique solution of the Cauchy problem (A.104)–(A.105) has the form

$$u(x,t) = \frac{1}{(4i\pi t)^{N/2}} \int_{\mathbb{R}^N} \frac{e^{\frac{i|x-y|^2}{4t} - i|y|^2}}{(1+|y|^2)^m}\, dy. \tag{A.106}$$

First, we can immediately verify that the integral in (A.106) is defined for $(x,t) \neq (0,1/4)$. Now let $t = 1/4$ and $x \neq 0$. Then we have the following formula:

$$u\left(x, \frac{1}{4}\right) = \frac{1}{(4i\pi t)^{N/2}} \int_{\mathbb{R}^N} e^{-2i(x,y)} \frac{dy}{(1+|y|^2)^m} = c_1 \frac{K_{N/2-m}(|x|)}{|x|^{N/2-m}}. \tag{A.107}$$

The Macdonald function $K_\nu(|x|)$ has the following asymptotic behavior as $x \to 0$:

$$K_0(|x|) \sim \log|x| \qquad \text{as } x \to 0,$$
$$K_\nu(|x|) \sim \frac{c}{|x|^\nu} \qquad \text{as } x \to 0,\ \nu \neq 0.$$

Thus there exists a sequence $\{(x_n, t_n)\}$ such that

$$(x_n, t_n) \to (0, 1/4), \qquad \lim_{n\to+\infty} |u(x_n, t_n)| = +\infty.$$

Example A.5 (**"hyperbolic" linear Schrödinger equation**). Consider the following Cauchy problem:

$$i\frac{\partial u}{\partial t} + \frac{\partial^2 u}{\partial x_1^2} - \frac{\partial^2 u}{\partial x_2^2} = 0, \quad (x,t) \in \mathbb{R}^2 \times \mathbb{R}^1_+, \quad u(x,0) = u_0(x). \tag{A.108}$$

This equation is the linearized equation of surface gravity waves on "deep" water.

The fundamental solution of the operator

$$i\frac{\partial}{\partial t} + \frac{\partial^2}{\partial x_1^2} - \frac{\partial^2}{\partial x_2^2}$$

has the following explicit form:

$$\mathcal{E}(x,t) = \frac{\theta(t)}{4i\pi t} e^{\frac{x_1^2 - x_2^2}{4t}}, \quad \theta(t) = \begin{cases} 1 & \text{for } t \geq 0, \\ 0 & \text{for } t < 0, \end{cases} \quad x = (x_1, x_2). \tag{A.109}$$

Similarly to Example A.1, we take the following initial function:

$$u_0(x) = \frac{e^{-i(x_1^2 - x_2^2)}}{(1 + x_1^2)^m (1 + x_2^2)^p}, \quad \frac{1}{4} < m, p \leq \frac{1}{2}. \tag{A.110}$$

Thus $u_0(x) \in \mathbb{C}^\infty(\mathbb{R}^2) \cap \mathbb{L}^\infty(\mathbb{R}^2) \cap \mathbb{L}^2(\mathbb{R}^2)$. Let $t > 0$ and $t = 1/4$. Then the solution of the Cauchy problem (A.108) with the initial function (A.110) has the following form:

$$u(x,t) = \frac{c_0}{t} e^{\frac{x_1^2 - x_2^2}{4t}} I(x,t), \tag{A.111}$$

where

$$\begin{aligned} I(x,t) &= \int_{\mathbb{R}^2} e^{-i\alpha(x_1 y_1 - x_2 y_2)} \frac{e^{i\beta(y_1^2 - y_2^2)}}{(1 + y_1^2)^m (1 + y_2^2)^p}\, dy \\ &= \int_{-\infty}^{+\infty} e^{-i\alpha x_1 y_1} \frac{e^{i\beta y_1^2}}{(1 + y_1^2)^m}\, dy_1 \int_{-\infty}^{+\infty} e^{i\alpha x_2 y_2} \frac{e^{-i\beta y_2^2}}{(1 + y_2^2)^p}\, dy_2 \end{aligned} \tag{A.112}$$

and

$$\alpha = \frac{1}{2t}, \quad \beta = \frac{1}{4t} - 1.$$

From the explicit form (A.111) of the solution of the Cauchy problem (A.108), taking into account (A.112), we detect the blow-up for a finite time at the points $(x_1, x_2, t) = (0, x_2, 1/4)$ and $(x_1, x_2, t) = (x_1, 0, 1/4)$.

Bibliography

[1] A. Al'shin, M. O. Korpusov, and A. G. Sveshnikov, *Blow-Up in Nonlinear Sobolev-Type Equations*, De Gruyter Ser. Nonlin. Anal. Appl., **15**, De Gruyter, Berlin–New York (2011).

[2] J. Ball, Remarks on blow-up and nonexistence theorems for nonlinear evolution equations. *Quart. J. Math. Oxford Ser.*, **28**, 473–486 (1977).

[3] R. Bellman, *Stability Theory of Differential Equations*, McGraw-Hill, New York–Toronto–London (1953).

[4] H. Berestycki and P. L. Lions. Existence of stationary states in nonlinear scalar field equations. In: *Bifurcation Phenomena in Mathematical Physics and Related Topics* (1976), pp. 269–292.

[5] J. L. Bona and J. C. Saut, Dispersive blow-up of solutions of generalized Korteveg–de Vries equations. *J. Differ. Equ.*, **103**, 3–57 (1993).

[6] J. L. Bona and J. C. Saut, Dispersive blow-up, II. Schrödinger-type equations, optical and oceanic rogue waves. *Chin. Ann. Math., Ser. B*, **31**, No. 6, 793–818 (2010).

[7] J. L. Bona and F. B. Weissler, Blow-up of spatially periodic complex-valued solutions of nonlinear dispersive equations. *Indiana Univ. Math. J.*, **50**, No. 2, 759–782 (2001).

[8] J. L. Bona, G. Ponce, J. K. Saut, and M. Tom, A model system for strong interaction between internal solitary waves. *Commun. Math. Phys.*, **143**, No. 2, 287–313 (1992).

[9] J. Bourgain and W. Wang, Construction of blow-up solutions for nonlinear Schrödinger equation with critical nonlinearity. *Ann. Sc. Norm. Super. Pisa, Cl. Sci.*, **25**, Nos. 1–2, 197–215 (1997).

[10] M. M. Cavalcanti, V. N. D. Cavalcanti, and J. A. Soriano, Exponential decay for the solution of semilinear viscoelastic wave equations with localized damping. *Electron. J. Differ. Equ.*, **2002**, No. 44, 1–14 (2002).

[11] P. J. Chen and M. E. Gurtin, On a theory of heat conduction involving two temperatures. *Z. Angew. Math. Phys.*, **19**, 614–627 (1968).

[12] A. Constantin and J. Escher, Global existence and blow-up for a shallow water equation. *Ann. Sc. Norm. Super. Pisa*, **23**, 303–328 (1998).

[13] G. G. Elenin, N. V. Zmitrenko, S. P. Kurdyumov, A. N. Mikhailov, and A. A. Samarskii. Inertia of heat and dissipative structures. In: *Numeric Simulation of Hydrodynamic Instability*, Keldysh Institute of Applied Mathematics, Moscow (1980), pp. 5–27, in Russian.

[14] L. E. Elsgolts, *Differential Equations and Calculus of Variations*, Nauka, Moscow (1969), in Russian.

[15] R. Emden, *Gaskugeln*, Teubner, Leipzig (1907).

[16] A. Erdélyi, W. Magnus, F. Oberhettinger, and F. G. Tricomi, *Higher Transcendental Functions*, Vol. 1, Bateman Manuscript Project, McGraw-Hill, New York (1953).

[17] V. I. Erofeev, V. V. Kazhaev, and N. P. Semerikova, *Waves in Rods. Dispersion. Dissipation. Nonlinearity*, Fizmatlit, Moscow (2002), in Russian.

[18] L. C. Evans, *Partial Differential Equations*, Grad. Stud. Math., **19**, Am. Math. Soc., Providence, Rhode Island (2010).

[19] H. Feder, Curvature measure. *Trans. Am. Math. Soc.*, **93**, 418–490 (1959).

[20] W. Fleming and R. Rishel, An integral formula for total gradient variation. *Arch. Math.*, **11**, 218–220 (1960).

[21] A. Friedman, *Partial Differential Equations of Parabolic Type*, Prentice-Hall, Englewood Cliffs, New Jersey (1964).

[22] A. Friedman, Remarks on nonlinear parabolic equations. *Proc. Symp. Appl. Math.*, **13**, 3–23 (1965).

https://doi.org/10.1515/9783110602074-013

[23] H. Fujita, On the blowing up solutions of the Cauchy problem for $u_t = \Delta u + u^{1+\alpha}$. *J. Fac. Sci., Univ. Tokyo, Sect. 1A, Math.*, **13**, 109–124 (1966).

[24] S. A. Gabov, *Introduction to the Theory of Nonlinear Waves*, Moscow State Univ., Moscow (1988), in Russian.

[25] E. I. Galakhov, On differential inequalities with point singularities on the boundary. *Tr. Mat. Inst. Steklova*, **260**, 119–129 (2008).

[26] V. A. Galaktionov, Approximate self-similar solutions of equations of heat conduction type. *Differ. Uravn.*, **16**, No. 9, 1660–1676 (1980).

[27] V. A. Galaktionov and S. I. Pokhozhaev, Blow-up, critical exponents and asymptotic spectra for nonlinear hyperbolic equations. Math. Preprint 00/10, Univ. of Bath (2000).

[28] V. A. Galaktionov, E. Mitidieri, and S. I. Pokhozhaev, On global solutions and blow-up for Kuramoto–Sivashinsky-type models, and well-posed Burnett equations. *Nonlinear Anal., Theory Methods Appl.*, **8**, 2930–2952 (2009).

[29] V. Georgiev and G. Todorova, Existence of solutions of the wave equations with nonlinear damping and source terms. *J. Differ. Equ.*, **109**, 295–308 (1994).

[30] J. Ginibre and G. Velo, On a class of nonlinear Schrödinger equations, I. The Cauchy problem, general case. *J. Funct. Anal.*, **32**, 1–32 (1979).

[31] R. T. Glassey, Blow-up theorems for nonlinear wave equations. *Math. Z.*, **132**, 183–203 (1972).

[32] R. T. Glassey, On the blowing up of solutions to the Cauchy problem for nonlinear Schrödinger equations. *J. Math. Phys.*, **18**, 1794–1797 (1977).

[33] A. Haraux and E. Zuazua, Decay estimates for some semilinear damped hyperbolic problems. *Arch. Ration. Mech. Anal.*, **150**, 191–206 (1988).

[34] P. Hartman, *Ordinary Differential Equations*, Wiley, New York–London–Sydney (1964).

[35] J. Hay, On necessary conditions for the existence of local solutions to singular nonlinear ordinary differential equations and inequalities. *Mat. Zametki*, **72**, No. 6, 924–935 (2002).

[36] K. Jörgens, *Nonlinear Wave Equations*, Univ. of Colorado (1970).

[37] V. K. Kalantarov and O. A. Ladyzhenskaya, The occurrence of collapse for quasilinear equations of parabolic and hyperbolic type. *J. Sov. Math.*, **10**, 53–70 (1978).

[38] S. Kaplan, On the growth of solutions of quasilinear parabolic equations. *Commun. Pure Appl. Math.*, **16**, 305–330 (1963).

[39] T. Kato, Blow-up of solutions of some nonlinear hyperbolic equations. *Commun. Pure Appl. Math.*, **32**, 501–505 (1980).

[40] H. Kauderer, *Nichtlineare Mechanik*, Springer-Verlag, Berlin (1958).

[41] J. B. Keller, On solutions of nonlinear wave equations. *Commun. Pure Appl. Math.*, **10**, 523–530 (1957).

[42] I. T. Kiguradze and T. A. Chanturia, *Asymptotic Properties of Solutions to Nonautonomous Ordinary Differential Equations*, Nauka, Moscow (1990), in Russian.

[43] V. A. Kondratiev and V. S. Samovol, On some asymptotic properties of solutions to Emden–Fowler-type equations. *Differ. Uravn.*, **17**, No. 4, 749–750 (1981).

[44] M. O. Korpusov, *Blow-Up in Nonclassical Wave Equations*, URSS, Moscow (2010), in Russian.

[45] M. O. Korpusov, *Blow-Up in Nonclassical Nonlocal Equations*, URSS, Moscow (2011), in Russian.

[46] M. O. Korpusov, *Blow-Up in Nonlinear Wave Equations with Positive Energy*, URSS, Moscow (2012), in Russian.

[47] M. O. Korpusov, *Blow-Up in Nonlinear Parabolic and Pseudo-Parabolic Equations with Double Nonlinearities*, URSS, Moscow (2012), in Russian.

[48] M. O. Korpusov, *Blow-Up in Nonlinear System of Mixed Type*, URSS, Moscow (2012), in Russian.

[49] M. O. Korpusov and A. A. Panin, Local solvability and blow-up of solutions of the Benjamen–Bona–Mahony–Burgers equation with nonlocal boundary conditions. *Teor. Mat. Fiz.*, **175**, No. 2, 159–172 (2013).

[50] M. O. Korpusov, A. G. Sveshnikov, and E. V. Yushkov, Blow-up of solutions of nonlinear equations of Kadomtsev–Petviashvili and Zakharov–Kuznetsov types. *Izv. Ross. Akad. Nauk, Ser. Mat.*, **78**, No. 3, 79–110 (2014).

[51] A. I. Kozhanov, Initial-boundary-value problem for generalized Boussinesq-type equations with nonlinear sources. *Mat. Zametki*, **65**, No. 1, 70–75 (1999).

[52] S. P. Kurdyumov, G. G. Malinetskii, Yu. A. Poveshchenko, Yu. P. Popov, and A. A. Samarskii, Interaction of dissipative heat structures in nonlinear media. *Dokl. Akad. Nauk SSSR*, **251**, No. 4, 836–839 (1980).

[53] M. V. Kuzelev and A. A. Rukhadze, *Method of the Wave Theory with Dispersion*, Fizmatlit, Moscow (2007), in Russian.

[54] O. A. Ladyzhenskaya and N. N. Ural'tseva, *Linear and Quasilinear Elliptic Equations*, Nauka, Moscow (1973), in Russian.

[55] O. A. Ladyzhenskaya, V. A. Solonnikov, and N. N. Ural'tseva, *Linear and Quasilinear Parabolic Equations*, Nauka, Moscow (1967), in Russian.

[56] L. D. Landau and E. M. Lifshits, *Electrodynamics of Continuous Media*, Nauka, Moscow (1992), in Russian.

[57] H. A. Levine, Some nonexistence and instability theorems for solutions of formally parabolic equations of the form $Pu_t = -Au + F(u)$. *Arch. Ration. Mech. Anal.*, **51**, 371–386 (1973).

[58] H. A. Levine, Instability and nonexistence of global solutions of nonlinear wave equation of the form $Pu_{tt} = Au + F(u)$. *Trans. Am. Math. Soc.*, **192**, 1–21 (1974).

[59] H. A. Levine, An estimate for the best constant in a Sobolev inequality involving three integral norms. *Ann. Mat. Pura Appl.*, **124**, 181–197 (1980).

[60] H. A. Levine, Some additional remarks on the nonexistence of global solutions to nonlinear wave equations. *SIAM J. Math. Anal.*, **137**, 341–361 (1997).

[61] H. A. Levine and L. E. Payne, A nonexistence theorem for heat equation with a nonlinear boundary condition and for the porous medium equation. *J. Differ. Equ.*, **16**, 319–334 (1974).

[62] M.-R. Li and L.-Y. Tsai, On a system of nonlinear wave equations. *Taiwan. J. Math.*, **4**, 557–573 (2003).

[63] M.-R. Li and L.-Y. Tsai, Existence and nonexistence of global solutions of some system of semilinear wave equations. *Nonlinear Anal.*, **54**, No. 8, 1397–1415 (2003).

[64] K. Li and Q. Zhang, Existence and nonexistence of global solutions for the equation of dislocation. *J. Differ. Equ.*, **10**, No. 1, 5–21 (1998).

[65] F. Linares, A higher order modified Korteweg–de Vries equation. *Mat. Apl. Comput.*, **14**, No. 3, 253–267 (1995).

[66] J. L. Lions, *Quelques Méthodes de Résolution des Problèmes aux Limites Nonlinéaires*, Dunod, Gautier-Villars, Paris (1969).

[67] A. E. H. Love, *A Treatise on the Mathematical Theory of Elasticity*, Dover, New York (1944).

[68] V. G. Makhankov, Dynamics of classical solutions in integrable systems. *Phys. Rep.*, **35**, 1–128 (1978).

[69] K. McLeod and J. Serrin, Uniqueness of solutions of semilinear Poisson equations. *Proc. Natl. Acad. Sci. Math.*, **10**, 64–74 (1941).

[70] L. A. Medeiros and G. P. Menzala, On a mixed problem for a class of nonlinear Klein–Gordon equations. *Acta Math. Hung.*, **52**, 61–69 (1988).

[71] L. A. Medeiros and M. M. Miranda, Weak solutions for a system of nonlinear Klein–Gordon equations. *Ann. Mat. Pura Appl.*, **146**, 173–183 (1987).

[72] F. Merle and P. Raphael, The blow-up dynamic and upper bound on the blow-up rate for critical nonlinear Schrödinger equation. *Ann. Math.*, **161**, 157–222 (2005).

[73] S. A. Messaoudi, Blow-up of positive-initial-energy solutions of a nonlinear viscoelastic hyperbolic equation. *J. Math. Anal. Appl.*, **320**, 902–915 (2006).

[74] E. L. Mitidieri and S. I. Pokhozhaev, A priori estimates and blow-up of solutions to nonlinear partial differential equations and inequalities. *Tr. Mat. Inst. Steklova*, **234** (2001), in Russian.

[75] B. V. Sz. Nagy, Über Integralgleichungen zwischen einer Funktion und ihrer Ableitung. *Acta Sci. Math.*, **10**, 64–74 (1941).

[76] P. I. Naumkin and I. A. Shishmarev, *Nonlinear Nonlocal Equations in the Theory of Waves*, Transl. Math. Monogr., **133**, Am. Math. Soc., Providence, Rhode Island (1994).

[77] Z. Nehari, On a nonlinear differential equation arising in nuclear physics. *Proc. R. Ir. Acad., Sci.*, **62**, 117–135 (1963).

[78] A. F. Nikiforov and V. B. Uvarov, *Special Functions of Mathematical Physics*, Nauka, Moscow (1984), in Russian.

[79] A. A. Panin, Local solvability and blow-up of solution for the Rosenau–Burgers with various boundary conditions. *Teor. Mat. Fiz.*, **177**, No. 1, 93–110 (2013).

[80] A. A. Panin. On a nonextendable solution of the integral Volterra–Hammerstein equation. In: *Proc. IV Int. School-Seminar "Nonlinear Analysis and Extremal Problems"*, Irkutsk, June 22–28, 2014, Matrosov Institute for System Dynamics and Control Theory of Siberian Branch of Russian Academy of Sciences, Irkutsk (2014), pp. 44–45, in Russian.

[81] L. E. Payne, Uniqueness criteria for steady state solutions of Navier–Stokes equations. *Atti Simp. Inter. sulle Appl. dell' Anal. alla Fis. Mat., Cagliari-Sassari*, **28**, 130–153 (1964).

[82] L. E. Payne and D. H. Sattinger, Saddle points and instability of nonlinear hyperbolic equations. *Isr. J. Math.*, **22**, Nos. 3–4, 273–303 (1975).

[83] G. Perelman, On the formation of singularities in solutions of the critical nonlinear Schrödinger equation. *Ann. Henri Poincaré*, **2**, 605–673 (2001).

[84] S. I. Pokhozhaev, On eigenfunctions of the equation $\Delta u + \lambda f(u) = 0$. *Dokl. Akad. Nauk SSSR*, **165**, No. 1, 36–39 (1965).

[85] S. I. Pokhozhaev, Blow-up of sign-alternating solutions of quasilinear parabolic equations. *Tr. Mat. Inst. Steklova*, **269**, 215–224 (2010).

[86] S. I. Pokhozhaev, Blow-up of sign-alternating solutions of a quasilinear parabolic equation. *Differ. Uravn.*, **47**, No. 3, 376–384 (2011).

[87] S. I. Pokhozhaev, On the nonexistence of global solutions for some initial-boundary-value problems for the Korteweg–de Vries equation. *Differ. Equ.*, **47**, No. 4, 488–493 (2011).

[88] S. I. Pokhozhaev, Weighted identities for the solutions of generalized Korteweg–de Vries equations. *Math. Notes*, **89**, No. 3, 382–396 (2011).

[89] S. I. Pokhozhaev, On the absence of global solutions of the Korteweg–de Vries equation. *J. Math. Sci.*, **190**, No. 1, 147–156 (2013).

[90] M. H. Protter and H. F. Weinberger, *Maximum Principles in Differential Equations*, Prentice-Hall, Englewood Cliff, New Jersey (1967).

[91] M. Reed, *Abstract Nonlinear Wave Equations*, Springer-Verlag, Berlin (1976).

[92] G. H. Ryder, Boundary-value problems for a class of nonlinear differential equations. *Pac. J. Math.*, **22**, 477–503 (1967).

[93] A. A. Samarskii, G. G. Elenin, N. V. Zmitrenko, S. P. Kurdyumov, and A. P. Mikhailov, Burning of nonlinear media in the form of complex structures. *Dokl. Akad. Nauk SSSR*, **237**, No. 6, 1330–1333 (1977).

[94] A. A. Samarskii, V. A. Galaktionov, S. P. Kurdyumov, and A. P. Mikhailov, *Blow-Up in Quasilinear Parabolic Equations*, Walter de Gruyter, Berlin (1995).

[95] G. Sansone, *Equazioni Differenziali nel Campo Reale*, Zanichelli, Bologna (1949).

[96] D. H. Sattinger, Stability of nonlinear hyperbolic equations. *Arch. Ration. Mech. Anal.*, **28**, 226–244 (1968).
[97] I. Segal, Nonlinear semigroups. *Ann. Math.*, **78**, 339–364 (1963).
[98] I. Segal, Nonlinear partial differential equations in quantum field theory. *Proc. Symp. Appl. Math.*, **17**, 210–226 (1965).
[99] W. A. Strauss, Existence of solitary waves in higher dimensions. *Commun. Math. Phys.*, **55**, 149–162 (1977).
[100] M. Struwe, *Variational Methods. Applications to Nonlinear Partial Differential Equations and Hamiltonian Systems*, Springer-Verlag, Berlin (2007).
[101] G. Todorova, The Cauchy problem for nonlinear wave equations with nonlinear damping and source terms. *Nonlinear Anal., Theory Methods Appl.*, **41**, Nos. 7–8, 891–905 (2000).
[102] M. Tom, *Global well-posedness, local smoothing and dispersive blow-up of some nonlinear dispersive equations*, Ph.D. thesis. The Pennsylvania State Univ. (1990).
[103] M. Tsutsumi, Existence and Nonexistence of Global Solutions for Nonlinear Parabolic Equations, manuscript. DOI: 10.2977/prims/1195193108.
[104] M. Tsutsumi, Nonexistence of Global Solutions to Nonlinear Schrödinger Equations, unpublished manuscript.
[105] V. N. Vlasov and A. Weinstein, Sur le problème de Cauchy pour l'équation de Poisson et l'équation des ondes. *C. R. Acad. Sci. Paris*, **234**, 2584–2585 (1952).
[106] Y. Wang, Nonexistence of global solutions of a class of coupled nonlinear Klein–Gordon equations with nonnegative potentials and arbitrary initial energy. *IMA J. Appl. Math.*, **24**, No. 3, 392–415 (2009). Oxford Univ. Press.
[107] A. Weinstein, Nonlinear Schrödinger equations and sharp interpolation estimates. *Commun. Math. Phys.*, **87**, 567–576 (1983).
[108] E. V. Yushkov, Blow-up of solutions of a Korteweg–de Vries-type equation. *Theor. Math. Phys.*, **172**, No. 1, 932–938 (2012).
[109] E. V. Yushkov, Blow-up in Korteweg–de Vries-type systems. *Theor. Math. Phys.*, **173**, No. 2, 1498–1506 (2012).
[110] E. V. Yushkov, Blow-up in systems with nonlinear viscosity. *Mat. Zametki*, **95**, No. 4, 615–629 (2014).
[111] V. E. Zakharov, Stochastization of one-dimensional chains of nonlinear operators. *Zh. Èksp. Teor. Fiz.*, **65**, 219–228 (1973).
[112] Ya. B. Zel'dovich, S. I. Blinnikov, and N. I. Shakura, *Physical Foundations of the Structure and Evolution of Stars*, Moscow State Univ., Moscow (1981), in Russian.
[113] N. M. Zmitrenko, S. P. Kurdyumov, A. P. Mikhailov, and A. A. Samarskii, *Appearance of Structures in Nonlinear Media and Nonstationary Dynamics of Hyperbolic Growth*, Keldysh Institute of Applied Mathematics, Moscow (1976), in Russian.

Index

De Gruyter Series in Nonlinear Analysis and Applications

Volume 26
Saïd Abbas, Mouffak Benchohra, John R. Graef, Johnny Henderson
Implicit Fractional Differential and Integral Equations. Existence and Stability, 2018
ISBN 978-3-11-055313-0, e-ISBN (PDF) 978-3-11-055381-9,
e-ISBN (EPUB) 978-3-11-055318-5

Volume 25
Luboš Pick, Alois Kufner, Oldřich John, Svatopluk Fucík
Function Spaces. Volume 2, 2018
ISBN 978-3-11-027373-1, e-ISBN (PDF) 978-3-11-032747-2,
e-ISBN (EPUB) 978-3-11-038221-1

Volume 24
Alexander A. Kovalevsky, Igor I. Skrypnik, Andrey E. Shishkov
Singular Solutions of Nonlinear Elliptic and Parabolic Equations, 2016
ISBN 978-3-11-031548-6, e-ISBN (PDF) 978-3-11-033224-7,
e-ISBN (EPUB) 978-3-11-039008-7

Volume 23/2
Sergey G. Glebov, Oleg M. Kiselev, Nikolai N. Tarkhanov
Nonlinear Equations with Small Parameter. Volume 2: Partial Differential Equations, 2018
ISBN 978-3-11-053383-5, e-ISBN (PDF) 978-3-11-053497-9,
e-ISBN (EPUB) 978-3-11-053390-3

Volume 23/1
Sergey G. Glebov, Oleg M. Kiselev, Nikolai N. Tarkhanov
Nonlinear Equations with Small Parameter. Volume 1: Ordinary Differential Equations, 2017
ISBN 978-3-11-033554-5, e-ISBN (PDF) 978-3-11-033568-2,
e-ISBN (EPUB) 978-3-11-038272-3

Volume 22
Miroslav Bácak
Convex Analysis and Optimization in Hadamard Spaces, 2014
ISBN 978-3-11-036103-2, e-ISBN (PDF) 978-3-11-036162-9,
e-ISBN (EPUB) 978-3-11-039108-4

Volume 21
Moshe Marcus, Laurent Véron
Nonlinear Second Order Elliptic Equations Involving Measures, 2013
ISBN 978-3-11-030515-9, e-ISBN (PDF) 978-3-11-030531-9

Volume 20
John R. Graef, Johnny Henderson, Abdelghani Ouahab
Impulsive Differential Inclusions: A Fixed Point Approach, 2013
ISBN 978-3-11-029361-6, e-ISBN (PDF) 978-3-11-029531-3

Volume 19
Petr Hájek, Michal Johanis
Smooth Analysis in Banach Spaces, 2014
ISBN 978-3-11-025898-1, e-ISBN (PDF) 978-3-11-025899-8,
e-ISBN (EPUB) 978-3-11-039199-2

Volume 18
Smaïl Djebali, Lech Górniewicz, Abdelghani Ouahab
Solution Sets for Differential Equations and Inclusions, 2012
ISBN 978-3-11-029344-9, e-ISBN (PDF) 978-3-11-029356-2, Set-ISBN 978-3-11-029357-9

Volume 17
Jürgen Appell, Józef Banas, Nelson José Merentes Díaz
Bounded Variation and Around, 2013
ISBN 978-3-11-026507-1, e-ISBN (PDF) 978-3-11-026511-8, Set-ISBN 978-3-11-026624-5

Volume 16
Martin Väth
Topological Analysis: From the Basics to the Triple Degree for Nonlinear Fredholm Inclusions, 2012
ISBN 978-3-11-027722-7, e-ISBN (PDF) 978-3-11-027733-3, Set-ISBN 978-3-11-027734-0

Volume 15
Alexander B. Al'shin, Maxim O. Korpusov, Alexey G. Sveshnikov
Blow-up in Nonlinear Sobolev Type Equations, 2011
ISBN 978-3-11-025527-0, e-ISBN (PDF) 978-3-11-025529-4, Set-ISBN 978-3-11-218918-4

Volume 14
Luboš Pick, Alois Kufner, Oldrich John, Svatopluk Fucík
Function Spaces: Volume 1, 2012
ISBN 978-3-11-025041-1, e-ISBN (PDF) 978-3-11-025042-8, Set-ISBN 978-3-11-218901-6

Volume 13
Peter Kosmol, Dieter Müller-Wichards
Optimization in Function Spaces: With Stability Considerations in Orlicz Spaces, 2011
ISBN 978-3-11-025020-6, e-ISBN (PDF) 978-3-11-025021-3, Set-ISBN 978-3-11-214606-4

Volume 12
Victor G. Zvyagin, Dmitry A. Vorotnikov
Topological Approximation Methods for Evolutionary Problems of Nonlinear Hydrodynamics, 2008
ISBN 978-3-11-020222-9, e-ISBN (PDF) 978-3-11-020828-3, Set-ISBN 978-3-11-916747-5

Volume 11
Stanislav V. Emelyanov, Sergey K. Korovin, Nikolai A. Bobylev, Alexander V. Bulatov
Homotopy of Extremal Problems: Theory and Applications, 2007
ISBN 978-3-11-018942-1, e-ISBN (PDF) 978-3-11-089301-4, Set-ISBN 978-3-11-185292-8

Volume 10
Jürgen Appell, Espedito De Pascale, Alfonso Vignoli
Nonlinear Spectral Theory, 2004
ISBN 978-3-11-018143-2, e-ISBN (PDF) 978-3-11-019926-0, Set-ISBN 978-3-11-916047-6

Volume 9
Carlo Bardaro, Julian Musielak, Gianluca Vinti
Nonlinear Integral Operators and Applications, 2003
ISBN 978-3-11-017551-6, e-ISBN (PDF) 978-3-11-019927-7, Set-ISBN 978-3-11-916519-8

Volume 8
Jorge Ize, Alfonso Vignoli
Equivariant Degree Theory, 2003
ISBN 978-3-11-017550-9, e-ISBN (PDF) 978-3-11-020002-7, Set-ISBN 978-3-11-916004-9

Volume 7
Mikhail Kamenskii, Valeri Obukhovskii, Pietro Zecca
Condensing Multivalued Maps and Semilinear Differential Inclusions in Banach Spaces, 2001
ISBN 978-3-11-016989-8, e-ISBN (PDF) 978-3-11-087089-3, Set-ISBN 978-3-11-188280-2

Volume 6
Wu, Jianhong
Introduction to Neural Dynamics and Signal Transmission Delay, 2001
ISBN 978-3-11-016988-1, e-ISBN (PDF) 978-3-11-087997-1, Set-ISBN 978-3-11-179117-3

Volume 5
Pavel Drábek, Alois Kufner, Francesco Nicolosi
Quasilinear Elliptic Equations with Degenerations and Singularities, 1997
ISBN 978-3-11-015490-0, e-ISBN (PDF) 978-3-11-080477-5, Set-ISBN 978-3-11-184978-2

Volume 4
Tomas Roubicek
Relaxation in Optimization Theory and Variational Calculus, 1997
ISBN 978-3-11-014542-7, e-ISBN (PDF) 978-3-11-081191-9, Set-ISBN 978-3-11-174517-6

Volume 3
Thomas Runst, Winfried Sickel
Sobolev Spaces of Fractional Order, Nemytskij Operators, and Nonlinear Partial Differential Equations, 1996
ISBN 978-3-11-015113-8, e-ISBN (PDF) 978-3-11-081241-1, Set-ISBN 978-3-11-176620-1

Volume 2
Nikolai A. Bobylev, Yuri M. Burman, Sergey K. Korovin
Approximation Procedures in Nonlinear Oscillation Theory, 1994
ISBN 978-3-11-014132-0, e-ISBN (PDF) 978-3-11-088571-2, Set-ISBN 978-3-11-177704-7

Volume 1
Klaus Deimling,
Multivalued Differential Equations, 1992
ISBN 978-3-11-013212-0, e-ISBN (PDF) 978-3-11-087422-8, Set-ISBN 978-3-11-178559-2

www.degruyter.com

Bei Fragen zur Produktsicherheit wenden Sie sich bitte an:
If you have any questions regarding product safety,
please contact:

Walter de Gruyter GmbH
Genthiner Straße 13
10785 Berlin
productsafety@degruyterbrill.com